"Listen To The Stars"

A Comprehensive Reference & Sketch Book on the 110 Messier Objects

JJ Evans

I gaze upon the night so clear
and for a moment the stars seem near,

And no better view, I feel for awhile
Until I remember my Granddaughters smile

"...Enjoy The Views."

- JJ Evans

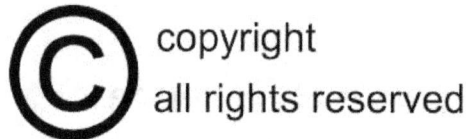

Introduction

To my fellow amateur, professional and soon to be beginner astronomers…
Thank you for choosing this Messier Astronomy Sketch Book, "Listen To The Stars".
Now you may be asking yourself, 'Why THAT title'? Well, more on that later. As you will soon see, this book is more than just sketch pages.

Like many of you, I enjoy the many aspects of this fascinating & rewarding hobby/profession. Besides the visual element, there are other "spin-offs" like, imaging, astro art, software, chart reading & DIY telescope making to name a few.

One such topic, sketching, has a huge following of people. Forums such as Astronomy Forum even has awards for sketching certain amounts of Messier Objects. If sketching is not your thing, they have a visual only section where all that is required is to list the objects you've seen. The Astronomy League also has numerous programs that will garnish you an award, or two, depending on which program suits you.

But no matter what sparks your interest, you've probably accumulated many pages of charts, drawings and notes. I know I did.
When I began sketching, none of the templates on the Internet gave me everything I was looking for. So, I customized my own sketch templates to include key information that reads more like a reference page than just a sketch sheet. I even included a photo of the Messier object to help verify what I saw. Additionally, there is a globe to indicate exactly where in the sky the object was located.

But I decided to take it a step further…

I decided that I wanted to make a Messier Book that, when completed, would be more of a reference or better yet, a display book that would look great on any table. So I decided to add much more to this "sketch book".

- History & Facts Pages adjacent to every sketch page. This allows you to read about the object while doing the optional sketching.
- Object Log - photos of every Messier Object with information and check off boxes.
- Atlas Info - more advanced information
- Personal Viewing Note Pages -
- Messier Object By Constellation
- Messier Objects by Season
- Messier Marathon Info

The book ended up becoming an excellent reference and sketch book. It can be used at the scope or the info can be added later at your leisure. It becomes a culmination of facts, history and sketches that nurtures that creative outlet for your viewing sessions.

But, again, I decided to take it a step further…

As mentioned, one of the fascinating areas of astronomy is reading all the interesting history about the object. I remember, looking at one of my favorite Messier's, M13 in Hercules. The next morning, I would read all the facts about this incredible object. But something was lacking.
The next night, while again viewing M13, I read the facts *at the scope*. This, as you know, requires a red light as to not interfere with your night vision. This can be difficult at times. Add to this, going from eye piece to paper to eye piece to paper to…well you get the idea. Although it added greatly to the session, something was still missing.

Ah…why just READ about the stars, why not "Listen To The Stars". Wouldn't it be fun to listen to the history and facts while never taking your eye from the scope or binoculars?

Think about it… You were able to hunt down an elusive object or your revisiting a favorite. You turn on your audio player and now you can listen to all of the interesting history of this famous object you are viewing…while never taking your attention away from the eye piece.
To hear facts, such as, "…You are looking at a stellar nursery where stars are born.
Or, "…It contains only a dozen or so variable stars and is estimated to be 13.7 billion years old." Another example, "…there is enough dust/debris, in the nebula, to create 30-40k earths!"
And maybe this one will capture your attention… "It took the light you are looking at over 65 million years to reach your telescope. It left that object while dinosaurs were still roaming the Earth."

There is something very astonishing to read that an object you are looking at is over a BILLION years old. It will give you the sense of viewing "live" history. It will fully immerse you into the view.

So…"Listen To The Stars" audio book was born. This book is the accompaniment to the audio book. It is not required that you purchase both although each does compliment the other quite well.
This book can give you excellent information in locating your object as well as document and sketch your findings. The audio book is meant to just sit back, relax, listen to the history and just 'Enjoy The Views'.

You will be happy to know that a portion of the proceeds will be going to the Children's Hospital of Philadelphia. The Children's Hospital of Philadelphia is the nation's first hospital devoted exclusively to the care of children. Since their start in 1855, CHOP has been the birthplace for many dramatic firsts in pediatric medicine. The Hospital has fostered medical discoveries and innovations that have improved pediatric healthcare and saved countless children's lives. Today, families facing complex conditions come to CHOP from all over the world, and their care and innovation has repeatedly earned them a spot on the *U.S. News & World Report's* Honor Roll of the nation's best children's hospitals.

Contents

What is a Messier Object?

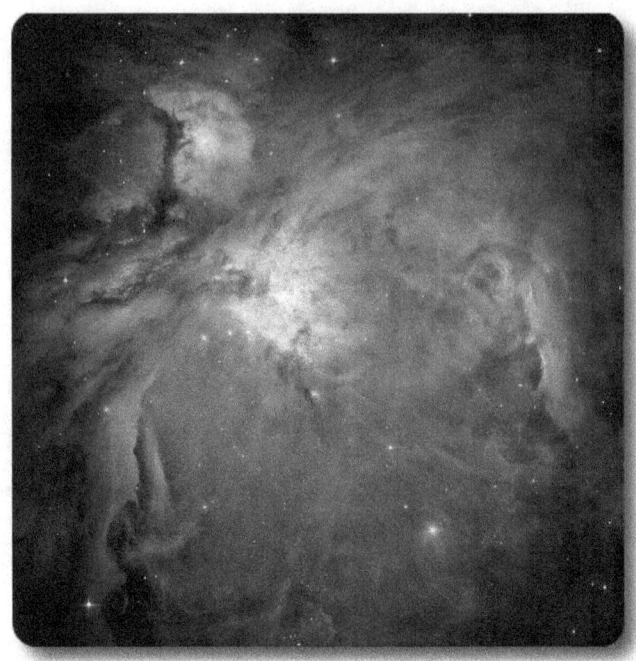

A Messier Object is one of the 110 deep-sky objects catalogued by Charles Messier in the 18th century.

They are mostly large, fuzzy objects that Messier wanted to distinguish from comets.

They are not stars or planets, but galaxies, clusters and nebulae. Messier considered these objects "nuisances" that got in the way of searching for comets.

Today's amateur astronomer considers them some of the most interesting objects to search for and the most beautiful to look at. This log will help you find and catalogue all the Messier objects.

You may do this over years, as I have, or in a single, glorious night in March — a "Messier Marathon" — if you have the stamina and star-hopping abilities.

Clear (and Dark) Skies!

The Messier objects

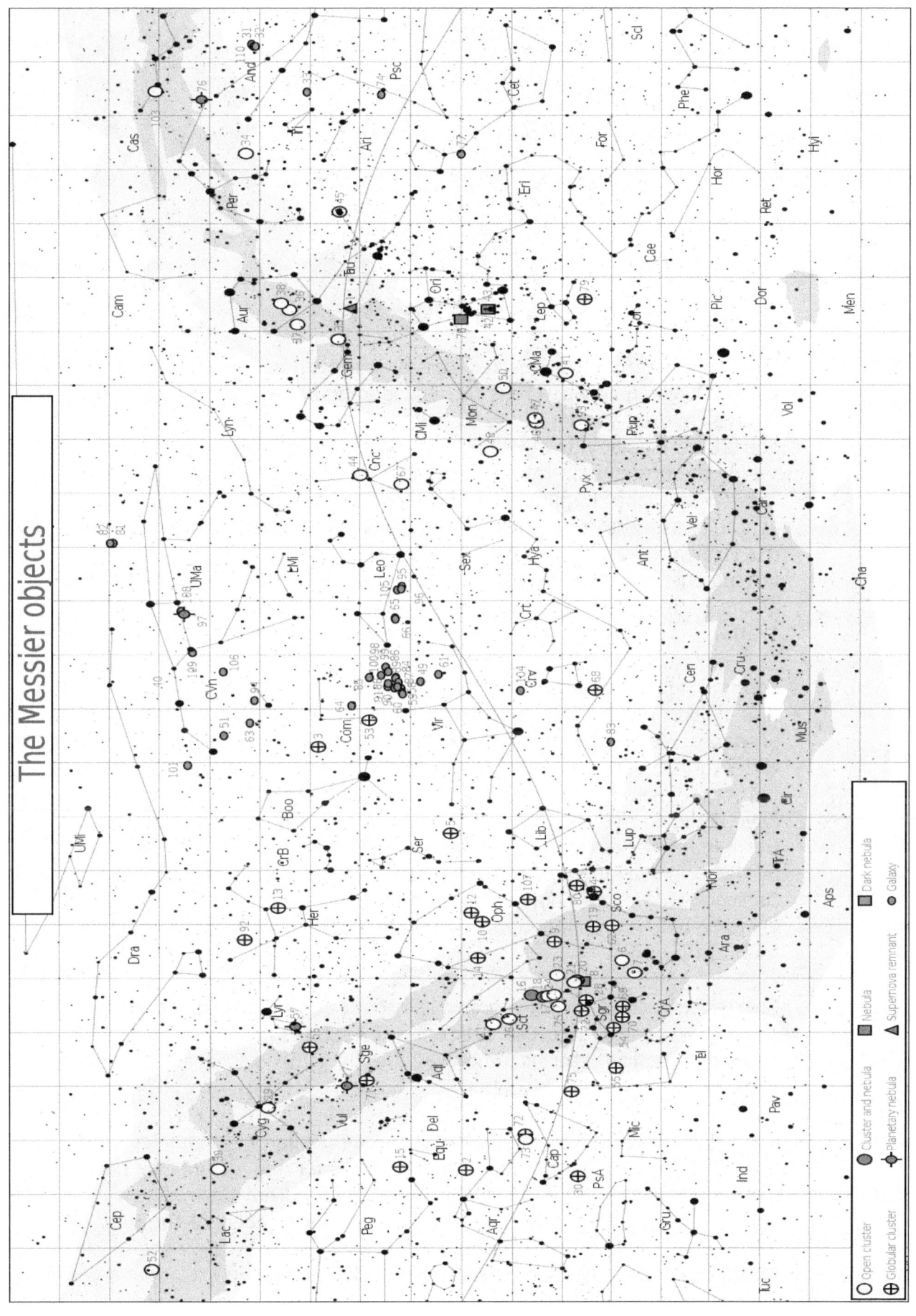

Legend:
- ○ Open cluster
- ⊕ Globular cluster
- ◉ Cluster and nebula
- ✦ Planetary nebula
- ■ Nebula
- ▲ Supernova remnant
- ▧ Dark nebula
- ◉ Galaxy

CHARLES MESSIER (JUNE 26, 1730 - APRIL 12, 1817)

The French astronomer Charles Messier was born in Lorraine on June 26, 1730. When he was 11 his father died and being the tenth of twelve children Messier consequently had little opportunity for education. As a boy he developed an avid interest in astronomy after seeing the brilliant six-tailed comet of 1744. Even though Messier came from a poor family and had limited schooling he was hired at the age of 21 as a draftsman by Joseph-Nicholas de l'Isle, Astronomer to the French Navy. Messier soon learned to use astronomical instruments and became a skilled observer. He was promoted to clerk at the Marine Observatory at the Hotel de Cluny in Paris by the mid-1750's.

Edmund Halley had predicted that the comet of 1682 would return in late 1758 or early 1759. Using charts that de l'Isle had incorrectly prepared, Messier began searching for the comet with a small reflector. On January 21, 1759 he located the comet but de l'Isle initially refused to let Messier announce his discovery. (As fate would have it the comet was first sighted on Christmas Night of 1758 by a German farmer and amateur astronomer named Palitzch.) Undaunted by the embarrassment of the late announcement, Messier from that time onward devoted himself to searching for comets. In the coming years he held a near monopoly on comet discoveries claiming to have found 21 by 1798.

Charles Messier used over a dozen telescopes during his career but his favorite was a 7.5 inch 104x Gregorian reflector. Later when the apochromatic refractor became available he utilized several 3.5 inch 120x apochromatics.

On August 28, 1758 Messier found by chance a small nebulous (cloudy) object in the constellation of Taurus while observing a comet he had discovered two weeks earlier. This object, a supernova remnant known today as the Crab Nebula (M1), was later to become the first entry on a list of comet-like objects that eventually became the most famous catalog of galaxies, nebulae and star clusters in astronomy.

Messier became the chief astronomer of the Marine Observatory in 1759. He was elected to the Royal Society of London in 1764 and the Paris Academy of Sciences in 1770.

During a seven-month period of searching for comets in 1764 Messier added 38 objects to his list including M13 (the great globular cluster in Hercules), the Swan nebula (M17) in Sagittarius and the Andromeda galaxy (M31). In January of the following year he logged M41, the open cluster southwest of Sirius. Messier determined the positions of the Orion Nebula (M42 and M43), the Beehive cluster (M44) and the Pleiades (M45) on March 4, 1769.

Messier also began compiling reports of discoveries by other astronomers. In fact, only 17 of the 45 objects in the first installment of Messier's catalog published in 1774 were discovered by Messier himself. (The first object that Messier is given credit for discovering is the globular cluster M3 which was first located on May 3, 1764.) By 1780 the number of objects in his catalog had increased to 80.

Because of Messier's undeniable success as a comet hunter King Louis XV of France nicknamed him the "Ferret of Comets". But Messier was no mathematician and relied on his aristocratic friend Bouchart de Saron who was the President of the French Assembly to compute the orbits of his comets. Messier also collaborated with the younger astronomer Pierre Francois Andre Mechain who was a successful comet hunter as well. During 1780 and 1781 he discovered some 32 new nebulous objects and reported their positions to Messier. Mechain was the first astronomer to note the profusion of nebulae in Coma Berenices and Virgo. Messier recorded 9 new nebulae in this region in a single night on March 18, 1781.

On April 13, 1781 Messier added the one hundredth object to the catalog. Three subsequent observations by Mechain were included hastily without verification and what was to be the final revision of the catalog was published in 1781. Forty of the 103 objects listed had been discovered by Messier and 27 by Mechain. In November of 1781 Messier suffered a severe fall and further work on his catalog was ended.

During his convalescence Messier was provided for by President de Saron and members of the Academie Royale. After a year of recuperating Messier made preparations to study the transit of Mercury and began observing the newly discovered planet Uranus as well as searching for more comets.

The French Revolution was a disastrous period for Messier and his compatriots. In 1794 Messier lost his Academie pension and naval salary and the navy stopped paying the rent on his observatory. President de Saron, the talented mathematician who was one of the first men to realize that William Herschel's Uranus was in fact a planet, was guillotined a few days after computing the orbit of a comet that Messier had discovered the previous year.

Mechain lost his estate and all of his savings. With the coming of Napoleon Bonaparte the lives of Messier and Mechain improved greatly. Mechain was made the director of the Paris Observatory and both he and Messier were admitted to the new Academy of Sciences and the Bureau of Longitudes. Messier received the cross of the Legion of Honor from Napoleon himself.

Messier made his last discovery in 1798. He continued to observe until he suffered a debilitating stroke. Two years later on April 12, 1817 he died at the age of 86.

In the twentieth century 7 objects known to have been logged by Messier were added to the Messier Catalog. M110, the final entry, was added in 1967.

Today it is known that M40 is merely a binary star and M73 is only an asterism. M102 is thought to be a duplication of M101 but NGC 5866 is often accepted as being M102. The true identity of M91 is also questionable. Because of an error in their coordinates M47 and M48 were at one time deemed to be "lost" Messier objects.

Charles Messier was limited as a scientist but he was an astute observational astronomer who studied sunspots, eclipses and occultations in addition to discovering many comets and nebulous objects. He was so totally dedicated to astronomy that when his wife lay dying it was with the greatest reluctance that he left his telescope to be at her side. Messier's lasting legacy to amateur astronomy, the Messier Catalog, includes most of the best deep sky objects visible in the northern hemisphere.

The Complete Messier Object Log

"Astronomy, as nothing else can do, teaches men humility".

— Arthur C. Clarke

The Complete Messier Object Log

Messier # M001

NGC 1 9 5 2

Seen? ☐

Date seen

Type Supernova remnant
Name (if any) Crab Nebula
Constellation Taurus

Magnitude
8.4

Messier # M002

NGC 7 0 8 9

Seen? ☐

Date seen

Type Globular Cluster
Name (if any) N/A
Constellation Aquarius

Magnitude
6.5

Messier # M003

NGC 5 2 7 2

Seen? ☐

Date seen

Type Globular Cluster
Name (if any) N/A
Constellation Canes Venatici

Magnitude
6.4

Messier # M004

NGC 6 1 2 1

Seen? ☐

Date seen

Type Globular Cluster
Name (if any) N/A
Constellation Scorpio

Magnitude
5.9

Messier # M005

NGC 5 9 0 4

Seen? ☐

Date seen

Type Globular Cluster
Name (if any) N/A
Constellation Serpens Caput

Magnitude
5.8

Messier # M006

NGC 6 4 0 5

Seen? ☐

Date seen

Type Open Cluster
Name (if any) Butterfly Cluster
Constellation Scorpius

Magnitude
4.2

Messier # M007

NGC 6 4 7 5

Seen? ☐

Date seen

Type Open Cluster
Name (if any) Ptolemy's Cluster
Constellation Scorpius

Magnitude
3.3

The Complete Messier Object Log

Messier # M008

NGC 6 5 2 3

Seen? ☐

Date seen

Type **Diffuse Nebula**

Name (if any) **Lagoon Nebula**

Constellation **Sagittarius**

Magnitude
5.0

Messier # M009

NGC 6 3 3 3

Seen? ☐

Date seen

Type **Globular Cluster**

Name (if any) **N/A**

Constellation **Ophiuchus**

Magnitude
7.9

Messier # M010

NGC 6 2 5 4

Seen? ☐

Date seen

Type **Globular Cluster**

Name (if any) **N/A**

Constellation **Ophiuchus**

Magnitude
6.6

Messier # M011

NGC 6 7 0 5

Seen? ☐

Date seen

Type **Open Cluster**

Name (if any) **Wild Duck Cluster**

Constellation **Scutum**

Magnitude
5.8

Messier # M012

NGC 6 2 1 8

Seen? ☐

Date seen

Type **Globular Cluster**

Name (if any) **N/A**

Constellation **Ophiuchus**

Magnitude
6.6

Messier # M013

NGC 6 2 0 5

Seen? ☐

Date seen

Type **Globular Cluster**

Name (if any) **N/A**

Constellation **Hercules**

Magnitude
5.9

Messier # M014

NGC 6 4 0 2

Seen? ☐

Date seen

Type **Globular Cluster**

Name (if any) **N/A**

Constellation **Ophiuchus**

Magnitude
7.6

The Complete Messier Object Log

Messier # M015

NGC 7 0 7 8

Seen? ☐

Date seen

Type Globular Cluster

Name (if any) N/A

Constellation Pegasus

Magnitude
6.4

Messier # M016

NGC 6 6 1 1

Seen? ☐

Date seen

Type Open Cluster

Name (if any) Eagle Nebula Cluster

Constellation Serpens Claudia

Magnitude
6.5

Messier # M017

NGC 6 6 1 8

Seen? ☐

Date seen

Type Diffuse   Nebula

Name (if any) Omega,   Swan, Horseshoe Nebula

Constellation Sagittarius

Magnitude
7.0

Messier # M018

NGC 6 6 1 3

Seen? ☐

Date seen

Type Open Cluster

Name (if any) N/A

Constellation Sagittarius

Magnitude
8.0

Messier # M019

NGC 6 2 7 3

Seen? ☐

Date seen

Type Globular Cluster

Name (if any) N/A

Constellation Ophiuchus

Magnitude
8.5

Messier # M020

NGC 6 5 1 4

Seen? ☐

Date seen

Type Diffuse Nebula

Name (if any) Trifid Nebula

Constellation Sagittarius

Magnitude
5.0

Messier # M021

NGC 6 5 3 1

Seen? ☐

Date seen

Type Open Cluster

Name (if any) N/A

Constellation Sagittarius

Magnitude
7.0

The Complete Messier Object Log

Messier # M022

NGC 6 6 5 6

Seen? ☐

Date seen

Type Globular Cluster
Name (if any) N/A
Constellation Sagittarius

Magnitude
6.5

Messier # M023

NGC 6 4 9 4

Seen? ☐

Date seen

Type Open Cluster
Name (if any) N/A
Constellation Sagittarius

Magnitude
6.0

Messier # M024

NGC 6 6 0 3

Seen? ☐

Date seen

Type Star Cloud
Name (if any) Milky Way Patch
Constellation Sagittarius

Magnitude
11.5

Messier # M025

NGC IC4725

Seen? ☐

Date seen

Type Open Cluster
Name (if any) N/A
Constellation Sagittarius

Magnitude
4.9

Messier # M026

NGC 6 6 9 4

Seen? ☐

Date seen

Type Open Cluster
Name (if any) N/A
Constellation Scutum

Magnitude
9.5

Messier # M027

NGC 6 8 5 3

Seen? ☐

Date seen

Type Planetary Nebula
Name (if any) Dumbbell Nebula
Constellation Vulpecula

Magnitude
7.5

Messier # M028

NGC 6 6 2 6

Seen? ☐

Date seen

Type Globular Cluster
Name (if any) N/A
Constellation Sagittarius

Magnitude
8.5

The Complete Messier Object Log

Messier # M029

NGC 6 9 1 3

Seen? ☐

Date seen

Type Open Cluster

Name (if any) N/A

Constellation Cygnus

Magnitude
9.0

Messier # M030

NGC 7 0 9 9

Seen? ☐

Date seen

Type Globular Cluster

Name (if any) N/A

Constellation Capricornus

Magnitude
8.5

Messier # M031

NGC 0 2 2 4

Seen? ☐

Date seen

Type Spiral Galaxy

Name (if any) Andromeda Galaxy

Constellation Andromeda

Magnitude
4.5

Messier # M032

NGC 0 2 2 1

Seen? ☐

Date seen

Type Elliptical Galaxy

Name (if any) Satellite of M31

Constellation Andromeda

Magnitude
10.0

Messier # M033

NGC 0 5 9 8

Seen? ☐

Date seen

Type Spiral Galaxy

Name (if any) Triangulum or Pinwheel Galaxy

Constellation Triangulum

Magnitude
7.0

Messier # M034

NGC 1 0 3 9

Seen? ☐

Date seen

Type Open Cluster

Name (if any) N/A

Constellation Perseus

Magnitude
6.0

Messier # M035

NGC 2 1 6 8

Seen? ☐

Date seen

Type Open Cluster

Name (if any) N/A

Constellation Gemini

Magnitude
5.5

The Complete Messier Object Log

Messier # M036

NGC 1 9 6 0

Seen? ☐

Date seen

Type **Open Cluster**

Name (if any) N/A

Constellation **Auriga**

Magnitude

6.5

Messier # M037

NGC 2 0 9 9

Seen? ☐

Date seen

Type **Open Cluster**

Name (if any) N/A

Constellation **Auriga**

Magnitude

6.0

Messier # M038

NGC 1 9 2 2

Seen? ☐

Date seen

Type **Open Cluster**

Name (if any) N/A

Constellation **Auriga**

Magnitude

7.0

Messier # M039

NGC 7 0 9 2

Seen? ☐

Date seen

Type **Open Cluster**

Name (if any) N/A

Constellation **Cygnus**

Magnitude

5.5

Messier # M040

NGC 7 0 9 2

Seen? ☐

Date seen

Type **Double Star**

Name (if any) Winecke 4

Constellation **Ursa Major**

Magnitude

9.0

Messier # M041

NGC 2 2 8 7

Seen? ☐

Date seen

Type **Open Cluster**

Name (if any) N/A

Constellation **Canis Major**

Magnitude

5.0

Messier # M042

NGC 1 9 7 6

Seen? ☐

Date seen

Type **Diffuse Nebula**

Name (if any) Great Orion Nebula

Constellation **Orion**

Magnitude

5.0

The Complete Messier Object Log

Messier # M043

NGC 1 9 8 2

Seen? ☐

Date seen

Type Diffuse Nebula

Name (if any) de Mairan's Nebula

Constellation Orion

Magnitude
7.0

Messier # M044

NGC 2 6 3 2

Seen? ☐

Date seen

Type Open Cluster

Name (if any) Beehive Cluster (Praesepe)

Constellation Cancer

Magnitude
4.0

Messier # M045

NGC 1 4 3 2

Seen? ☐

Date seen

Type Open Cluster

Name (if any) Pleiades, Subaru, Seven Sisters

Constellation Taurus

Magnitude
1.4

Messier # M046

NGC 2 4 3 7

Seen? ☐

Date seen

Type Open Cluster

Name (if any) N/A

Constellation Puppis

Magnitude
6.5

Messier # M047

NGC 2 4 2 2

Seen? ☐

Date seen

Type Open Cluster

Name (if any) N/A

Constellation Puppis

Magnitude
4.5

Messier # M048

NGC 2 5 4 8

Seen? ☐

Date seen

Type Open Cluster

Name (if any) N/A

Constellation Hydra

Magnitude
5.5

Messier # M049

NGC 4 4 7 2

Seen? ☐

Date seen

Type Elliptical Galaxy

Name (if any) N/A

Constellation Virgo

Magnitude
10.0

The Complete Messier Object Log

Messier # M050

NGC 2 3 2 3

Seen? ☐

Date seen

Type Open Cluster
Name (if any) N/A
Constellation Monocerus

Magnitude
7.0

Messier # M051

NGC 5 1 9 4

Seen? ☐

Date seen

Type Spiral   Galaxy
Name (if any) Whirlpool Galaxy
Constellation Ursa   Major

Magnitude
8.0

Messier # M052

NGC 7 6 5 4

Seen? ☐

Date seen

Type Open Cluster
Name (if any) N/A
Constellation Cassiopeia

Magnitude
8.0

Messier # M053

NGC 5 0 2 4

Seen? ☐

Date seen

Type Globular Cluster
Name (if any) N/A
Constellation Coma Berenices

Magnitude
8.5

Messier # M054

NGC 6 7 1 5

Seen? ☐

Date seen

Type Globular Cluster
Name (if any) N/A
Constellation Sagittarius

Magnitude
8.5

Messier # M055

NGC 6 8 0 9

Seen? ☐

Date seen

Type Globular Cluster
Name (if any) N/A
Constellation Sagittarius

Magnitude
7.0

Messier # M056

NGC 6 7 7 9

Seen? ☐

Date seen

Type Globular Cluster
Name (if any) N/A
Constellation Lyra

Magnitude
9.5

The Complete Messier Object Log

Messier # M057   NGC 6 7 2 0

Seen? ☐

Date seen

Type Planetary Nebula

Name (if any) Ring Nebula

Magnitude

Constellation Lyra

9.5

Messier # M058   NGC 4 5 7 9

Seen? ☐

Date seen

Type Spiral Galaxy

Name (if any) N/A

Magnitude

Constellation Virgo

11.0

Messier # M059   NGC 4 6 2 1

Seen? ☐

Date seen

Type Elliptical Galaxy

Name (if any) N/A

Magnitude

Constellation Virgo

11.5

Messier # M060   NGC 4 6 4 9

Seen? ☐

Date seen

Type Elliptical Galaxy

Name (if any) N/A

Magnitude

Constellation Virgo

10.5

Messier # M061   NGC 4 3 0 3

Seen? ☐

Date seen

Type Spiral Galaxy

Name (if any) N/A

Magnitude

Constellation Virgo

10.5

Messier # M062   NGC 6 2 6 6

Seen? ☐

Date seen

Type Globular Cluster

Name (if any) N/A

Magnitude

Constellation Ophiuchus

8.0

Messier # M063   NGC 5 0 5 5

Seen? ☐

Date seen

Type Spiral Galaxy

Name (if any) Sunflower Galaxy

Magnitude

Constellation Canes Venatici

8.5

The Complete Messier Object Log

Messier # M064   NGC 4 8 2 6   *Seen?* ☐
Date seen

Type **Spiral   Galaxy**
Name (if any) **Blackeye Galaxy**   *Magnitude*
Constellation **Coma   Berenices**   9.0

Messier # M065   NGC 3 6 2 3   *Seen?* ☐
Date seen

Type **Spiral Galaxy**
Name (if any) **N/A**   *Magnitude*
Constellation **Leo**   10.5

Messier # M066   NGC 3 6 2 7   *Seen?* ☐
Date seen

Type **Spiral Galaxy**
Name (if any) **N/A**   *Magnitude*
Constellation **Leo**   10.0

Messier # M067   NGC 2 6 2 8   *Seen?* ☐
Date seen

Type **Open Cluster**
Name (if any) **N/A**   *Magnitude*
Constellation **Cancer**   7.5

Messier # M068   NGC 4 5 9 0   *Seen?* ☐
Date seen

Type **Globular Cluster**
Name (if any) **N/A**   *Magnitude*
Constellation **Hydra**   9.0

Messier # M069   NGC 6 6 3 7   *Seen?* ☐
Date seen

Type **Globular Cluster**
Name (if any) **N/A**   *Magnitude*
Constellation **Sagittarius**   9.0

Messier # M070   NGC 6 6 8 1   *Seen?* ☐
Date seen

Type **Globular Cluster**
Name (if any) **N/A**   *Magnitude*
Constellation **Sagittarius**   9.0

The Complete Messier Object Log

Messier # M071

NGC 6 8 3 8

Seen? ☐

Date seen

Type Globular Cluster
Name (if any) N/A
Constellation Sagittarius

Magnitude
8.5

Messier # M072

NGC 6 9 8 1

Seen? ☐

Date seen

Type Globular Cluster
Name (if any) N/A
Constellation Aquarius

Magnitude
10.0

Messier # M073

NGC 6 9 9 4

Seen? ☐

Date seen

Type Group/Asterism
Name (if any) N/A
Constellation Aquarius

Magnitude
9.0

Messier # M074

NGC 0 6 2 8

Seen? ☐

Date seen

Type Spiral Galaxy
Name (if any) N/A
Constellation Pisces

Magnitude
10.5

Messier # M075

NGC 6 8 6 4

Seen? ☐

Date seen

Type Globular Cluster
Name (if any) N/A
Constellation Sagittarius

Magnitude
9.5

Messier # M076

NGC 0 6 5 0

Seen? ☐

Date seen

Type Planetary Nebula
Name (if any) Little Dumbbell, Cork, Butterfly Nebula
Constellation Perseus

Magnitude
12.0

Messier # M077

NGC 1 0 6 8

Seen? ☐

Date seen

Type Spiral Galaxy
Name (if any) N/A
Constellation Cetus

Magnitude
10.5

The Complete Messier Object Log

Messier # M078

NGC 2 0 6 8

Seen? ☐

Date seen

Type Diffuse Nebula

Name (if any) N/A

Constellation Orion

Magnitude
8.0

Messier # M079

NGC 1 9 0 4

Seen? ☐

Date seen

Type Globular Cluster

Name (if any) N/A

Constellation Lepus

Magnitude
8.5

Messier # M080

NGC 6 0 9 3

Seen? ☐

Date seen

Type Globular Cluster

Name (if any) N/A

Constellation Scorpius

Magnitude
8.5

Messier # M081

NGC 3 0 3 1

Seen? ☐

Date seen

Type Spiral   Galaxy

Name (if any) Bode's Galaxy - 1/2 of Double Galaxy

Constellation Ursa   Major

Magnitude
8.5

Messier # M082

NGC 3 0 3 4

Seen? ☐

Date seen

Type Irregular Galaxy

Name (if any) Cigar Galaxy - 1/2 of Double Galaxy

Constellation Ursa Major

Magnitude
9.5

Messier # M083

NGC 5 2 3 6

Seen? ☐

Date seen

Type Spiral   Galaxy

Name (if any) Small   Pinwheel Galaxy

Constellation Hydra

Magnitude
8.5

Messier # M084

NGC 4 3 7 4

Seen? ☐

Date seen

Type Lenticular (S0) Galaxy

Name (if any) N/A

Constellation Virgo

Magnitude
11.0

The Complete Messier Object Log

Messier # M085

NGC 4 3 8 2

Seen? ☐

Date seen

Type Lenticular (S0) Galaxy

Name (if any) N/A

Constellation Coma Berenices

Magnitude

10.5

Messier # M086

NGC 4 4 0 6

Seen? ☐

Date seen

Type Lenticular (S0) Galaxy

Name (if any) N/A

Constellation Virgo

Magnitude

11.0

Messier # M087

NGC 4 4 8 6

Seen? ☐

Date seen

Type Elliptical Galaxy

Name (if any) Virgo A

Constellation Virgo

Magnitude

11.0

Messier # M088

NGC 4 5 0 1

Seen? ☐

Date seen

Type Spiral Galaxy

Name (if any) N/A

Constellation Coma Berenices

Magnitude

11.0

Messier # M089

NGC 4 5 5 2

Seen? ☐

Date seen

Type Elliptical Galaxy

Name (if any) N/A

Constellation Virgo

Magnitude

11.5

Messier # M090

NGC 4 5 6 9

Seen? ☐

Date seen

Type Spiral Galaxy

Name (if any) N/A

Constellation Virgo

Magnitude

11.0

Messier # M091

NGC 4 5 4 8

Seen? ☐

Date seen

Type Spiral Galaxy

Name (if any) N/A

Constellation Coma Berenices

Magnitude

11.5

The Complete Messier Object Log

Messier # M092

NGC 6 3 4 1

Seen? ☐

Date seen

Type Globular Cluster
Name (if any) N/A
Constellation Hercules

Magnitude
7.5

Messier # M093

NGC 2 4 4 7

Seen? ☐

Date seen

Type Open Cluster
Name (if any) N/A
Constellation Puppis

Magnitude
6.5

Messier # M094

NGC 4 7 3 6

Seen? ☐

Date seen

Type Spiral Galaxy
Name (if any) N/A
Constellation Canes Venatici

Magnitude
9.5

Messier # M095

NGC 3 3 5 1

Seen? ☐

Date seen

Type Spiral Galaxy
Name (if any) N/A
Constellation Leo

Magnitude
11.0

Messier # M096

NGC 3 3 6 8

Seen? ☐

Date seen

Type Spiral Galaxy
Name (if any) N/A
Constellation Leo

Magnitude
10.5

Messier # M097

NGC 3 5 8 7

Seen? ☐

Date seen

Type Planetary Nebula
Name (if any) Owl   Nebula
Constellation Ursa   Major

Magnitude
12.0

Messier # M098

NGC 4 1 9 2

Seen? ☐

Date seen

Type Spiral Galaxy
Name (if any) N/A
Constellation Coma Berenices

Magnitude
11.0

The Complete Messier Object Log

Messier # M099   NGC 4 2 5 4   *Seen?* ☐

Date seen

Type Spiral Galaxy

Name (if any) N/A   *Magnitude*

Constellation Coma Berenices   10.5

Messier # M100   NGC 4 3 2 1   *Seen?* ☐

Date seen

Type Spiral Galaxy

Name (if any) N/A   *Magnitude*

Constellation Coma Berenices   10.5

Messier # M101   NGC 5 4 5 7   *Seen?* ☐

Date seen

Type Spiral   Galaxy

Name (if any) Pinwheel Galaxy   *Magnitude*

Constellation Ursa   Major   8.5

Messier # M102   NGC 5 8 6 6   *Seen?* ☐

Date seen

Type Lenticular (S0)   Galaxy

Name (if any) Spindle Galaxy   *Magnitude*

Constellation Draco   10.5

Messier # M103   NGC 0 5 8 1   *Seen?* ☐

Date seen

Type Open Cluster

Name (if any) N/A   *Magnitude*

Constellation Cassiopeia   7.0

Messier # M104   NGC 4 5 9 4   *Seen?* ☐

Date seen

Type Spiral Galaxy

Name (if any) Sombrero Galaxy   *Magnitude*

Constellation Virgo   9.5

Messier # M105   NGC 3 3 7 9   *Seen?* ☐

Date seen

Type Elliptical Galaxy

Name (if any) N/A   *Magnitude*

Constellation Leo   11.0

The Complete Messier Object Log

Messier # M106

NGC 4 2 5 8

Seen? ☐

Date seen

Type Spiral Galaxy
Name (if any) N/A
Constellation Ursa Major

Magnitude
9.5

Messier # M107

NGC 6 1 7 1

Seen? ☐

Date seen

Type Globular Cluster
Name (if any) N/A
Constellation Ophiuchus

Magnitude
10.0

Messier # M108

NGC 3 5 5 6

Seen? ☐

Date seen

Type Spiral Galaxy
Name (if any) N/A
Constellation Ursa Major

Magnitude
11.0

Messier # M109

NGC 3 9 9 2

Seen? ☐

Date seen

Type Spiral Galaxy
Name (if any) N/A
Constellation Ursa Major

Magnitude
11.0

Messier # M110

NGC 0 2 0 5

Seen? ☐

Date seen

Type Elliptical Galaxy
Name (if any) Satellite of M31
Constellation Andromeda

Magnitude
10.0

The Complete Messier
Facts & Sketch Pages

"Astronomy compels the soul to look upward

and leads us from this world to another."

- Plato

Messier 1 - M1 - Crab Nebula (Supernova Remnant)

The first item in Messier's catalogue is the famous Crab Nebula, a remnant of a supernova explosion, observed and recorded by Chinese and Arab astronomers' way back in 1054. The supernova itself was brilliant, easily visible in daylight with a peak brightness of magnitude –7. Today, it still remains one of the brightest natural stellar events ever recorded. Roll on 950+ years; the initial explosive has long since faded but the resulting aftermath - the nebula - still remains visible. At magnitude +8.4 and with an apparent size of 7x5 arc minutes, it's a relatively easy target under dark skies where it can be spotted with just a pair of binoculars.

M1 is located in the zodiacal constellation of Taurus "the Bull". It is not difficult to find as it is positioned only 1 degree northwest of magnitude +3.0 star Zeta Tauri (ζ Tau). The nebula itself was first observed by John Bevis in 1731. Charles Messier observed it in September 1758 and the appearance of this dying remnant inspired him to begin compiling a list of nebulae that could possibly be mistaken for comets. The list eventually became his famous catalogue. The following century, William Parsons, the Third Earl of Rosse made a drawing of M1 (around 1844) and christened it the "Crab Nebula" due to its wispy filamentary structure. It's best seen from the Northern Hemisphere during the months of November, December and January.

When viewed through 10x50 binoculars, M1 appears as a reasonably sized oblong patch of light that is more pronounced when using averted vision. Larger 20x80mm binoculars or a small 80mm (3.1-inch) telescope show some variations in brightness around the edges, especially on nights of good seeing. A 150mm (6-inch) or 200mm (8-inch) telescope displays the Crab Nebula as a large, oval, diffuse patch of light with finer streaks visible. A 300mm (12-inch) or larger amateur telescope will reveal more intricate details including twists, filaments and a surface brightness that changes over the complete face of the nebula. A tantalizing view that hints towards Lord Rosse's famous "Crab" structure.

M1 is located about 6,500 light years from Earth and is the only supernova remnant in Messier's catalogue. It has a radius of 5.5 light-years. At the centre of the nebula is a pulsar - the remains of the original star - that is believed to be about 30 km in diameter, emitting pulses of radiation every 33 milliseconds.

M1 Data Table

Messier	1
NGC	1952
Name	Crab Nebula
Object Type	Supernova Remnant
Constellation	Taurus
Distance (kly)	6.5
Apparent Mag.	8.4
RA (J2000)	05h 34m 32s
DEC (J2000)	22d 00m 52s
Apparent Size (arcmins)	7.0 x 4.8
Radius (light years)	5.5
Other Designation	Sharpless 244

Messier # 001

Date:	Time:

Site:		
Temp:	Wind:	Hum:
Clouds:	Moon:	
Scope:		
EP:	Mag:	
NELM:	See/Trans:	
Type:	# Stars:	
Mag:	Age:	
Const:		

Notes:

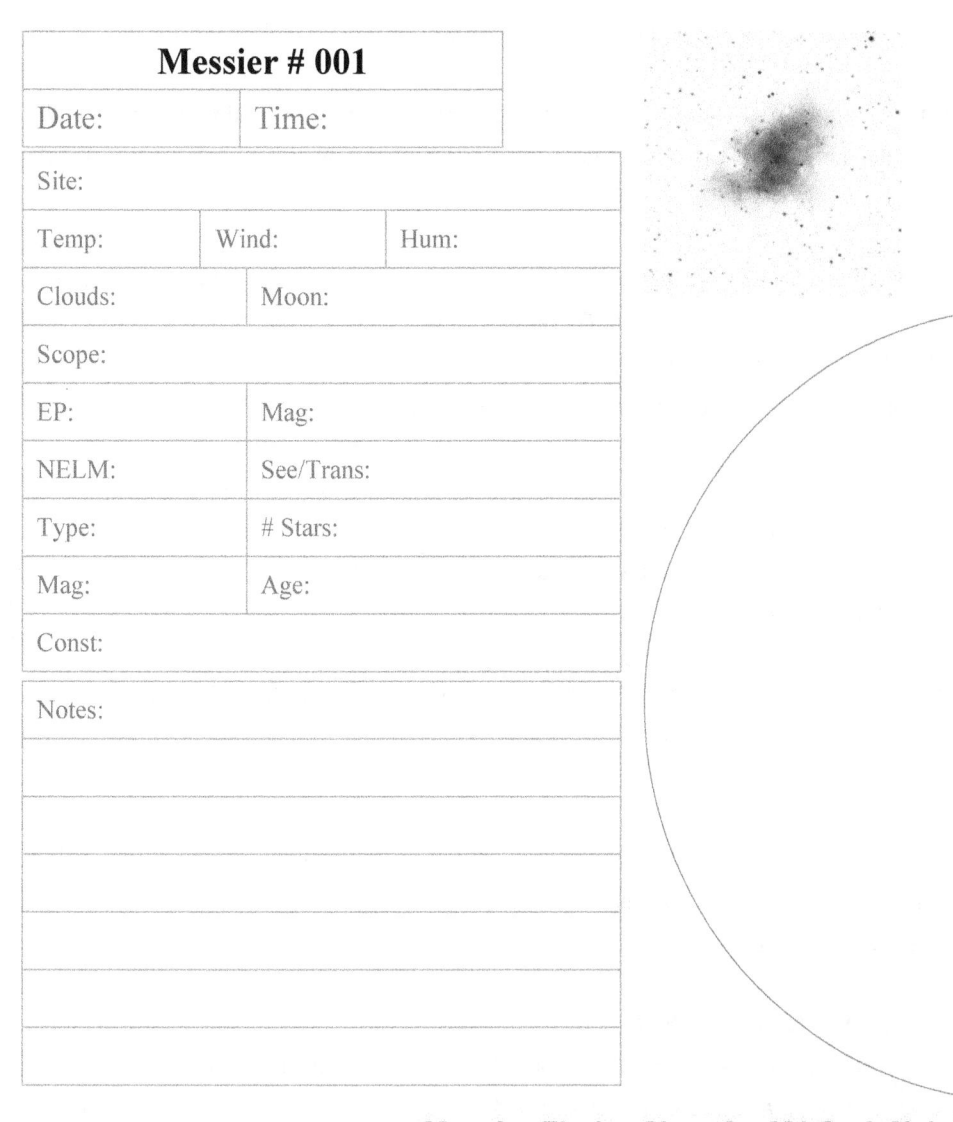

Messier Finder Chart for M1 Crab Nebula

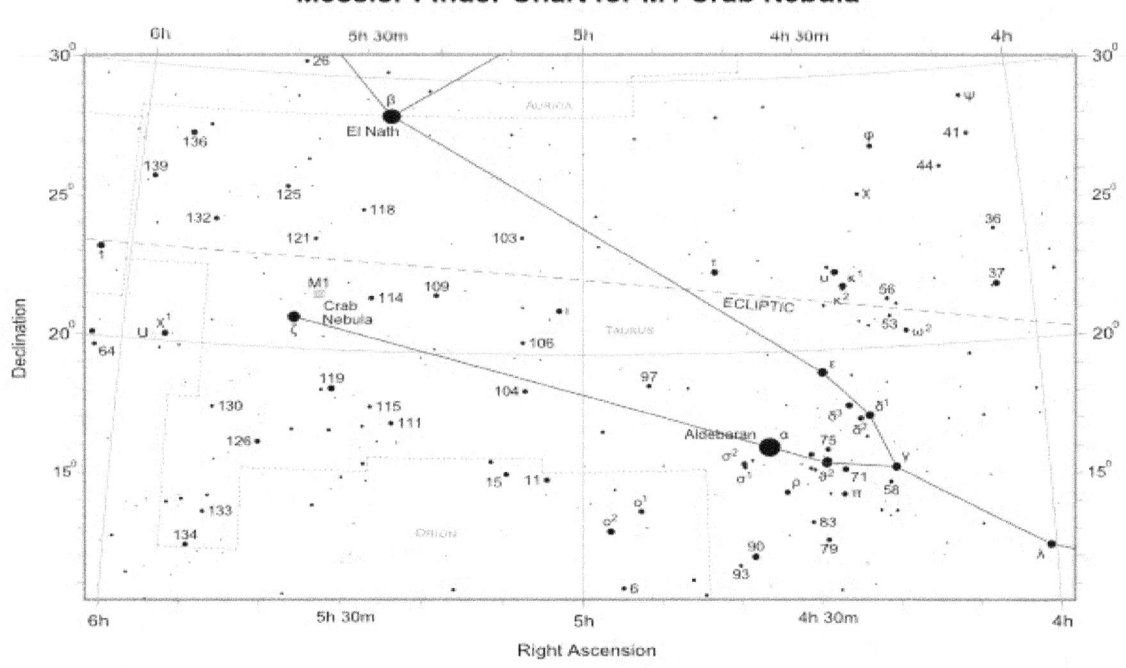

Messier 2 - M2 - Globular Cluster

M2 is a rich compact globular cluster at the edge of naked eye visibility that is located in the constellation of Aquarius. It's an easy binocular object (mag. +6.3), but requires a dark site with extremely good seeing conditions to be spotted with the naked eye.

M2 was discovered on September 11, 1746 by French astronomer Jean-Dominique Maraldi while observing a comet with Jacques Cassini, the son of legendary Italian astronomer Giovanni Cassini. Charles Messier then rediscovered the cluster exactly 14 years later on the September 11, 1760. He described it, like many of his other objects, "as a nebula without any stars associated with it". William Herschel was the first to resolve individual stars in the cluster, in 1783.

Finding M2 can be initially challenging as it's located in a relatively sparsely populated area of sky. The key is to locate the two brightest stars in Aquarius, Sadalsuud (β Aqr - mag. +2.90) and Sadalmelik (α Aqr - mag. +2.95). Sadalsuud is positioned 10 degrees southwest of Sadalmelik. Located 5 degrees north of Sadalsuud is M2. Together the two stars and the globular cluster form a large right-angled triangle.

The finder chart below shows the position of M2. Since M2 straddles the celestial equator it's visible from all over the World, although it appears higher in the sky when viewed from tropical latitudes. The globular is best seen between July and October.

Through 7x50 or 10x50 binoculars, M2 appears obviously non-stellar, like an out of focus star. A small 80mm (3.1-inch) telescope at high powers reveal a bright centre surrounded by a small fuzzy halo in an otherwise barren field of view. Close by is a pair of 6th/7th magnitude white stars that point almost directly towards M2.

A medium size 200mm (8-inch) telescope will resolve some of the outer stars, of which the brightest are of apparent magnitude +13.1. Higher magnifications certainly help with a larger 250mm (10-inch) scope resolving many more stars. The globular doesn't quite have the wow factor of M13 and the other great southern clusters, but it's still an impressive sight. Even larger telescopes of the order of 300mm (12-inch) or greater reveal many stars scattered across the complete face of M2. The globular also appears elliptical in shape with a peculiar dark lane that crosses the north-east edge visible.

As Messier globulars go, M2 is relatively distant at 37,500 light-years. However, it's intrinsically large, therefore despite the distance it still covers a respectable 16 arcminutes of apparent sky when viewed from Earth. The cluster is estimated to be 13,000 Million years old with about 150,000 stars in total.

M2 Data Table

Messier	2		DEC (J2000)	-00d 49m 23s
NGC	7089		Apparent Size (arcmins)	16 x 16
Object Type	Globular cluster		Radius (light years)	87
Constellation	Aquarius		Age (years)	13,000M
Distance (kly)	37.5		Number of Stars	150,000
Apparent Mag.	6.3			
RA (J2000)	21h 33m 29s			

Messier # 002

Date:		Time:	
Site:			
Temp:	Wind:		Hum:
Clouds:		Moon:	
Scope:			
EP:		Mag:	
NELM:		See/Trans:	
Type:		# Stars:	
Mag:		Age:	
Const:			

Notes:

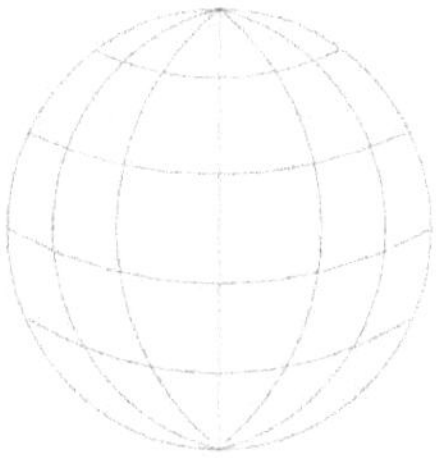

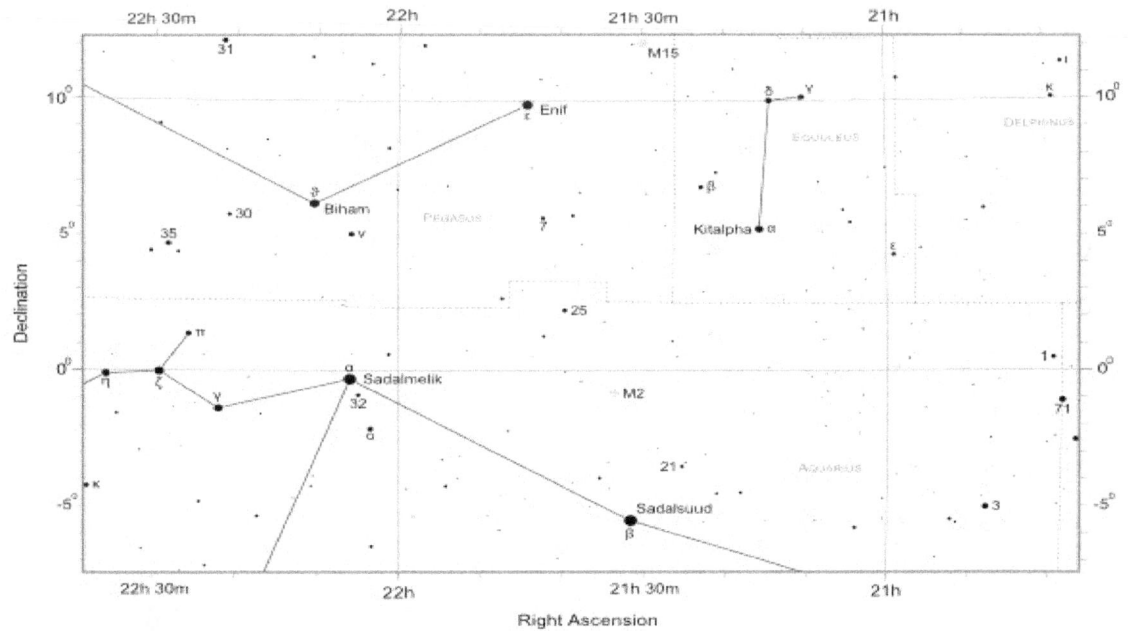

Messier Finder Chart for M2
Also shown M15

Messier 3 - M3 - Globular Cluster

M3 is a fine globular cluster located in the northern constellation of Canes Venatici. It's widely considered by amateur astronomers to be one of the best examples of its type in the northern section of the sky, beaten only by M13 "the Great Hercules Cluster". With an apparent magnitude of +6.2, M3 is beyond naked eye visibility except from an extremely dark site. However, it's easily visible with binoculars, appearing distinctly non-stellar unlike the stars surrounding it. The cluster is best seen from northern latitudes during the months of March, April and May.

M3 lies in the southern part of Canes Venatici, practically on the border with Boötes. It's located 12 degrees northwest of orange giant star Arcturus (α Boo - mag. -0.05) and about halfway along an imaginary line connecting Arcturus with Cor Caroli (α CVn - mag. +2.9). The area of sky surrounding M3 is rather barren, but when observed through binoculars a number of 6th and 7th magnitude stars are visible along with the cluster itself. One such 6th magnitude star is located just 0.5 degrees southwest of M3.

Popular 7x50 and 10x50 binoculars show M3 as a hazy spot, not unlike an out of focus star. When viewed under good conditions through a 80mm (3.1 inch) telescope or 20x80 binoculars it appears as a large diffuse ball with a hint of graininess. Medium size scopes of 150mm (6-inch) or 200mm (8-inch) aperture reveal the globular as a large and bright diffuse ball of mottled light with a noticeably brighter core. Under dark skies a number of stars are resolvable towards the outer edge, with averted vision revealing many more. Through larger telescopes of the order of 300mm (12-inch) or greater, M3 is a spectacular sight covering much - if not all - of the field of view. The component stars are scattered in every direction and under black skies it's resolvable to the core. The apparent size is 18 arc minutes.

M3 was discovered by Charles Messier on May 3, 1764 and first resolved into stars by William Herschel around 1784. It is located about 33,900 light-years from Earth and is a large globular cluster, estimated to contain at least 500,000 stars. With an estimated age of about 8 billion years, M3 is young. It also contains a large number of variable stars; at least 274 exist - more than any other globular - of which 133 have been determined to be of the RR Lyrae type.

M3 Data Table

Messier	3	**(arcmins)**	
NGC	5272	**Radius (light years)**	90
Object Type	Globular cluster	**Age (years)**	8,000M
Constellation	Canes Venatici	**Number of Stars**	>500,000
Distance (kly)	33.9	**Notable Features**	A much studied globular cluster. It contains at least 274 variables stars.
Apparent Mag.	6.2		
RA (J2000)	13h 42m 11s		
DEC (J2000)	28d 22m 32s		
Apparent Size	18 x 18		

Messier # 003

Date:	Time:

Site:

Temp:	Wind:	Hum:

Clouds:	Moon:

Scope:

EP:	Mag:
NELM:	See/Trans:
Type:	# Stars:
Mag:	Age:
Const:	

Notes:

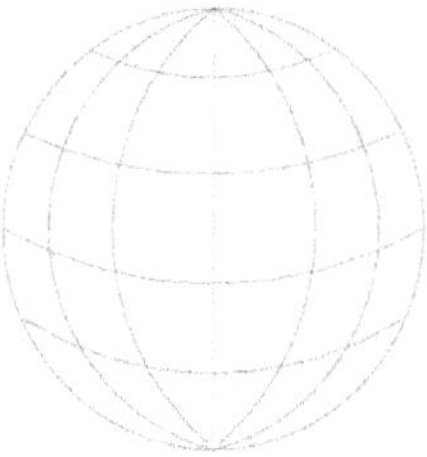

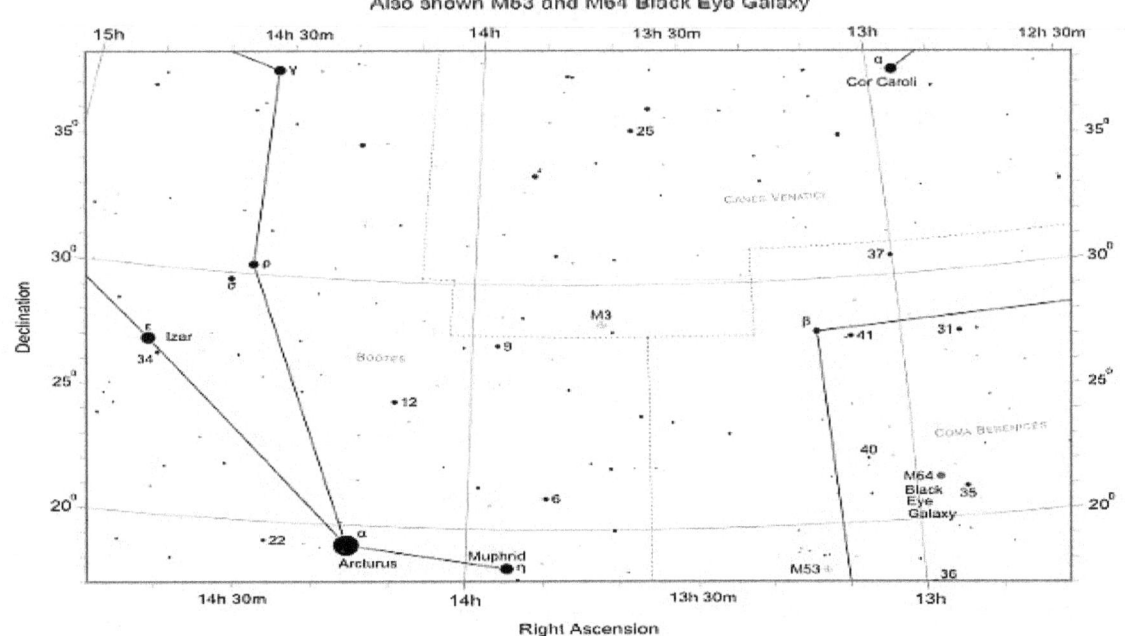

Messier Finder Chart for M3
Also shown M53 and M64 Black Eye Galaxy

Declination

Right Ascension

Star magnitudes: -1 0 1 2 3 4 5 6 7

Double star   Variable stars   Galaxy   Globular cluster

Messier 4 - M4 - Globular Cluster

M4 is a large apparent size magnificent globular cluster that's located amongst the rich Milky Way star fields of the constellation of Scorpius, "the Scorpion". With an apparent magnitude of +5.6, it can be spotted with the naked eye from dark skies. Even the simplest optical aid, reveals it as an obvious non-stellar fuzzy object. In addition, not only is M4 bright but it's also one of the easiest globulars to find, located just 1.3 degrees west of the brightest star in Scorpius, striking red giant Antares (α Sco – mag. +1.0).

M4 was discovered by Philippe Loys de Chéseaux in 1746, who listed it as number 19 in his catalog. It was also included in Lacaille's catalog as Lacaille I.9 and subsequently added by Charles Messier to his famous list on May 8th, 1764. M4 holds the distinction of being the first globular cluster ever to be resolved into stars by a telescope. Messier himself achieved this task, describing it "as a cluster of very small faint stars". Incidentally this was the only globular cluster that he could manage to resolve with his modest instruments.

The finder chart below shows the position of M4 in Scorpius. The constellation is best seen from tropical and southern hemisphere latitudes where it appears high in the sky during May, June and July. At least part of the constellation can be seen from most northern hemisphere latitudes, although the constellation never rises particularly high above the southern horizon.

On close inspection to the naked eye under dark skies, M4 appears as a slightly unusual star. Through 10x50 binoculars, the centre of the cluster is relatively bright, surrounded by a thin halo of light that hints at resolution. In reality, M4 is an easy globular to resolve in small scopes due to its large size (26 arc minutes) and loose unconcentrated structure. On nights of good seeing, a 100mm (4-inch) scope at high magnifications resolves the entire face of the cluster into a multitude of stars. Medium to large aperture telescopes reveal a mass of stars including a strange central bar type structure from roughly below left to above right of center that was first noted way back in 1783 by Sir William Herschel. The bar consists of 11th magnitude stars.

The reason why M4 is bright and large is simply due to its distance. At only 7,200 light years it's a stone's throw away, making it one of the closest globular clusters to our Solar System. Currently the only other real contenders for the title of closest globular are NGC 6397 in the far southern constellation of Ara "The Altar", which is about the same distance as M4 and the recently discovered (in 2006) FSR 1767 globular at 4,900 light years.

M4 Data Table

Messier	4	**Apparent Size (arcmins)**	26 x 26
NGC	6121	**Radius (light years)**	27
Object Type	Globular cluster	**Age (years)**	12,200M
Constellation	Scorpius	**Number of Stars**	>20,000
Distance (kly)	7.2	**Notable Feature**	Located only 1.3 degrees west of Antares
Apparent Mag.	5.6		
RA (J2000)	16h 23m 35s		
DEC (J2000)	-26d 31m 32s		

Messier # 004

Date:	Time:

Site:

Temp:	Wind:	Hum:

Clouds:	Moon:

Scope:

EP:	Mag:
NELM:	See/Trans:
Type:	# Stars:
Mag:	Age:

Const:

Notes:

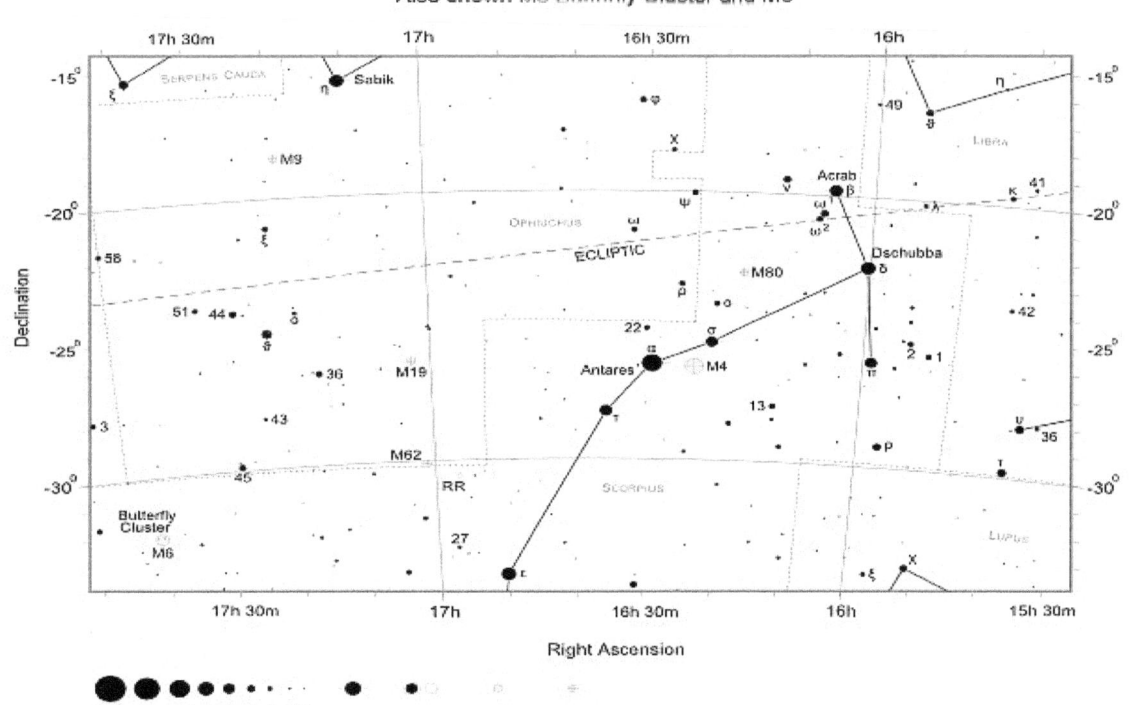

Messier Finder Chart for M4, M19, M62 and M80
Also shown M6 Butterfly Cluster and M9

Declination

Right Ascension

Star magnitudes -1 0 1 2 3 4 5 6 7

Double star   Variable stars   Open cluster   Globular cluster

Messier 5 - M5 - Globular Cluster

M5 is a superb globular cluster that's located close to the celestial equator in the constellation of Serpens (Caput). Shining at magnitude +5.7, it's visible to the naked eye under dark skies, appearing as a faint "star". The cluster is large, covering 23 arc-minutes of apparent sky. At a distance of 24,500 light-years from Earth this corresponds to a spatial diameter of 160 light-years, making M5 one of the largest globulars in terms of both apparent and actual size.

It was discovered by Gottfried Kirch and his wife Maria Margarethe on the May 5, 1702. At the time, the couple were observing a comet when they stumbled across the cluster and noted it as a star with nebulosity. Charles Messier found it independently 62 years later on the May 23, 1764, describing it as a round nebula which "doesn't contain any stars". The first person to resolve the cluster into stars was William Herschel, who counted 200 of them using his 40-foot (12.2 meter) focal length reflector in 1791.

M5 is located 23 degrees southeast of orange giant star Arcturus (α Boo) which at mag. −0.04 is the fourth brightest star. It's positioned 4 degrees east of 110 Her (mag. +4.4) and next to a small triangle of 6th magnitude stars. The northern tip of the triangle is double star 5 Ser (mag. +5.0) with M5 located just northwest of this star.

Under very good viewing conditions, M5 can be just about glimpsed with the naked eye as a faint point of light. With binoculars, it's easily visible as small fuzzy patch. A small 80mm (3.1-inch) telescope reveals a bright glowing core wrapped inside a much fainter halo of nebulosity. A 100mm (4-inch) telescope under excellent skies will start to resolve individual stars, the brightest of which are 11th magnitude. A 150mm (6-inch) scope at high power resolves more of the outer edges of M5. Through larger scopes it's a spectacular sight with thousands of stars coming into view, radiating like chains or legs outwards from the bright centre. The cluster appears noticeably elongated with a compact and somewhat clumpy core. Overall, M5 is a beautiful globular cluster that's best seen during the months of March, April and May.

With an estimated age of 13 billion years, M5 is one of the oldest Milky Way globular clusters. It's estimated to contain at least 500,000 stars. A total of 105 variable stars have been observed of which 97 belong to the RR Lyrae type. Also a dwarf nova has been observed in M5.

M5 Data Table

Messier	5
NGC	5904
Object Type	Globular cluster
Constellation	Serpens
Distance (kly)	24.5
Apparent Mag.	5.7
RA (J2000)	15h 18m 34s
DEC (J2000)	02d 04m 58s

Apparent Size (arcmins)	23 x 23
Radius (light years)	80
Age (years)	13,000M
Number of Stars	>500,000
Notable Feature	A dwarf nova has been observed in this globular

Messier # 005

Date:	Time:
Site:	

Temp:	Wind:	Hum:

Clouds:	Moon:
Scope:	
EP:	Mag:
NELM:	See/Trans:
Type:	# Stars:
Mag:	Age:
Const:	
Notes:	

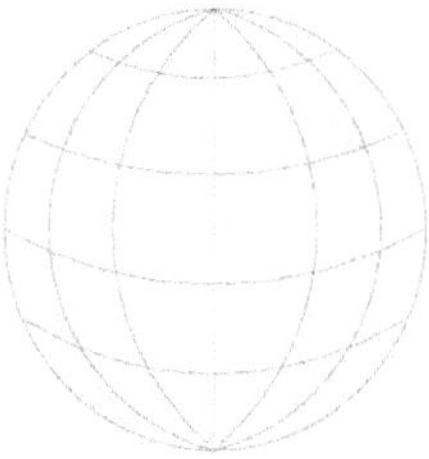

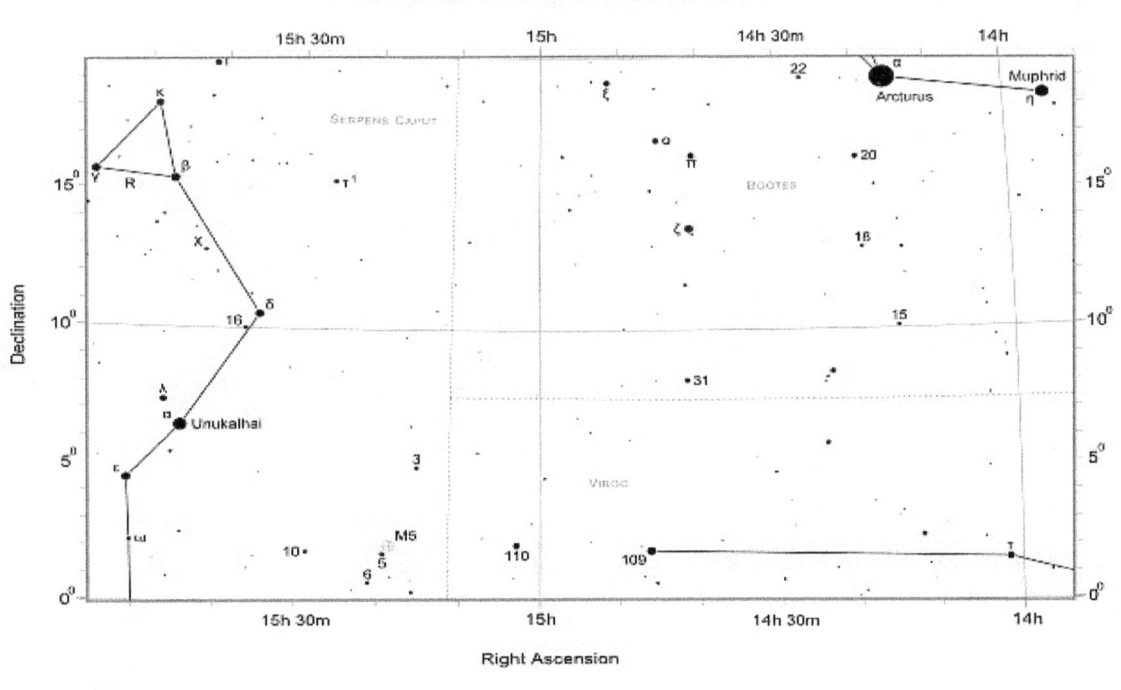

Messier Finder Chart for M5

Star magnitudes -1 0 1 2 3 4 5 6 7

Double star   Variable stars   Globular cluster

Messier 6 - M6 - The Butterfly Cluster (Open Cluster)

M6 is a superb bright naked eye open cluster in the constellation of Scorpius that's also known as the "Butterfly Cluster". This name was first coined by Robert Burnham who described it as a "charming group whose arrangement suggests the outline of a butterfly with open wings." At magnitude +4.2, it's one of the brightest open clusters in the Messier catalogue and a wonderful object for binocular and telescope owners. It covers 25 arc minutes of apparent sky and contains 80 stars. Located just a few degrees southeast of M6 - in this wonderfully rich area of the Milky Way - is an even brighter and larger open cluster, M7 "The Ptolemy Cluster".

Despite being visible to the naked eye, it's commonly believed that the first person to record the position of M6 was Giovanni Battista Hodierna in 1654. However, Robert Burnham proposed that Ptolemy might also have seen M6 with the naked eye while observing M7. Many years later, Charles Messier included both M6 and M7 in his catalogue on May 23, 1764.

M6 is located in eastern Scorpius. At the heart of the Scorpius is red supergiant star Antares (α Sco - mag. +1.0) the brightest star in the constellation and 16th brightest star in the nighttime sky. Follow the stars from Antares, curving in a southerly direction until arriving at lambda Sco (λ Sco - mag. +1.6) at the end of the tail. M6 is positioned 5 degrees north and 1.5 degrees east of this star.

From a dark site, M6 is an easy naked eye target appearing in the sky as a hazy patch that hints on resolution. With 7x50 or 10x50 binoculars the cluster is a wonderful sight. The brightest six stars are of approximately the same brightness and line up to form the beautiful butterfly shape. In a small 100mm (4-inch) telescope the cluster is awash with stars of various colors. Most of the bright stars in M6 are hot, blue B type stars but the brightest member is K type orange giant variable star, BM Sco (mag. +5.5 -> +7.0 - period ~850 days). The contrast between this star and the surrounding hot white/blue stars is striking, a wonderful sight.

M6 is 1,600 light years distant and has a spatial diameter of 12 light-years. The cluster is estimated to be 95 million years old. It's best seen during the months of June, July and August from the Southern Hemisphere, where it appears high in the sky. From northern temperate latitudes M6 never climbs particularly high above the southern horizon.

M6 Data Table

Messier	6	**DEC (J2000)**	-32d 15m 15s
NGC	6405	**Apparent Size (arcmins)**	25 x 25
Name	Butterfly Cluster	**Radius (light years)**	6
Object Type	Open cluster	**Age (years)**	95M
Constellation	Scorpius	**Number of Stars**	80
Distance (kly)	1.6	**Other Name**	Collinder 341
Apparent Mag.	4.2		
RA (J2000)	17h 40m 21s		

Messier # 006

Date:	Time:
Site:	
Temp:	Wind:	Hum:
Clouds:	Moon:
Scope:	
EP:	Mag:
NELM:	See/Trans:
Type:	# Stars:
Mag:	Age:
Const:	

Notes:

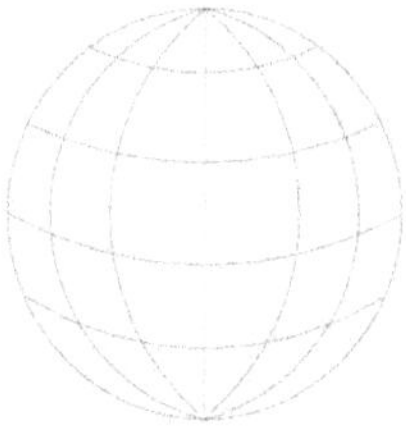

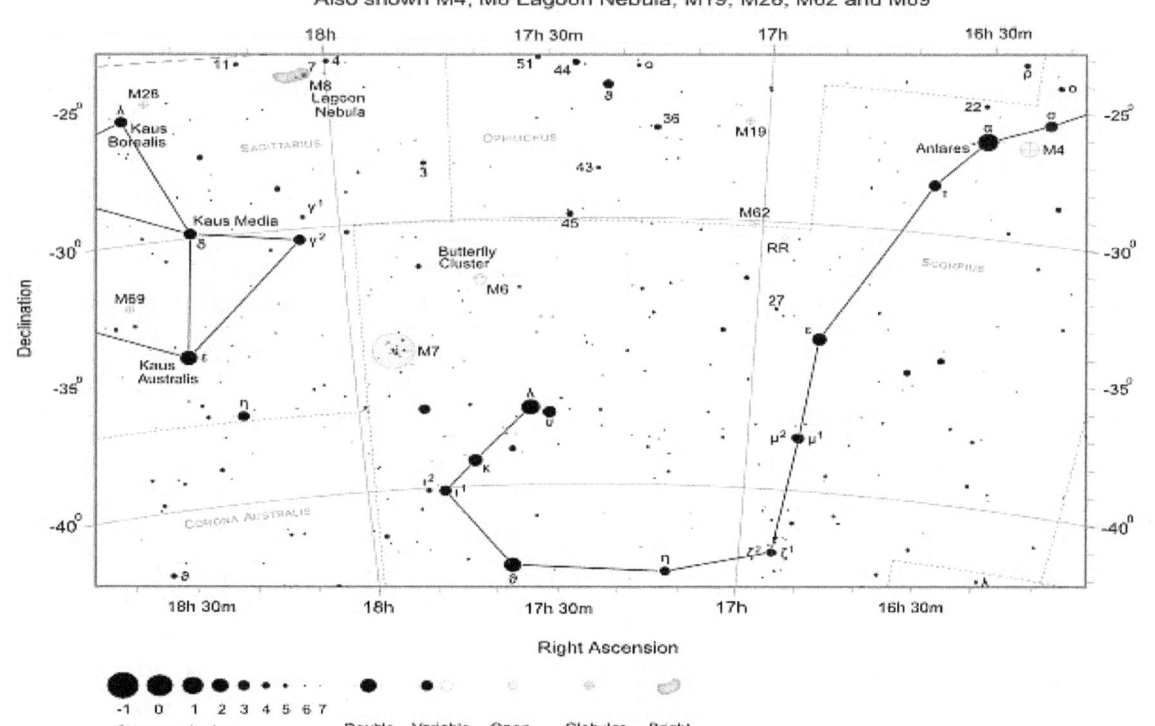

Messier Finder Chart for M6 Butterfly Cluster and M7
Also shown M4, M8 Lagoon Nebula, M19, M28, M62 and M69

Messier 7 - M7 - The Ptolemy Cluster (Open Cluster)

M7 is a large magnificent naked eye open cluster located in the constellation of Scorpius. It's one of the brightest open clusters and has been known since ancient times; Greek-Roman astronomer Ptolemy first recorded it in 130 AD. In recognition of this early observation, M7 is often referred to as "Ptolemy Cluster". Italian astronomer Giovanni Batista Hodierna observed 30 stars sometime before 1654 and Charles Messier adding it to his catalogue in 1764.

With a combined magnitude of +3.3, M7 is brightest and most obvious deep sky object in Scorpius. It's a giant group of 80 stars that has an apparent diameter of 80 arc minutes, almost 3x that of the full Moon. To the naked eye M7 appears as a very large hazy patch with its brightest stars just about resolvable. It's so bright that it's even noticeable under suburban skies. With a declination of -34.8 degrees, the cluster is the southernmost Messier object and is best seen from the Southern Hemisphere during the months of June, July and August. From most northern temperate locations it appears low down, at best climbing just a few degrees above the southern horizon. From northern locations above 56 degrees it never even rises.

M7 is located in eastern Scorpius, close to the Sagittarius border. It's positioned 4.75 degrees northeast of lambda Sco (λ Sco - mag. +1.6), the constellations second brightest star. Also known as Shaula, λ Sco is at the end of the Scorpions tail and part of the "stinger". The "Butterfly Cluster", M6 is located 4 degrees northwest of M7.

Due to its large size, M7 is best seen through binoculars or at low magnifications in a wide field telescope where it appears as a loose cluster of approximately twenty scattered stars. Averted vision resolves a few more fainter background stars. With a 200mm (8-inch) telescope even at low powers, M7 will often overflow the eyepiece field of view appearing as a very loose cluster of mainly white and blue/white stars. There is one notable exception, the brightest individual cluster member star shines at magnitude +5.6 and is an orange/yellow star of spectral type G8.

In the same field of view as M7 is the faint and distant globular cluster NGC 6453. It appears as an 11th magnitude spot of fuzzy nebulosity that spans only one arc minute across.

M7 is located about 800 years from Earth and has an actual diameter of 40 light years. It's estimated to be 220 Million years old.

M7 Data Table

Messier	7
NGC	6475
Name	The Ptolemy Cluster
Object Type	Open cluster
Constellation	Scorpius
Distance (kly)	0.8
Apparent Mag.	3.3
RA (J2000)	17h 53m 51s

DEC (J2000)	-34d 47m 34s
Apparent Size (arcmins)	80 x 80
Radius (light years)	20
Age (years)	220M
Number of Stars	80
Other Name	Collinder 354

Messier # 007

Date:	Time:
Site:	
Temp:	Wind:	Hum:
Clouds:	Moon:
Scope:	
EP:	Mag:
NELM:	See/Trans:
Type:	# Stars:
Mag:	Age:
Const:	

Notes:

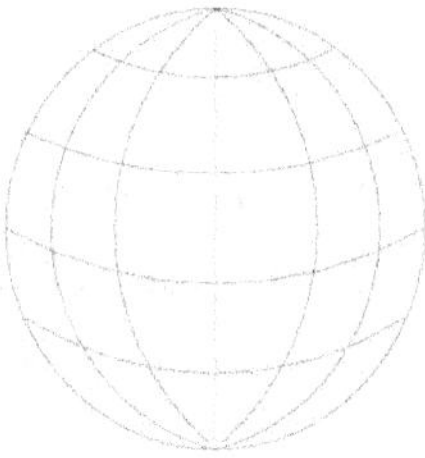

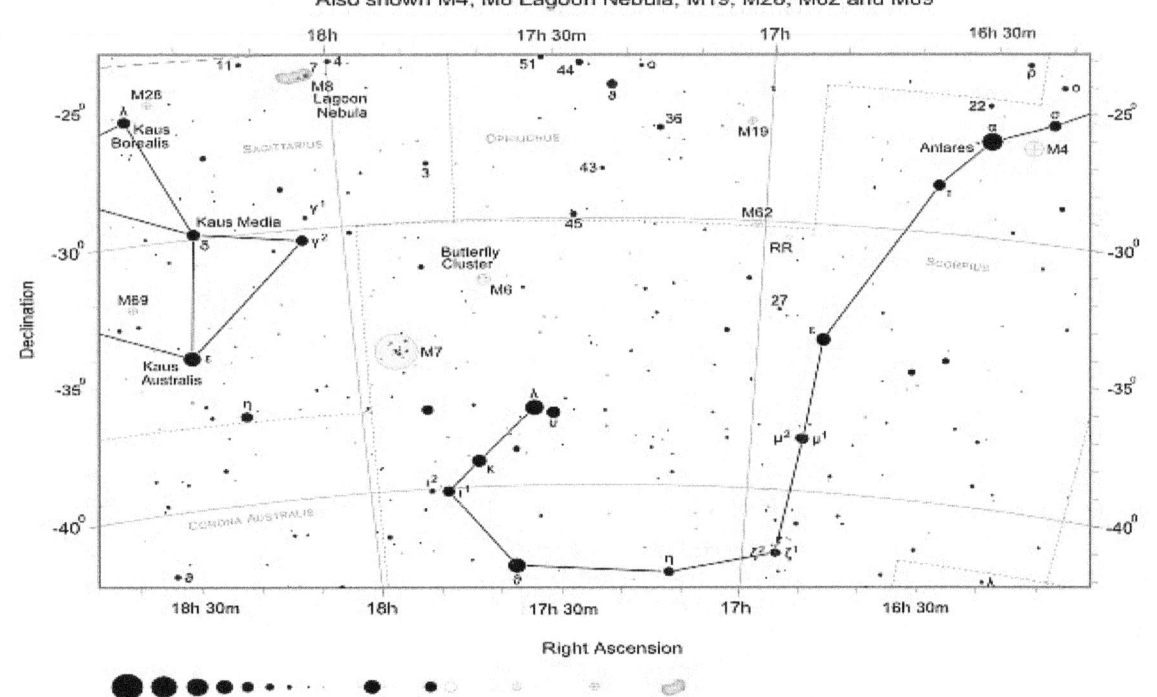

Messier Finder Chart for M6 Butterfly Cluster and M7
Also shown M4, M8 Lagoon Nebula, M19, M28, M62 and M69

Messier 8 - M8 - Lagoon Nebula (Emission Nebula)

M8 the Lagoon Nebula is a giant spectacular emission nebula in Sagittarius that's one of the brightest and finest star forming regions in the entire sky. With an apparent magnitude of +6.0, it's faintly visible to the naked eye and a wonderful sight through all types of optical instrument.

The nebula was discovered by Italian astronomer Giovanni Hodierna sometime before 1654. French astronomer Guillaume Le Gentil independently found it in 1747 before Charles Messier added the object to his catalogue on May 23, 1764. The distance to M8 is uncertain. It's currently estimated at 5,200 light-years although it might be as close as 4,100 light-years or as far away as 6,000 light-years. When seen from our perspective this is an extremely large object. It covers 90 by 40 arc minutes of apparent sky, which is many times larger than the full Moon and comparable in size to another celebrated star forming region, the Great Orion Nebula (M42). To locate M8, start by finding the bright teapot asterism of Sagittarius. The top three stars of the teapot are Kaus Borealis (λ Sgr - mag. +2.8), Kaus Media (δ Sgr - mag. +2.7) and φ Sgr (mag. +3.2). Now imagine a line connecting φ Sgr to Kaus Borealis and extend it in a westerly direction - curving slightly southwards - for about 6 degrees until arriving at M8. The Trifid Nebula (M20) is located 1.5 degrees north of M8.

The Lagoon Nebula is best seen from southern and equatorial regions during the months of June, July and August. When viewed through popular 7x50 or 10x50 binoculars, M8 appears as a cloud like, oblong shaped patch of diffuse light that's slight brighter towards the centre. A scattering of faint stars appear superimposed on the nebula. Small scopes of the order of 80mm (3.1-inch) aperture show the nebulosity is split in two distinct sections, separated by a dark dust band. Through a larger 200mm (8-inch) telescope, M8 is a wonderful sight. The cloud is better defined with a noticeable brighter core, intrigue twists and knots and numerous dark bands running through the centre. Many predominantly white stars are also visible. Positioned at the eastern edge of M8 is NGC 6530, an open cluster of bright stars. Like many nebulae when photographed or imaged M8 appears pink in color, but to the eye looking through binoculars or telescopes it's essential grey or at best has a green tint. This is a consequence of the poor color sensitivity of the retina at low light levels. Within the brightest part of the Lagoon Nebula is a feature known as the "Hourglass Nebula". It was discovered by John Herschel and occurs in a region where intensive star formation appears to be currently taking place. M8 also contains a number of Bok globules that are dark, collapsing clouds of dense dust and gas in which star formation can take place. They are known to be some of the coldest objects in the universe and have been and still are a subject of intense research. America astronomer E. E. Barnard catalogued many of these objects at the end of the 19th century.

M8 Data Table

Messier	8
NGC	6523
Name	Lagoon Nebula
Object Type	Emission Nebula
Constellation	Sagittarius
Distance (kly)	5.2
Apparent Mag.	6.0

RA (J2000)	18h 03m 41s
DEC (J2000)	-24d 22m 49s
Apparent Size (arcmins)	90 x 40
Radius (light years)	70 x 30
Other Name	Sharpless 25
Notable	Structure at centre is

Feature	known as the Hourglass Nebula (not to be confused with better known Hourglass Nebula in Musca)

Messier # 008

Date:		Time:	
Site:			
Temp:	Wind:		Hum:
Clouds:		Moon:	
Scope:			
EP:		Mag:	
NELM:		See/Trans:	
Type:		# Stars:	
Mag:		Age:	
Const:			

Notes:

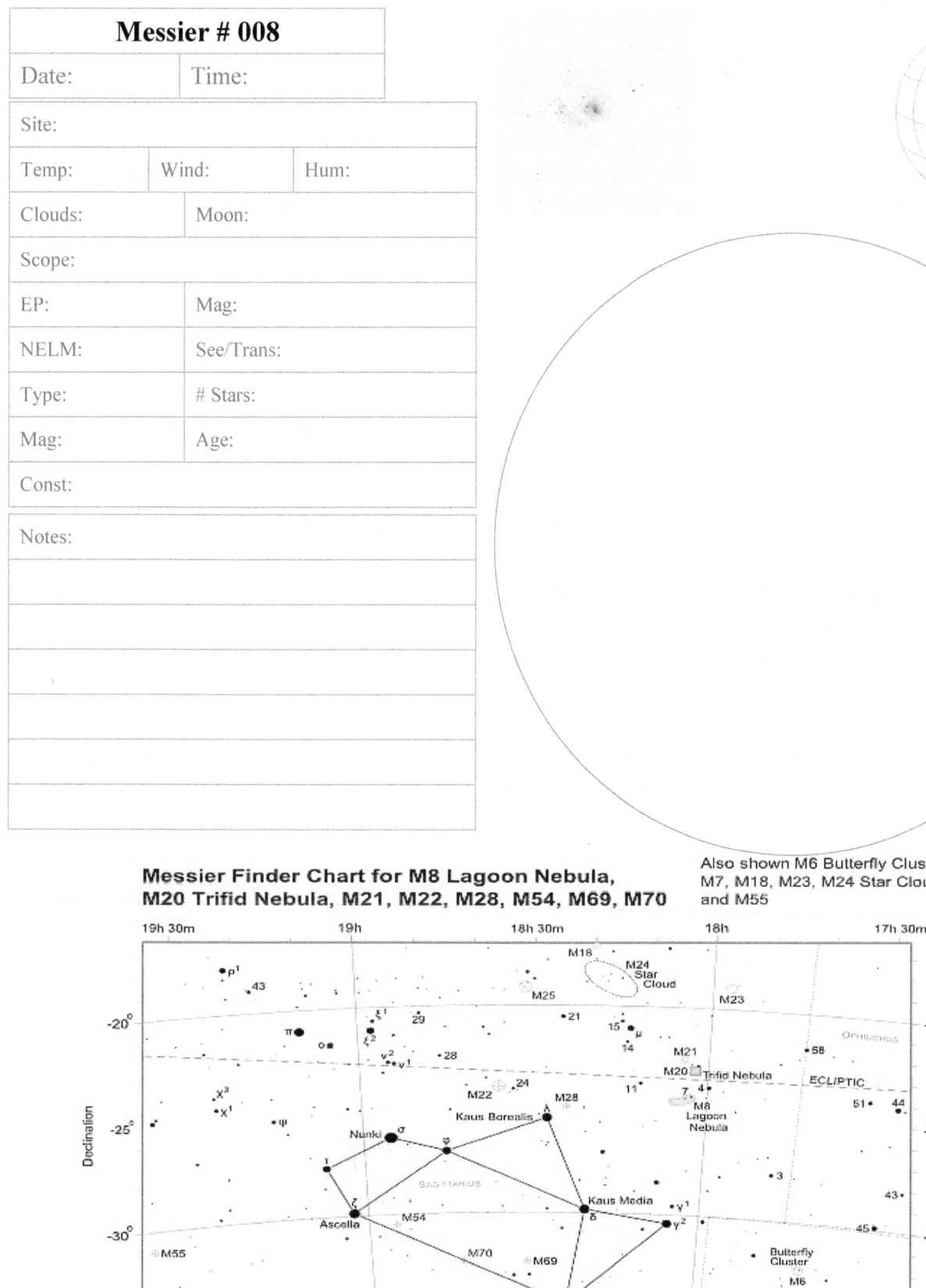

Messier Finder Chart for M8 Lagoon Nebula, M20 Trifid Nebula, M21, M22, M28, M54, M69, M70

Also shown M6 Butterfly Cluster, M7, M18, M23, M24 Star Cloud and M55

Messier 9 - M9 - Globular Cluster

M9 is a globular cluster located in the southern section of the large sprawling constellation of Ophiuchus. It was discovered by Charles Messier on May 28, 1764, who described it as a "nebula without star" of 3 arc minutes in diameter. With an apparent magnitude of +8.4, it's one of the fainter objects of its type in Messier's catalogue. Since not particularly bright M9 is a challenging object for binocular observers appearing at best as a slightly out of focus faint "star", which can be difficult to pick out against surrounding Milky Way stars. The cluster is much easier to spot with larger 15x70 or 20x80 binoculars but again not much detail is discernible.

M9 is located 25,800 light-years from Earth. At a distance of 5,500 light-years it's one of the nearer globular clusters to the center of the Milky Way Galaxy. The globular lies adjacent to a prominent dark nebula called Barnard 64, which significantly dims the light of the cluster due to intervening interstellar dust.

To find M9 start by locating Sabik (η Oph - mag. +2.4) the second brightest star in Ophiuchus. About 3 degrees southeast of Sabik is M9, which is best seen during the months of May, June and July.

When viewed through a small 80mm (3.1-inch) telescope, M9 appears as a small faint diffuse non-stellar patch of light. Even at high magnifications the cluster is not resolvable. A 200mm (8-inch) scope displays a concentrated core about 8 arc minutes in diameter surrounded by a small faint halo. At best a few stars are resolved around the edges, especially when using averted vision. Large amateur telescopes of aperture 300mm (12-inch) or greater resolve the cluster much better. Observers may be able to notice that M9 appears slightly little flattened due to the gravitational pull of our own galactic core. The brightest member stars are about magnitude +13.5.

In total, M9 has an apparent diameter of 12 arc minutes, which corresponds to an actual diameter of 90 light-years. It's estimated to be 12 billion years old and contains at least 100,000 stars. The globular is receding from us at a speed of 224 km/sec.

M9 Data Table

Messier	9
NGC	6333
Object Type	Globular cluster
Constellation	Ophiuchus
Distance (kly)	25.8
Apparent Mag.	8.4
RA (J2000)	17h 19m 12s
DEC (J2000)	-18d 30m 59s

Apparent Size (arcmins)	12 x 12
Radius (light years)	45
Age (years)	12,000M
Number of Stars	>100,000
Notable Feature	One of the closest globular clusters to the centre of the Milky Way

Messier # 009

Date:	Time:
Site:	
Temp:	Wind:	Hum:
Clouds:	Moon:
Scope:	
EP:	Mag:
NELM:	See/Trans:
Type:	# Stars:
Mag:	Age:
Const:	

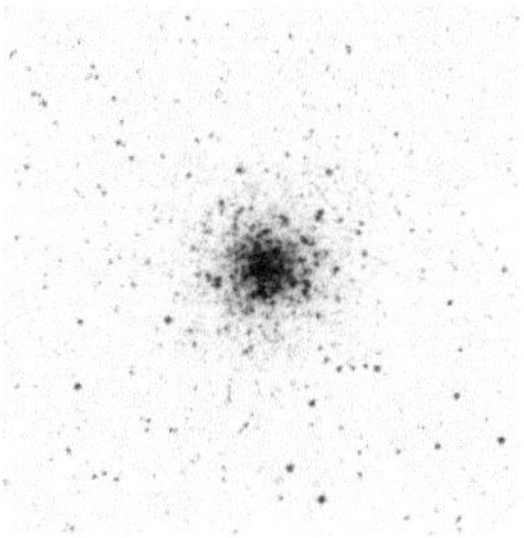

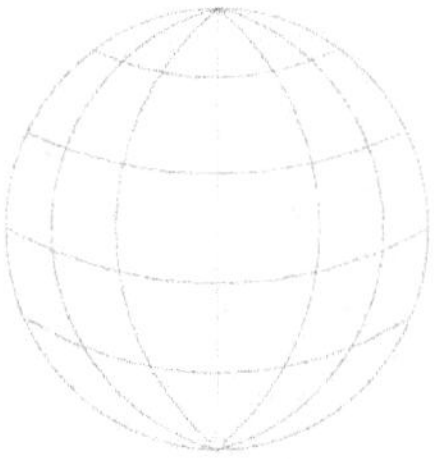

Notes:

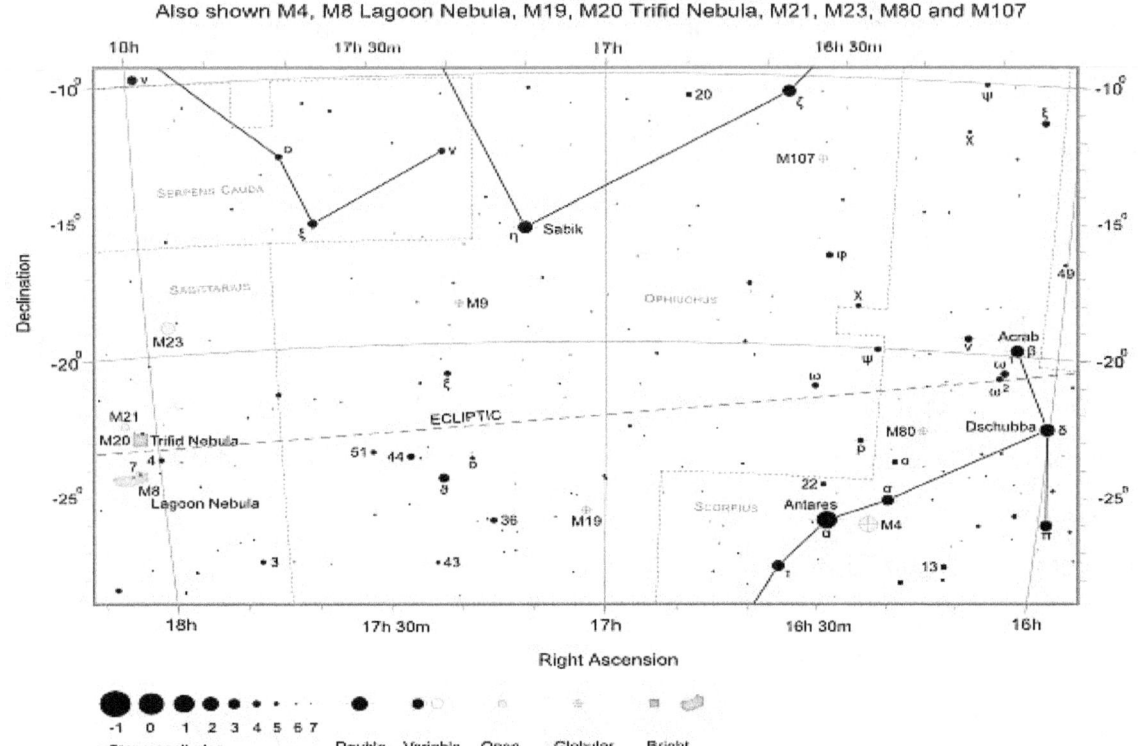

Messier Finder Chart for M9
Also shown M4, M8 Lagoon Nebula, M19, M20 Trifid Nebula, M21, M23, M80 and M107

Declination

Right Ascension

Star magnitudes
-1 0 1 2 3 4 5 6 7

Double star
Variable stars
Open cluster
Globular cluster
Bright nebulae

Messier 10 - M10 - Globular Cluster

M10 is a fine globular cluster that's located in the constellation of Ophiuchus. One of the largest constellations, Ophiuchus straddles the celestial equator and contains a host of globular clusters of which Messier catalogued seven of them. The brightest and best of them is M10 (mag. +6.6) which can be spotted with binoculars appearing like an out of focus fuzzy star.

Charles Messier discovered M10 on May 29, 1764, describing it as a "nebula without stars". Ten years later, German astronomer Johann Elert Bode noted it as a "very pale nebulous patch without stars". Both Messier and Bode used telescopes that suffered in quality and hence were unable to resolve the cluster. It was not until William Herschel using better and larger instruments was first able to spot individual member stars. He described it as a "beautiful cluster of extremely compressed stars". The best time of the year to observe M10 is during the months of May, June and July.

Locating M10 is not the easiest task as its near area of sky is devoid of bright stars. Start by locating Rasalhague (α Oph - mag +2.1) the brightest star in Ophiuchus. Join the stars of the constellation in a curve heading westwards and southwards until arriving at two close 3rd magnitude stars, Yed Prior (δ Oph - mag. +2.7) and Yed Posterior (ε Oph - mag. +3.2). M10 is located about 12 degrees east of Yed Prior with the star 30 Oph (mag. +4.8) one degree east of M10.

M10 makes a superb target for all telescopes. A small 80mm (3.1-inch) scope reveals an obvious non-stellar fuzzy ball of light spread over about 8 arc minutes. Through 150mm (6-inch) or 200mm (8-inch) instruments the globular appears dense with a large nebulous central core. A number of stars are resolvable especially in the outer halo that extends some 15-arc minutes in diameter. A larger 300mm (12-inch) telescope reveals many more stars and adds volume to the view. On good nights, it's possible to see stars across the entire face of the cluster. The fainter and smaller globular cluster M12 is located 3 degrees northwest of M10. In total M10 spans about 20 arc minutes of apparent sky.

M10 is 14,300 light years distant and is estimated to be 12,000 million years old. Although relatively close as far as globulars go, it's only 84 light years diameter and intrinsically small and therefore not as spectacular as might be expected. For example, other globulars such as M13 in Hercules are a much grander sight despite been much further away .It's estimated M10 contains 100,000 stars. However, an extremely low number are variable stars and to date only 4 have been discovered.

M10 Data Table

Messier	10	**DEC (J2000)**	-04d 05m 58s
NGC	6254	**Apparent Size (arcmins)**	20 x 20
Object Type	Globular cluster	**Radius (light years)**	42
Constellation	Ophiuchus	**Age (years)**	11,400M
Distance (kly)	14.3	**Number of Stars**	100,000
Apparent Mag.	6.6		
RA (J2000)	16h 57m 09s		

Messier # 010	
Date:	Time:
Site:	
Temp:	Wind:	Hum:
Clouds:	Moon:
Scope:	
EP:	Mag:
NELM:	See/Trans:
Type:	# Stars:
Mag:	Age:
Const:	
Notes:	

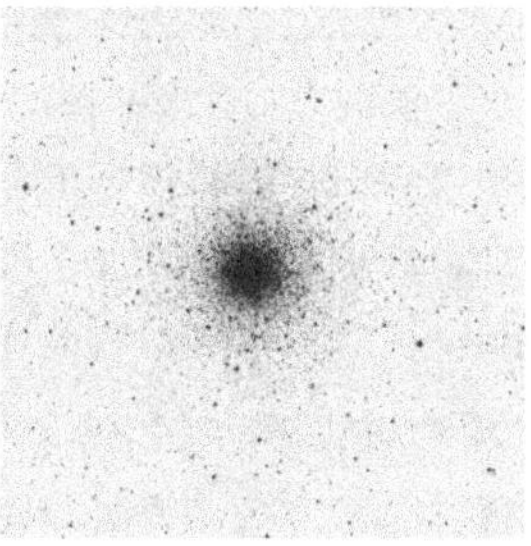

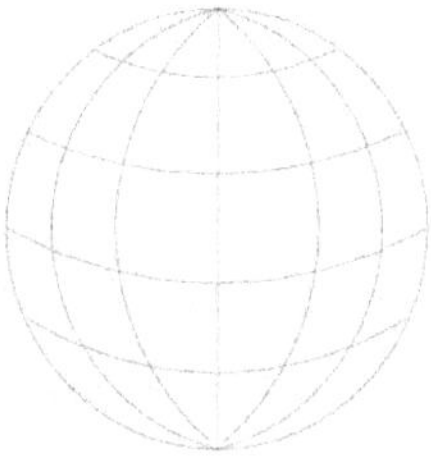

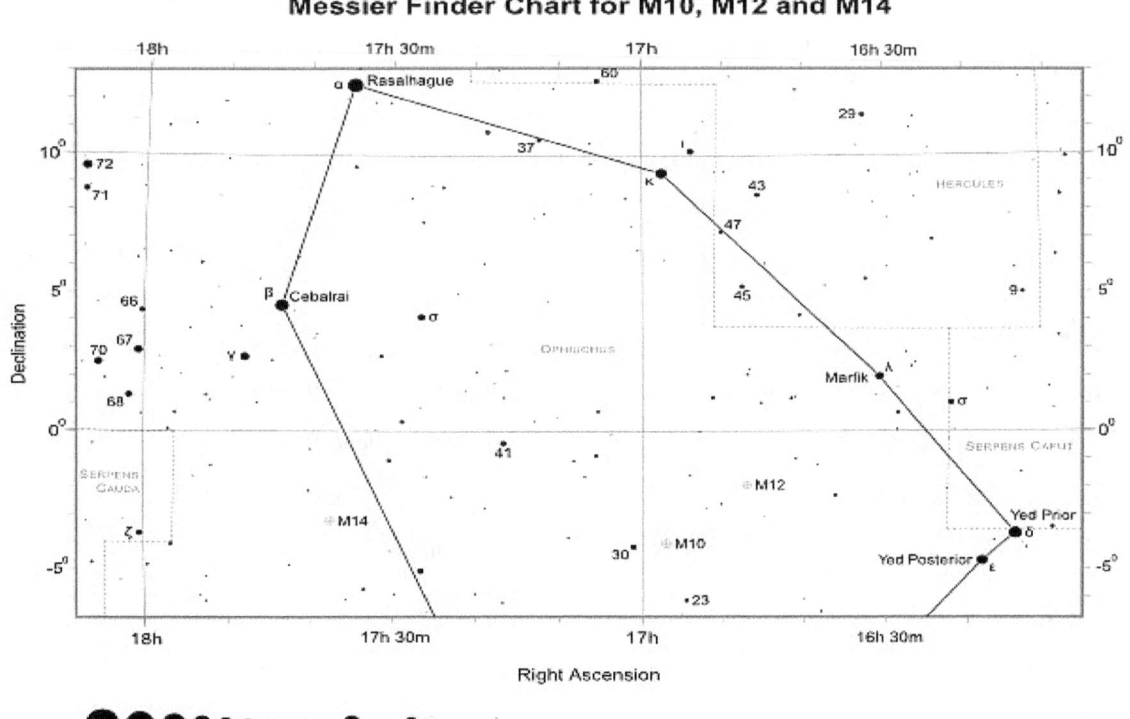

Messier Finder Chart for M10, M12 and M14

Star magnitudes -1 0 1 2 3 4 5 6 7

Double star   Variable stars   Globular cluster

Messier 11 - M11 - Wild Duck Cluster - Open Cluster

M11, the Wild Duck Cluster is a famous open cluster located in the constellation of Scutum. It's just beyond naked eye visibility but easily visible with binoculars and an outstanding telescope object. The brightest stars form a triangle that has been likened to a flock of flying ducks, hence the name "Wild Duck Cluster". Of all know open clusters, M11 is one of the richest and most compact with about 2900 members spread over a spatial diameter of 25 light-years. An observer located inside the group would have a fantastic night sky view, with several hundred first magnitude stars visible.M11 was discovered by German astronomer Gottfried Kirch of the Berlin observatory in 1681. English clergyman William Derham is believed to have been the first person to resolve it into stars (around 1733), with Charles Messier adding it to his catalogue on May 30, 1764.

The cluster is an easy target to find despite been located in the small and dim constellation of Scutum, whose brightest stars are of only 4th magnitude. The starting point on the way to the "Wild Duck Cluster" is to locate Altair (α Aql - mag. 0.8), the brightest star in Aquila and 12th brightest star in the night sky. Altair forms the southern corner of the famous "Summer triangle" along with first magnitude stars, Vega in Lyra and Deneb in Cygnus. To the southwest of Altair is mag. 3.4 star delta Aql (δ Aql). First imagine a line connecting Altair with δ Aql and then curve this line southwards for about the same distance again until you reach two 4th magnitude stars, λ Aql and 12 Aql. Located just over two degrees west of 12 Aql is the "Wild Duck cluster".M11 shines at magnitude +6.3 and is an easy binocular object; appearing as a diffuse fuzzy ball with a noticeably brighter center. Its appearance is not unlike that of a loose globular cluster and is often mistake as one. A 100mm (4-inch) telescope will resolve M11 into a swarm of mainly sparkling white stars, with one bright standout near the group's centre. When viewed through a larger 200mm (8-inch) scope, M11 is a spectacular sight with many hundreds of stars visible including some yellow and red stars. Since it's compact, covering 14 arc minutes in diameter, the magnification can be pushed up high to tease out more stars and details. In total, there are estimated to be at least 850 stars brighter than magnitude +17 in M11.The Wild Duck cluster is a tantalizing and fantastic rich open cluster when viewed through any type of optical instrument. It's best seen during the months of June, July, August and September. The name "The Wild Duck Cluster" was provided by British Admiral William Smyth who imagined the distinct V shape of the cluster as a flock of flying ducks. Since the cluster is one of the richest of its type, the density of stars in M11 has been the subject of discussion for decades. It appears that there are some 80 stars or so per cubic parsec in M11 and an observer inside this group would see several hundred first magnitude stars in his night sky.

This is a truly wonderful open cluster and a must see on all observers lists.

M11 Data Table

Messier	11		**DEC (J2000)**	-06d 16m 12s
NGC	6705		**Apparent Size (arcmins)**	14 x 14
Name	Wild Duck Cluster		**Radius (light years)**	12.5
Object Type	Open cluster		**Age (years)**	220M
Constellation	Scutum		**Number of Stars**	2900
Distance (kly)	6.2		**Other Name**	Collinder 391
Apparent Mag.	6.3			
RA (J2000)	18h 51m 06s			

Messier # 011

Date:	Time:
Site:	
Temp:	Wind:	Hum:
Clouds:	Moon:
Scope:	
EP:	Mag:
NELM:	See/Trans:
Type:	# Stars:
Mag:	Age:
Const:	
Notes:	

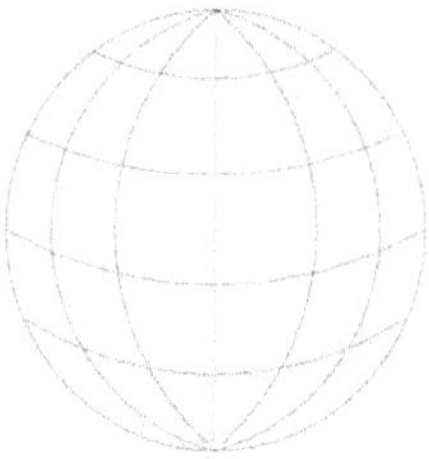

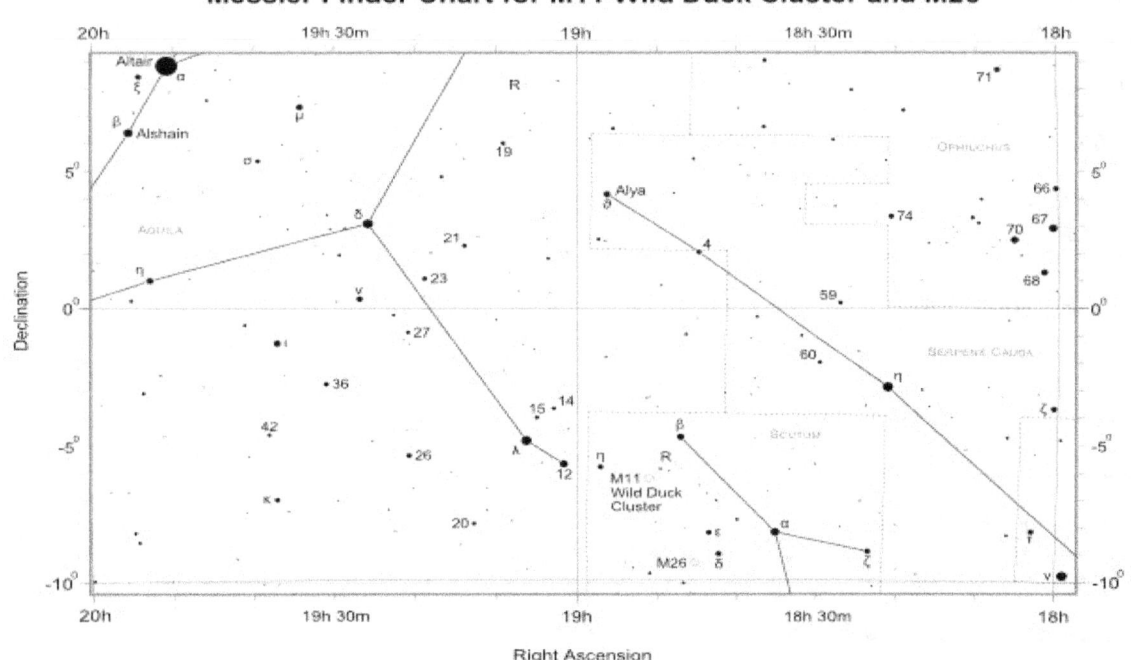

Messier Finder Chart for M11 Wild Duck Cluster and M26

Right Ascension

Declination

Star magnitudes
-1 0 1 2 3 4 5 6 7

Double star
Variable stars
Open cluster

Messier 12 - M12 - Globular Cluster

M12 is a magnitude +7.2 globular cluster in Ophiuchus that was discovered by Charles Messier on May 30, 1764. Messier was unable to resolve the cluster, describing it only as "nebula without stars". It was not until 1783 when William Herschel was the first to accomplish the task. Through good binoculars it appears as a faint hazy patch of light that's not well defined. Positioned nearby is the slightly brighter but similar looking globular M10. The two clusters are amongst the brightest of the seven Messier globulars located in Ophiuchus.

M12 is located in a barren area of sky that's devoid of bright stars and therefore finding it can require some patience. Start by locating Rasalhague (α Oph - mag +2.1) the brightest star in Ophiuchus. Join the stars of Ophiuchus in a curve heading westwards and southwards until arriving at two close together 3rd magnitude stars, Yed Prior (δ Oph - mag. +2.7) and Yed Posterior (ε Oph - mag. +3.2). M12 is located about 8 degrees northeast of these two stars. Positioned 3.25 degrees southeast of M12 is M10, with the star 30 Oph (mag. +4.8) located one degree east of M10.

The best time of the year to observe M12 is during the months of May, June and July.

M12 is not a particularly concentrated globular and was once believed to be an intermediate type object, something between a globular cluster and a dense open cluster (M11 for example). A small 80mm (3.1-inch) telescope reveals a fuzzy mottled ball that's obviously non-stellar. At medium to high powers, a 200mm (8-inch) scope resolves the clusters brightest stars with at least twenty pinpoints of light scattered throughout the cluster. Also visible are lines and branches of stars extending outwards from the diffuse center core. The core itself is not particularly bright or well defined. In total the globular covers 16 arc minutes of apparent sky and on good nights, large amateur telescopes show stars across the entire face of the cluster.

M12 is located at a distance of 18,000 light years and is estimated to be 12.6 billion years old. It has a spatial diameter of 80 light years and contains 70,000 stars. The brightest member star is of apparent magnitude +12.0.

M12 Data Table

Messier	12
NGC	6218
Object Type	Globular cluster
Constellation	Ophiuchus
Distance (kly)	18
Apparent Mag.	7.2
RA (J2000)	16h 47m 14s
DEC (J2000)	-01d 56m 52s
Apparent Size (arcmins)	16 x 16
Radius (light years)	40
Age (years)	12,600M
Number of Stars	70,000

Messier # 012

Date:	Time:

Site:		
Temp:	Wind:	Hum:
Clouds:	Moon:	
Scope:		
EP:	Mag:	
NELM:	See/Trans:	
Type:	# Stars:	
Mag:	Age:	
Const:		

Notes:

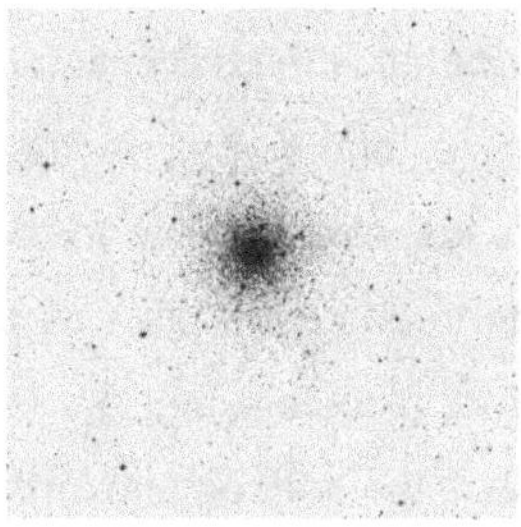

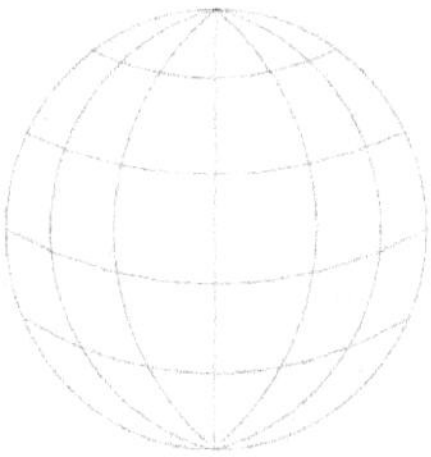

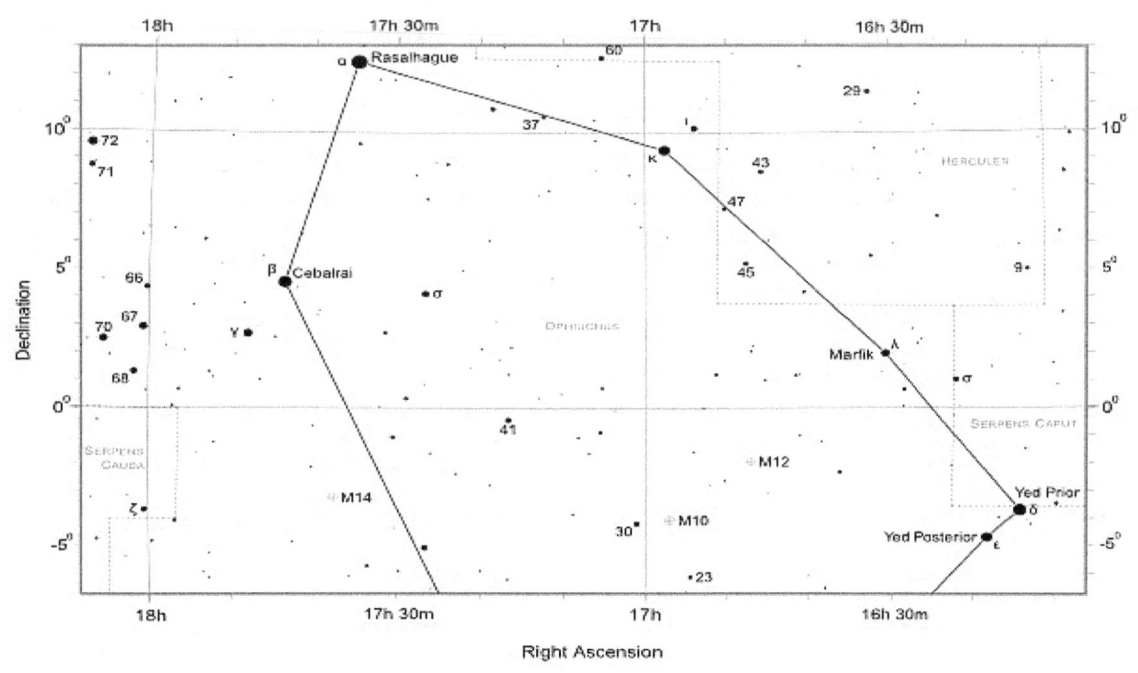

Messier Finder Chart for M10, M12 and M14

18h   17h 30m   17h   16h 30m

α Rasalhague

60

29

37

10°

72
71

ι

κ

43

HERCULES

47

Declination

5°

66

β Cebalrai

σ

45

9

67
70

γ

OPHIUCHUS

Marfik

λ

68

σ

0°

SERPENS CAPUT

41

SERPENS
CAUDA

M12

M14

ζ

Yed Prior
δ

30

M10

Yed Posterior
ε

-5°

23

18h   17h 30m   17h   16h 30m

Right Ascension

Star magnitudes: -1 0 1 2 3 4 5 6 7

Double star   Variable stars   Globular cluster

Messier 13 – M13 - Great Hercules Globular Cluster

M13 is a spectacular globular cluster and arguably the best example of its type in the northern section of the sky. It is the standout deep sky object in the constellation of Hercules and sometimes referred to as the Great Hercules Globular Cluster. At magnitude +5.8, M13 is an easy binocular target that's just about visible to the naked eye under dark skies. The popularity of M13 is mainly due to it declination; it lies at 36 degrees north and therefore well placed and often overhead during summer months for Northern Hemisphere observers. There are many other globulars that are larger and brighter than M13, but all are located in the southern section of the sky and either invisible or low on the horizon for North America, European and many Asian sky watchers. As a result, M13 is perhaps the most observed and studied globular cluster of all. The globular is easily found on the western side of the Hercules "Keystone" asterism, 2.5 degrees south of Eta Herculis (η Her) along a line connecting Eta Herculis (η Her) with Zeta Herculis (ζ Her). When viewed through 10x50 binoculars, it appears as bright fuzzy ball with a well-defined center that is obviously non-stellar but without resolution. It forms a right angled triangle with two nearby 7th magnitude stars. An 80 mm (3.1-inch) telescope shows M13 as a uniform extended hazy disk about 8 arc minutes across. At magnifications of about 100x the cluster appears like a zoomed in version of that of the binocular view. A 100mm (4-inch) telescope will resolve some of the outer stars with many more visible in 150mm (6-inch) and 200mm (8-inch) instruments. The brightest star in M13 is variable star V11 (apparent magnitude +11.95).In large amateur telescopes, M13 is truly sensational sight with the complete field awash with stars. When viewed through a 250mm (10-inch) telescope or larger, there are hundreds of stars visible against the dark background sky. The cluster appears 3-dimensional and breathtaking. In total, it has an apparent size of about 20 arc minutes, though visually it appears smaller, perhaps 12-13 arc minutes across. There are a couple of curious affects visible at a high magnification. Many of the outer stars seem to be arranged in long arcs weaving their way across the cluster face and the distribution of the bright stars is not even across the surface. This can result in an optical illusion of apparent voids or relatively barren areas interspersed across the cluster. Also visible at medium/high magnifications in telescopes of the order of 300mm (12-inch) or greater, are three dark dust lanes that form a Y shape towards the southeast of the core. This is known as the "propeller", first noticed by Bindon Stoney in the 1850s from Birr Castle in Ireland using the 72-inch reflector, the largest telescope in the world at the time.

M13 was discovered by the then Astronomer Royal, Sir Edmond Halley in 1714. He described it as, "a little patch, but shows itself to the naked eye, when the sky is serene and the Moon is absent". Fifty years later, Charles Messier catalogued it on June 1, 1764. It is located 25,100 light-years from Earth with a spatial diameter of 145 light-years and estimated to contain about 300,000 stars. M13 appears big and bright due to its close proximity, not because it is intrinsically luminous or large.

In 1974, M13 was chosen as the target for the Arecibo radio message. It was designed to communicate the existence of human life to hypothetical extraterrestrials that may live on a planet orbiting one of the thousands of stars in M13. The test was more of a technological demonstration than a realistic communications effort since the signal will take 25,100 years to arrive and by that time, M13 will have moved position and no longer in the correct location to receive the message.

M13 Data Table

Messier	13	**RA (J2000)**	16h 41m 41s
NGC	6205	**DEC (J2000)**	36d 27m 41s
Name	Great Hercules Globular Cluster	**Apparent Size (arcmins)**	20 x 20
Object Type	Globular Cluster	**Radius (light years)**	72.5
Constellation	Hercules	**Age (years)**	11,650M
Distance (kly)	25.1	**Number of Stars**	300,000
Apparent Mag.	5.8		

Messier # 013

Date:	Time:

Site:

Temp:	Wind:	Hum:

Clouds:	Moon:

Scope:

EP:	Mag:
NELM:	See/Trans:
Type:	# Stars:
Mag:	Age:
Const:	

Notes:

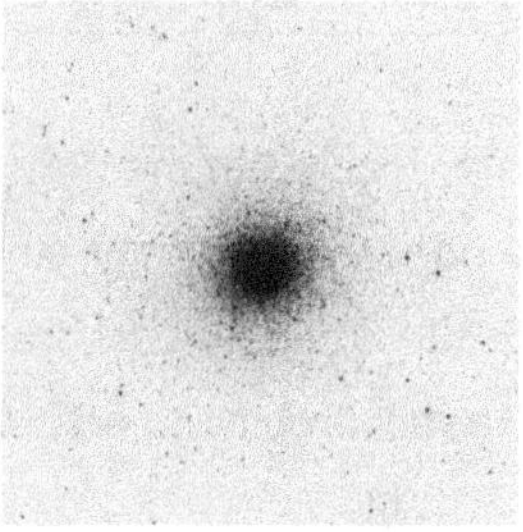

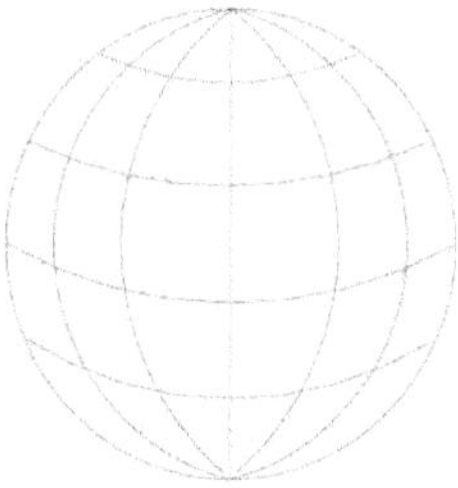

Messier Finder Chart for M13 Great Hercules Globular Cluster and M92

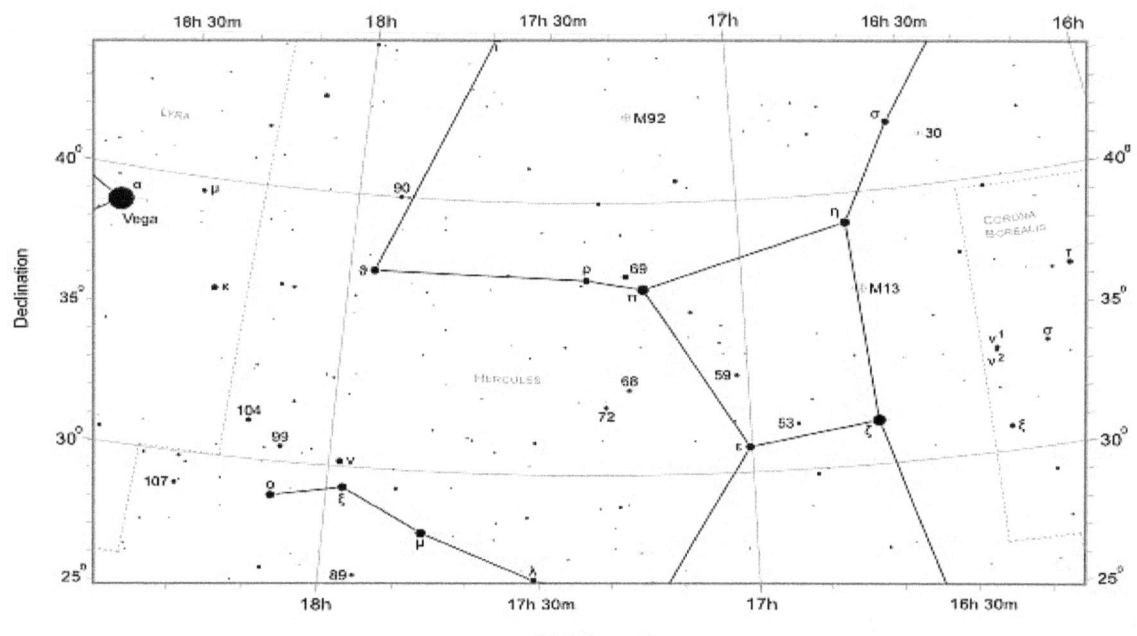

Declination

Right Ascension

Star magnitudes   -1 0 1 2 3 4 5 6 7

Double star   Variable stars   Globular cluster

Messier 14 - M14 - Globular Cluster

M14 is an eighth magnitude globular cluster located in the constellation Ophiuchus. It was discovery by Charles Messier on June 1, 1764, who described it as a "round nebula without stars". A few years later (1783), William Herschel became the first person to resolve M14 into stars. At 30,300 light-years from Earth this is one of the more distant globulars. However, since it's intrinsically bright M14 can be seen with binoculars, although at best appearing only as a out of focus faint "fuzzy star".

M14 is positioned in a rather barren area of sky and therefore not that easy to locate. The cluster is positioned 8 degrees south and a little west of giant orange star Cebalrai (β Oph - mag. +2.8) the fifth brightest star in Ophiuchus and about 11 degrees east of the brighter M10/M12 globular cluster pair.

The best time of the year to observe M14 is during the months of May, June and July.

Although M14 resembles a fainter dimmer version of M10 and M12, it's still quite impressive. Through a small 80mm (3.1-inch) scope, the cluster has a bright centre surrounded by a fuzzy outer halo. A larger 200mm (8-inch) scope displays the elliptical nature of the object with some graininess although no stars are resolvable. A telescope of 300mm (12 inches) aperture begins to resolve some of the individual stars, the brightest of which are of magnitude +14. The faint globular cluster NGC 6366 lies just over 3 degrees southwest of M14.

With an apparent magnitude of +7.9, M14 is more than half a magnitude fainter than M12 and over one magnitude fainter than M10. The reason why it's fainter is due to distance; M14 is almost twice as far away as M12 and more than twice as far away as M10. In reality, M14 is actually the largest of these three globulars but the distance factor wins out.

In total, M14 contains about 150,000 stars and has a spatial diameter of 100 light-years. It contains at least 70 variable stars, many of the W Virginis variety. In 1938, a nova that peaked at magnitude +9.2 appeared in M14.

M14 Data Table

Messier	14
NGC	6402
Object Type	Globular cluster
Constellation	Ophiuchus
Distance (kly)	30.3
Apparent Mag.	7.9
RA (J2000)	17h 37m 36s
DEC (J2000)	-03d 14m 46s
Apparent Size (arcmins)	11 x 11
Radius (light years)	50
Age (years)	13,000M
Number of Stars	150,000

Messier # 014

Date:	Time:
Site:	
Temp:	Wind:	Hum:
Clouds:	Moon:
Scope:	
EP:	Mag:
NELM:	See/Trans:
Type:	# Stars:
Mag:	Age:
Const:	
Notes:	

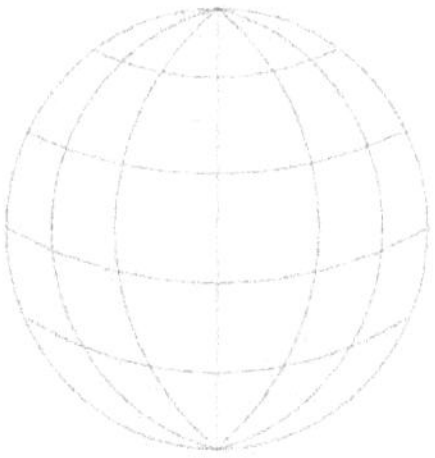

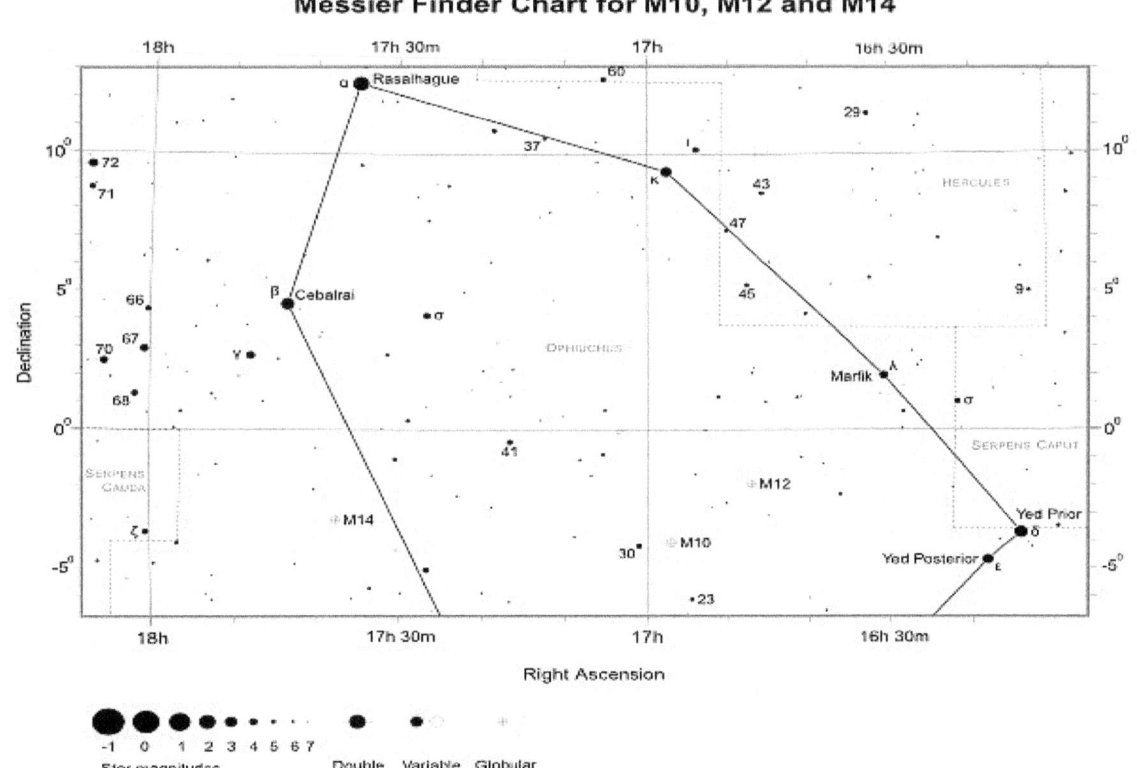

Messier Finder Chart for M10, M12 and M14

Right Ascension

Declination

Star magnitudes -1 0 1 2 3 4 5 6 7
Double star
Variable stars
Globular cluster

Messier 15 – M15 - Globular Cluster

M15 is one of the brightest and finest globular clusters in the northern section of the sky and the best deep-sky object in the constellation of Pegasus. It is only marginally fainter and smaller than M13, the finest northern globular cluster. M15 is relatively easy to find, located 4 degrees to the northwest of magnitude +2.4 star Enif (Epsilon Pegasi - ε Pegasi) and positioned on one edge of a right-angled triangle made up of stars of 6th, 7th and 8th magnitudes.

Approaching naked eye visibility under excellent conditions, this globular cluster is easily observed with binoculars or finder scopes, appearing as a magnitude +6.2 fuzzy star. A 100mm (4-inch) telescope at low power (40x) reveals a uniformly lit disk while high powers (>100x) hint at resolution of some of the outer stars. When viewed through a larger 200mm (8-inch) telescope, M15 appears as a large bright diffuse ball surrounding a dense compact centre region, with many individual stars resolved in the outer halo. The brightest of these stars are of magnitude +12.6. Larger telescopes do even better. A 300mm (12-inch) scope resolves many stars across the complete disk, creating a spectacular 3-dimensional effect. In total, the globular has an apparent diameter of 18 arc minutes, however in amateur telescopes it appears somewhat smaller, perhaps only 8 arc minutes visually.

M15 is one of the most densely packed globulars known in the Milky Way galaxy. Its core has undergone a contraction known as "core collapse" resulting in an enormous number of stars surrounding what may be a central black hole. It is also unusual in that it is one of only four known globulars (along with M22, NGC 6441 and Palomar 6) that contains a planetary nebula. The planetary nebula is named Pease 1.

The cluster was discovered by Italian born French astronomer Jean-Dominique Maraldi on September 7, 1746 and is located 33,600 light-years from Earth. With an age of at least 12.0 billion years, it's thought to be one of the oldest known Milky Way globular clusters.

M15 Data Table

Messier	15
NGC	7078
Object Type	Globular Cluster
Constellation	Pegasus
Distance (kly)	33.6
Apparent Mag.	6.2
RA (J2000)	21h 29m 58s
DEC (J2000)	12d 10m 00s
Apparent Size (arcmins)	18 x 18
Radius (light years)	88
Age (years)	12000M

Messier # 015

Date:	Time:

Site:		

Temp:	Wind:	Hum:

Clouds:	Moon:

Scope:	

EP:	Mag:

NELM:	See/Trans:

Type:	# Stars:

Mag:	Age:

Const:	

Notes:

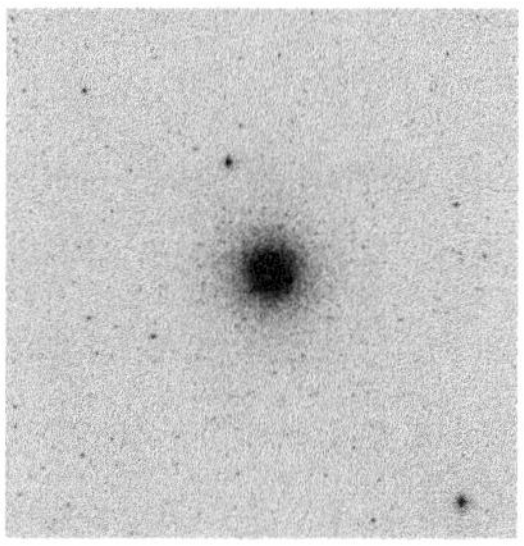

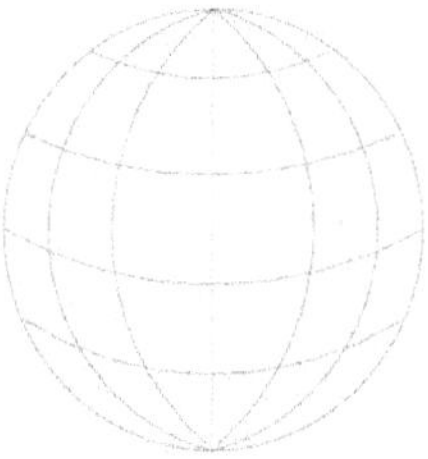

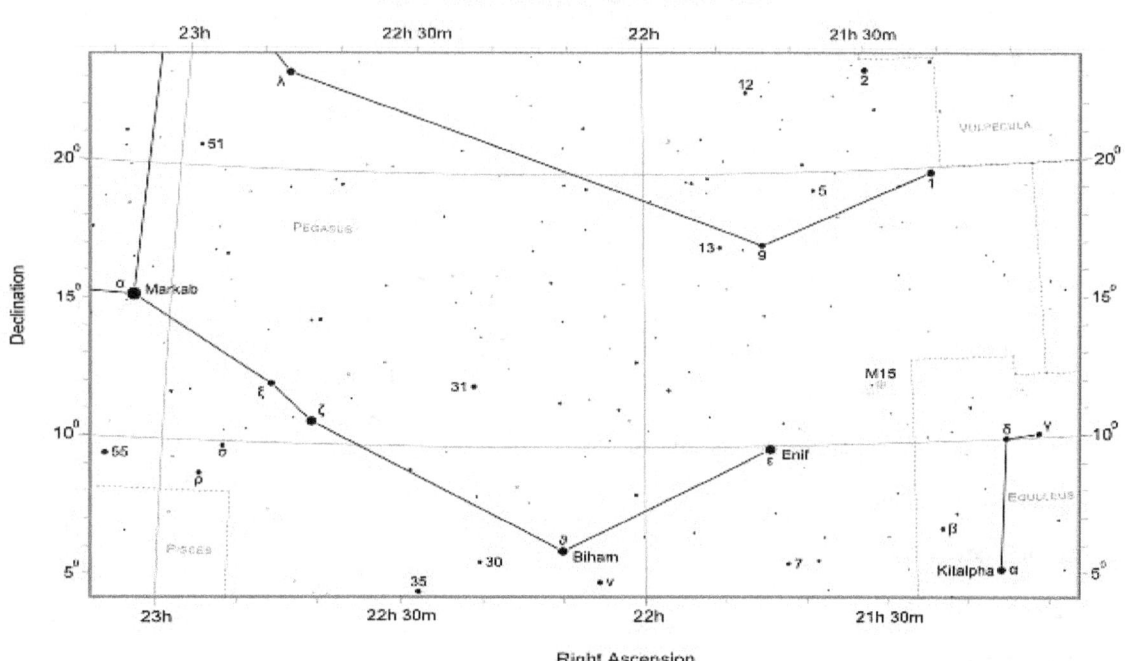

Messier Finder Chart for M15

Messier 16 - M16 - Eagle Nebula (Emission Nebula with Open Cluster)

M16 or the "Eagle Nebula" is a young open cluster of stars embedded within an extremely large cloud of interstellar gas and dust in the constellation of Serpens (Cauda). It's located 7,000 light years from Earth in the next inner spiral arm of the Milky Way. The emission part of the nebula - or H II region - is catalogued as IC 4703 and is an active star-forming region that's already created a significant cluster of young stars. The cluster itself lies at the heart of the Eagle Nebula and is known as NGC 6611. M16 was discovered by Philippe Loys de Chéseaux in 1745-6, but Charles Messier was the first to record the associated nebulosity on June 3, 1764.

The constellation of Serpens is faint but unique since its split into two separate sections. One half lies to the west of Ophiuchus and is named Serpens Caput and the other half, Serpens Cauda, lies on the eastern side of Ophiuchus. At the very southern tip of Serpens Cauda, close to the Scutum and Sagittarius border is M16. It can be found 2.5 degrees west of γ Sct (mag. +4.7) and a few degrees north of the Omega Nebula (M17), M18 and the Sagittarius Star Cloud (M24). This beautifully rich area of the sky is a delight to scan with binoculars. The Eagle Nebula was immortalized in 1995 when imaged several times by the Hubble Space Telescope. The resulting iconic photograph - titled the "Pillars of Creation" - showed three magnificent columns of interstellar gas and dust displayed in sensational detail.

M16 has an apparent magnitude of +6.2, which places it at the very edge of naked eye visibility but easily within the range of binoculars or small telescopes. A pair of 7x50 or 10x50 models will show a faint triangular shaped patch of light with the brightest cluster stars resolvable. Through a 100mm (4-inch) scope roughly 20 stars are revealed but spotting the emission nebula is much more challenging. Under dark skies, the nebula hints at visibility but due to its low surface brightness, medium to large sized amateur scopes are much better suited to the task. Through 10-inch (250-mm) scopes at low powers, the nebula appears wispy with subtle details visible including dark obscuring matter to the north along with many more stars. To spot the famous "Pillars of Creation", an instrument of at least 300mm (12-inch) aperture is recommended.

In total, the nebula part of M16 covers 65x50 arc minutes of apparent sky with the open cluster spanning 7 arc minutes. This corresponds to spatial diameters of 140x110 light-years and 15 light-years respectively. The open cluster is about 5.5 million years old and contains at least 450 stars of which the brightest member is of magnitude +8.2.M16 is best seen during the months of June, July and August.

M16 Data Table

Messier	16	DEC (J2000)	-13d 48m 26s
NGC	6611 (cluster)	Apparent Size (arcmins)	7.0 x 7.0 (cluster), 65 x 50 (nebula)
IC	4703 (nebula)		
Name	Eagle Nebula	Radius (light years)	7.5 (cluster), 70 x 55 (nebula)
Object Type	Emission nebula with open cluster		
Constellation	Serpens	Other Names	Collinder 375, Sharpless 49
Distance (kly)	7.0	Notable Feature	Subject of the famous Hubble Telescope "Pillars of Creation" photograph
Apparent Mag.	6.2		
RA (J2000)	18h 18m 48s		

Messier # 016

Date:	Time:

Site:		
Temp:	Wind:	Hum:
Clouds:	Moon:	
Scope:		
EP:	Mag:	
NELM:	See/Trans:	
Type:	# Stars:	
Mag:	Age:	
Const:		

Notes:

Messier Finder Chart for M16 Eagle Nebula, M17 Omega Nebula, M18, M23, M24 Star Cloud, M25

Also shown M8 Lagoon Nebula, M9, M20 Trifid Nebula. M21, M22 and M28

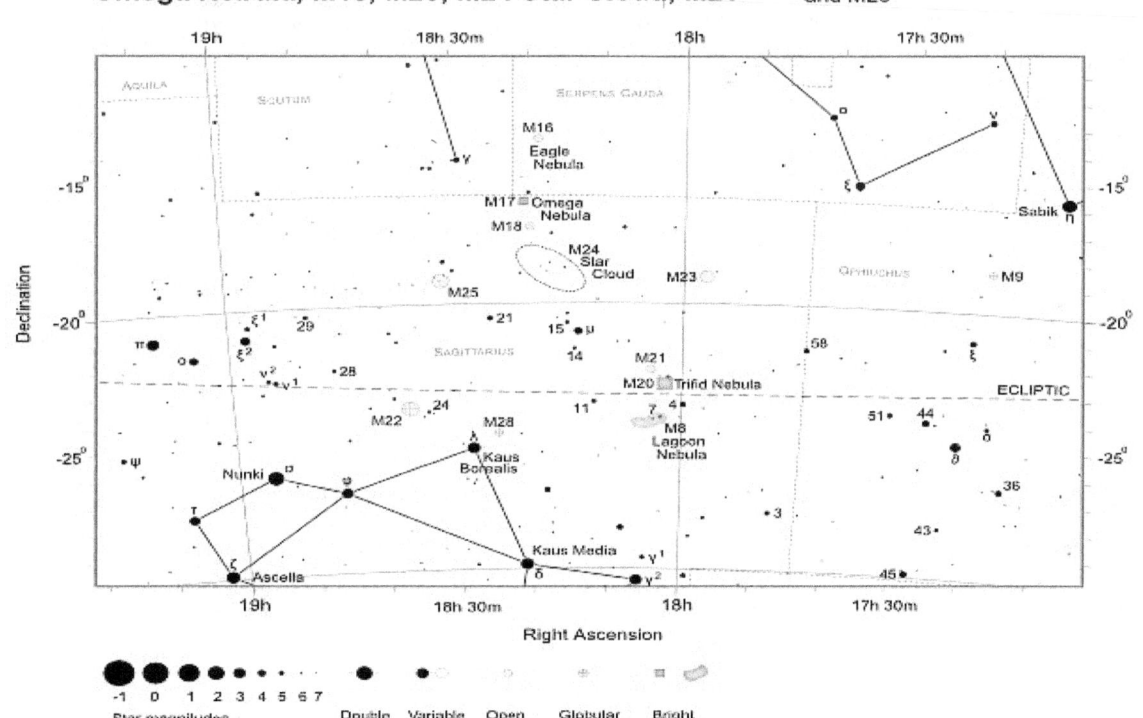

Messier 17 - M17 - The Omega Nebula (Emission Nebula with Open Cluster)

M17 also known as the Omega Nebula is a bright emission nebula located in the rich Milky Way star fields of Sagittarius. It's a H II star formation region that shines with an apparent magnitude of +6.0, placing it at the limit of naked eye visibility. Through binoculars, M17 appears as a diffuse patch of light that's oval shaped. In the same field of view to the south are open cluster M18 (mag. +7.5) and the very large Sagittarius Star Cloud (M24 - mag. +4.6).The Omega Nebula is located 5,500 light-years from Earth. Embedded within it is an open cluster of at least 35 stars that provides the source of the glowing gas. In many similar nebulae such stars are relatively easily visible, but not so in the case of M17. They are hidden deep within the structure and are not obvious. In total, there are many hundreds of stars contained in the nebula itself.

M17 was discovered by Philippe Loys de Chéseaux sometime between 1745-46. Charles Messier independently rediscovered it on June 3, 1764. The nebula covers 20x15 arc minutes of apparent sky, which corresponds to an actual diameter of 32 light-years. It's also sometimes referred to as the Swan Nebula, Horseshoe Nebula, Checkmark Nebula or Lobster Nebula. The Omega Nebula is positioned at the very north of Sagittarius, close to the Serpens Cauda and Scutum constellation boundaries. It's located about 15 degrees north of the "teapot" asterism of Sagittarius. A couple of degrees south of M18 is M24 with M17 sandwiched between them. The star γ Sct (mag. +4.7) lies 2 degrees northeast of M17 with M16 the Eagle Nebula located 2.5 degrees west of this star. The objects are best seen from southern and equatorial regions during the months of June, July and August.

M17 is one of the skies brightest and easily observed emission nebula. It has a high surface brightness and can be seen with the naked eye from a dark site with good transparency. For most observers, this is probably not the case and a minimum pair of binoculars will be required. Popular 7x50 or 10x50 models show the omega-shaped bar as a hazy patch of light.

When viewed through a 100mm (4-inch) telescope, the nebula appears elongated with subtle differences in brightness visible. A larger 200mm (8-inch) telescope reveals twists, contours and a hook shaped extension at one end. Through even larger scopes wispy details curving through the structure of the nebula can be seen.

The Omega Nebula is a popular target for astro imagers and a rewarding object whatever type of optical instrument is being used. It's one of the best examples of its type in the entire sky.

M17 Data Table

Messier	17	**Apparent Mag.**	6.0
NGC	6618	**RA (J2000)**	18h 20m 47s
Name	Omega Nebula	**DEC (J2000)**	-16d 10m 18s
Object Type	Emission nebula with open cluster	**Apparent Size (arcmins)**	20 x 15
Constellation	Sagittarius	**Radius (light years)**	16
Distance (kly)	5.5	**Other Names**	Collinder 377

Messier # 017

Date:	Time:
Site:	

Temp:	Wind:	Hum:

Clouds:	Moon:
Scope:	
EP:	Mag:
NELM:	See/Trans:
Type:	# Stars:
Mag:	Age:
Const:	

Notes:

Messier Finder Chart for M16 Eagle Nebula, M17 Omega Nebula, M18, M23, M24 Star Cloud, M25

Also shown M8 Lagoon Nebula, M9, M20 Trifid Nebula, M21, M22 and M28

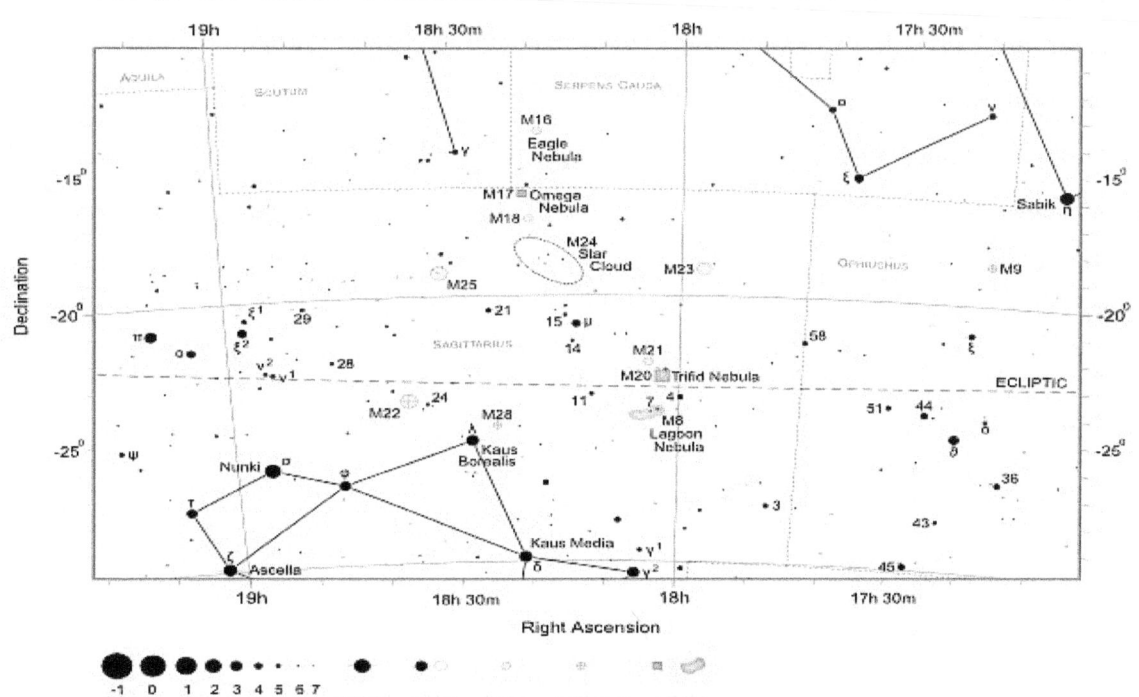

Messier 18 - M18 - Open Cluster

M18 is a small open star cluster located amongst the rich Milky Way star fields of Sagittarius. With an apparent magnitude of +7.5 it's easily visible with popular 7x50 or 10x50 binoculars, appearing as a somewhat dim hazy patch of light. The cluster was one of Charles Messier's original discoveries, which he catalogued on June 3, 1764.

M18 can be found not far from the "teapot" asterism of Sagittarius. It's positioned 8.5 degrees north and a little west of Kaus Borealis (λ Sgr - mag. +2.8), the teapot's top star. The surrounding area of sky is wonderfully rich for astronomers filled to the brim with numerous open clusters, globular cluster and nebulae. Two prominent examples are the Omega Nebula (M17) and the sprawling Sagittarius Star Cloud (M24), which are positioned one degree north and two degrees south of M18 respectively. All three objects are visible in the same binocular field of view.

M18 is best seen with binoculars and small scopes. A small 80mm (3.1-inch) refractor shows about 10 white / blue-white stars with fainter members visible with averted vision. They appear loosely arranged in an area that spans 9 arc minutes in diameter. In total M18 contains 20 stars. However, in larger scopes it's not as impressive due to its lack of concentration.

M18 is best seen from southern and equatorial regions during the months of June, July and August. The cluster is located 4,900 light-years from Earth, which corresponds to an actual diameter of 13 light-years. With an estimated age of 32 million years, it's a youthful star grouping.

M18 Data Table

Messier	18
NGC	6613
Object Type	Open cluster
Constellation	Sagittarius
Distance (kly)	4.9
Apparent Mag.	7.5
RA (J2000)	18h 19m 58s
DEC (J2000)	-17d 06m 07s
Apparent Size (arcmins)	9.0 x 9.0
Radius (light years)	6.5
Age (years)	32M
Number of Stars	20
Other Name	Collinder 376

Messier # 018

Date:	Time:

Site:		
Temp:	Wind:	Hum:

Clouds:	Moon:
Scope:	
EP:	Mag:
NELM:	See/Trans:
Type:	# Stars:
Mag:	Age:
Const:	

Notes:

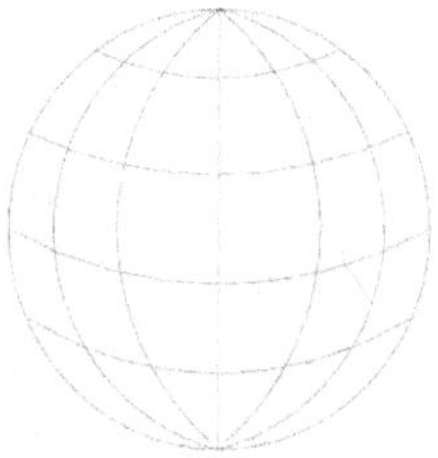

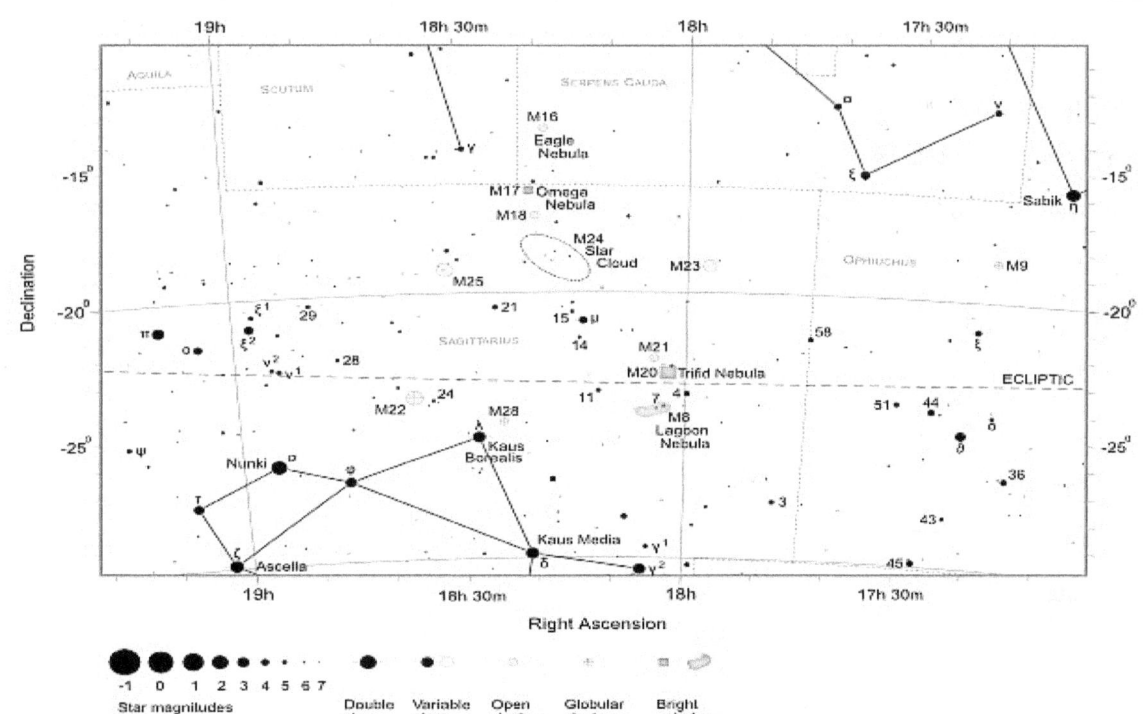

Messier Finder Chart for M16 Eagle Nebula, M17 Omega Nebula, M18, M23, M24 Star Cloud, M25

Also shown M8 Lagoon Nebula, M9, M20 Trifid Nebula, M21, M22 and M28

19h
18h 30m
18h
17h 30m
Aquila
Scutum
Serpens Cauda
M16
Eagle Nebula
M17 Omega Nebula
M18
M24 Star Cloud
M23
M25
Ophiuchus
M9
Sabik
Sagittarius
M21
M20 Trifid Nebula
ECLIPTIC
M22
M28
M8 Lagoon Nebula
Kaus Borealis
Nunki
Kaus Media
Ascella
-15°
-20°
-25°
Declination
Right Ascension
-1 0 1 2 3 4 5 6 7
Star magnitudes
Double star
Variable stars
Open cluster
Globular cluster
Bright nebulae

Messier 19 - M19 - Globular Cluster

M19 is a magnitude +7.2 globular cluster visible with binoculars that's located in the constellation of Ophiuchus. It's an intrinsically large object that's highly oblate in appearance, loosely packed and partly resolvable using medium sized amateur telescopes. The cluster was discovered on June 5, 1764 by Charles Messier and first resolved into stars by William Herschel in 1784.

Finding M19 is easy as its positioned 8 degrees east of Antares (α Sco; mag. +1.0) the brightest star in neighboring Scorpius. Located 4.5 degrees south of M19 is the slightly brighter Messier globular, M62(mag. +6.8).

M19 is best seen from tropical and southern hemisphere latitudes during May, June and July. However, from northern temperate based observers it never rises particularly high above the southern horizon.

When viewed through binoculars, M19 appears stellar like with a hint of fuzziness. A small 80mm (3.1-inch) telescope reveals an obviously diffuse object without a well-defined centre. On nights of good seeing and transparency a 200mm (8-inch) scope at medium to high magnifications will resolve the outer edges of M19. The oval shape of the cluster is noticeable with the long axis orientated in the north-south direction. When viewed through the largest of amateur scopes, countless more stars are visible across the face of M19. Visually it spans about 8 to 10 arc minutes of apparent sky.

M19 is located 28,700 light-years from Earth. In total, the cluster has a diameter of 17 arc minutes, which corresponds to an intrinsic diameter of 140 light-years. The brightest individual stars in M19 are of 14th magnitude with the cluster containing very few variable stars; only four RR Lyrae stars to date have been found.

It's estimated that M19 is 11.9 billion years old.

M19 Data Table

Messier	19
NGC	6273
Object Type	Globular cluster
Constellation	Ophiuchus
Distance (kly)	28.7
Apparent Mag.	7.2
RA (J2000)	17h 02m 38s
DEC (J2000)	-26d 16m 05s
Apparent Size (arcmins)	17 x 17
Radius (light years)	70
Age (years)	11,900M
Number of Stars	300,000

Messier # 019

Date:	Time:
Site:	
Temp:	Wind:	Hum:
Clouds:	Moon:
Scope:	
EP:	Mag:
NELM:	See/Trans:
Type:	# Stars:
Mag:	Age:
Const:	
Notes:	

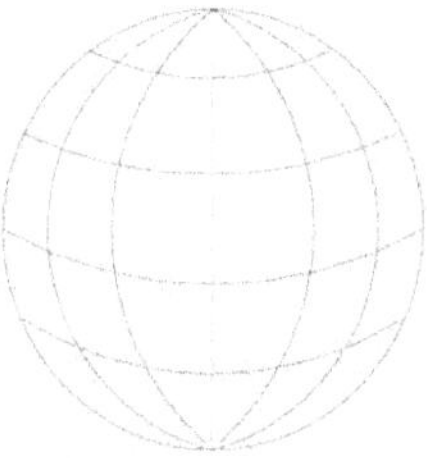

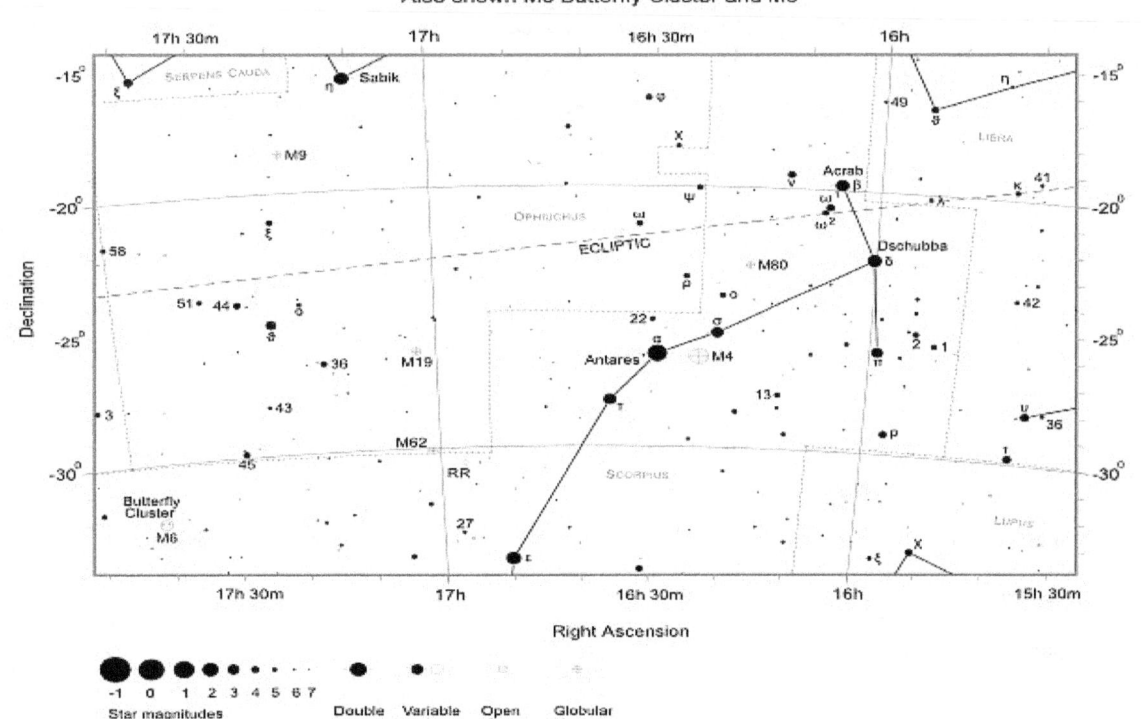

Messier Finder Chart for M4, M19, M62 and M80
Also shown M6 Butterfly Cluster and M9

Messier 20 - M20 - Trifid Nebula (Emission and Reflection Nebula)

M20 is the famous Trifid Nebula, a bright colorful emission and reflection nebula that's located in the constellation of Sagittarius. At magnitude +6.3, it's visible with binoculars. This remarkable object not only contains an emission and reflection nebula but also a dark nebula and an embedded open star cluster. When photographed or imaged, it looks spectacular with the emission nebula appearing red, the reflection nebula blue and mixed in between numerous dark lanes. The dark lanes appear to cut through the nebula splitting it into three prominent sections hence the popular name Trifid; meaning 'divided into three lobes'. The much larger and brighter Lagoon Nebula (M8) is located two degrees south of M20 with tightly packed open cluster M21 positioned 0.75 degrees northeast of M20.

Charles Messier discovered both M20 and M21 on June 5, 1764. He referred to M20 as an envelope of nebulosity. The surrounding area of sky is the richest part of the Milky Way; here you are looking towards the direction of galactic centre hence the abundance of stars, open clusters, globular clusters and nebulae. This wonderful region is perfect to scan with binoculars or small telescopes, especially at low magnifications.

To locate the Trifid, first focus on the bright familiar teapot asterism of Sagittarius. The top three stars of the teapot are Kaus Borealis (λ Sgr - mag. +2.8), Kaus Media (δ Sgr - mag. +2.7) and φ Sgr (mag. +3.2). Imagine a line connecting φ Sgr to Kaus Borealis and then extending it for just over 6 degrees to arrive at M20. The Trifid is best seen from southern and equatorial regions during the months of June, July and August.

At magnitude +6.3, M20 is a fine sight in 7x50 or 10x50 binoculars appearing as fuzzy circular diffuse shape that's just smaller than the diameter of the full Moon. A small 80mm (3.1-inch) telescope hints at dark lanes spreading from the center of the nebula especially when viewed under dark skies. With a 150mm (6-inch) or 200mm (8-inch) scope the Trifid becomes an exciting object. Its irregular shape is visible along with the trisecting dark lanes, exquisite twists/turns and under good conditions it's possible to notice hints of color. Of course, to really appreciate the full color beauty of this object an image or photograph is required.

M20 is located 5,200 light-years from Earth and has a spatial diameter of 42 light-years. The dark nebula that gives the Trifid its appearance was cataloged by E. E. Barnard as Barnard 85 (B85) and in 2005 the Spitzer Space Telescope discovered 30 embryonic stars and 120 newborn stars that had not been seen in visible light images.

M20 Data Table

Messier	20		**RA (J2000)**	18h 02m 21s
NGC	6514		**DEC (J2000)**	-23d 01m 38s
Name	Trifid Nebula		**Apparent Size (arcmins)**	28 x 28
Object Type	Emission and Reflection Nebula		**Radius (light years)**	21
Constellation	Sagittarius		**Other Name**	Collinder 360
Distance (kly)	5.2			
Apparent Mag.	6.3			

Messier # 020

Date:	Time:

Site:		
Temp:	Wind:	Hum:
Clouds:	Moon:	
Scope:		
EP:	Mag:	
NELM:	See/Trans:	
Type:	# Stars:	
Mag:	Age:	
Const:		

Notes:

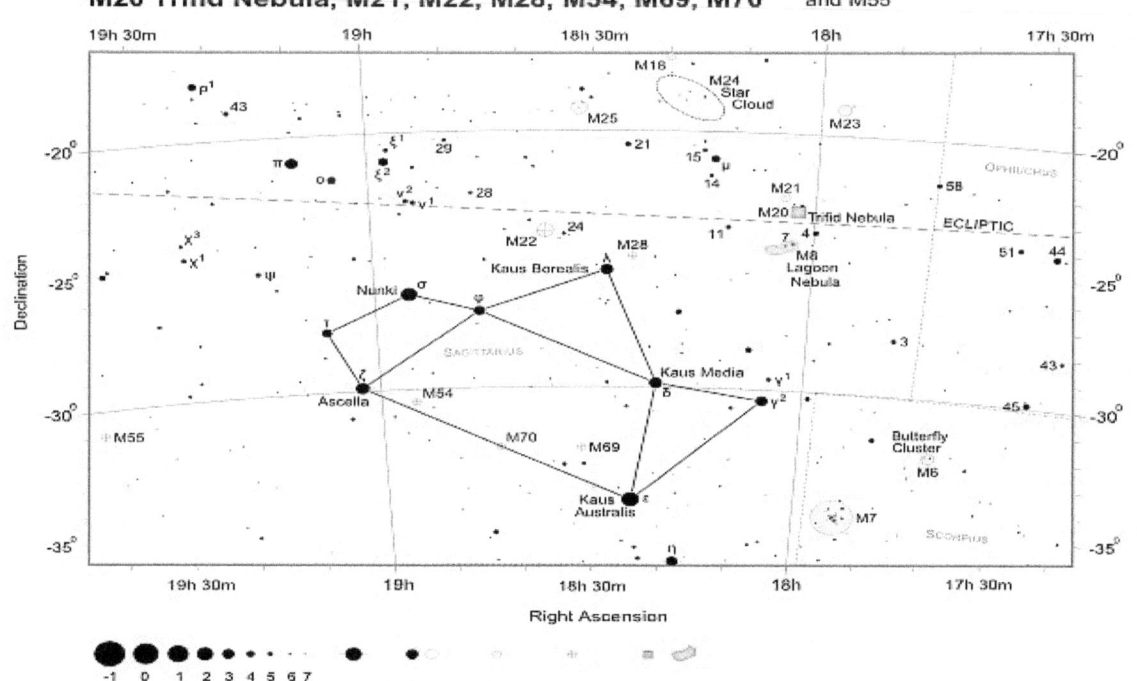

**Messier Finder Chart for M8 Lagoon Nebula,
M20 Trifid Nebula, M21, M22, M28, M54, M69, M70**

Also shown M6 Butterfly Cluster,
M7, M18, M23, M24 Star Cloud
and M55

Messier 21 - M21 - Open Cluster

M21 is a compact open cluster in the constellation of Sagittarius that's located less than one degree northeast of M20, the beautiful Trifid Nebula. The cluster contains 60 or so mainly white stars with a sprinkling of blue giant stars set in a tightly packed area covering about 13 arc minutes of sky. With an apparent magnitude of +6.5 it's a nice sight in binoculars and small telescopes and is compact enough that large scopes show it well, especially when viewed at low magnifications.

M21 was discovered - along with M20 - by Charles Messier on June 5, 1764. It's best seen from southern and equatorial regions during the months of June, July and August.

The ideal starting point to find M21 is teapot asterism that forms the core star grouping of Sagittarius. The top three stars of the teapot are Kaus Borealis (λ Sgr - mag. +2.8), Kaus Media (δ Sgr - mag. +2.7) and φ Sgr (mag. +3.2). Just over 6 degrees north of an imaginary line connecting φ Sgr and Kaus Borealis is M20, with M21 located 0.75 degrees northeast of M20.

M21 is a fine binocular object with the brightest stars resolvable. It appears compact and misty with a sprinkling of starlight especially when viewed with averted vision. Through a 80mm (3.1-inch) scope it's a wonderful sight with many stars visible. The cluster contains some 35 stars between the magnitudes of +8 and +12. A larger 150mm (6-inch) or 200mm (8-inch) scope reveals many stars tightly packed together.

M21 is a relatively young cluster at only 4.6 million years old. It's located 4250 light-years from Earth, which corresponds to an actual diameter of 16 light-years.

M21 Data Table

Messier	21
NGC	6531
Object Type	Open cluster
Constellation	Sagittarius
Distance (kly)	4.25
Apparent Mag.	6.5
RA (J2000)	18h 04m 13s
DEC (J2000)	-22d 29m 24s
Apparent Size (arcmins)	13 x 13
Radius (light years)	8.0
Age (years)	4.6M
Number of Stars	60

Messier # 021

Date:	Time:

Site:		
Temp:	Wind:	Hum:
Clouds:	Moon:	
Scope:		
EP:	Mag:	
NELM:	See/Trans:	
Type:	# Stars:	
Mag:	Age:	
Const:		

Notes:

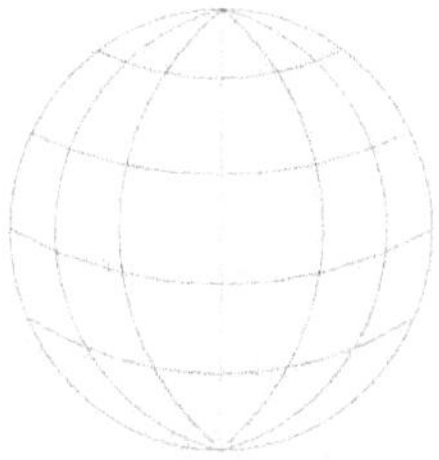

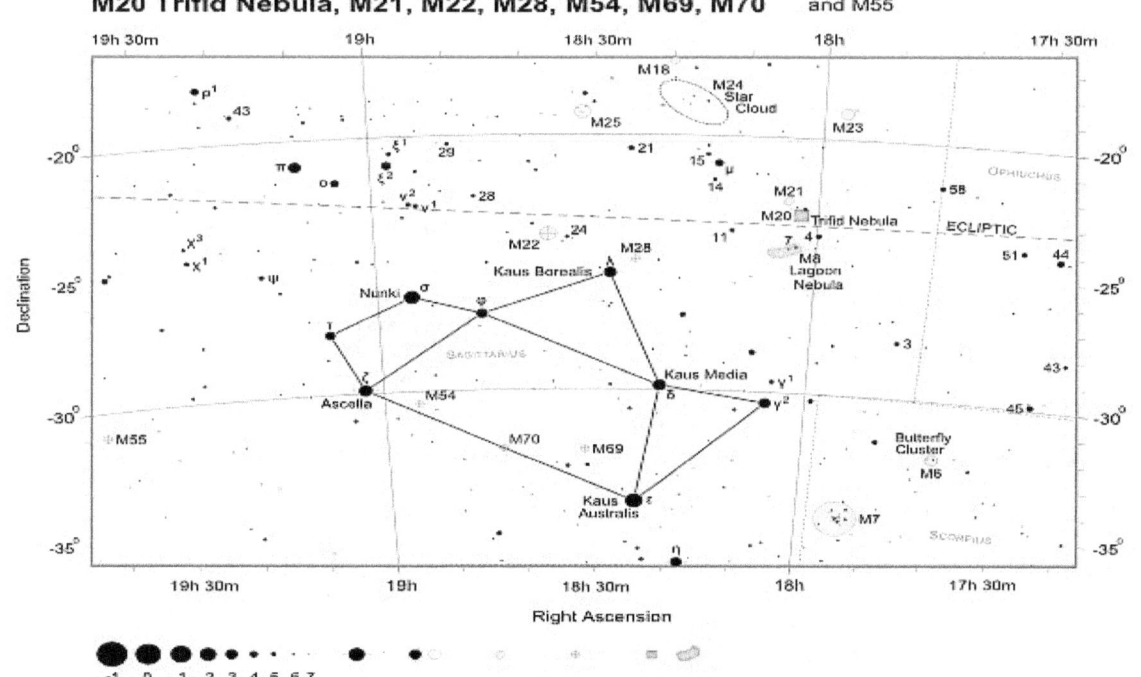

Messier Finder Chart for M8 Lagoon Nebula, M20 Trifid Nebula, M21, M22, M28, M54, M69, M70

Also shown M6 Butterfly Cluster, M7, M18, M23, M24 Star Cloud and M55

Messier 22 - M22 - Sagittarius Cluster (Globular Cluster)

M22 is a magnificent globular cluster located in the constellation of Sagittarius, that's one of the best objects of its type in the night sky. With a magnitude of +5.1, the cluster is visible to the naked eye under dark skies and is also the brightest globular in Messier's catalogue. Only the two great southern globulars Omega Centauri (NGC 5139) and 47 Tucanae (NGC 104) are more brilliant, however both are too far south in the sky to have been seen by Messier. With an apparent diameter extending 32 arc-minutes, M22 covers more sky than the Full Moon.

The main reason why M22 appears so large and bright is because it's close, only 10,400 light-years distant. It's also probably the first globular to have been discovered - by Abraham Ihle in 1665 - although it has been suggested that Hevelius may have seen it earlier. M22 was included in Edmund Halley's list of 6 objects published in 1715 and then catalogued by Charles Messier on the June 5, 1764.M22 is an easy object to locate as it's positioned 2.5 degrees northeast of Kaus Borealis (λ Sag - mag. +2.8), the top star of the "teapot" asterism of Sagittarius. The globular is best seen from southern and equatorial regions during the months of June, July and August. From northern temperate locations it never rises particularly high above the southern horizon.

M22 is a cluster for all types of optical instruments. Through 7x50 or 10x50 binoculars it appears as a diffuse ball of light that's obviously non-stellar with no discernible details visible. A small 80mm (3.1-inch) telescope displays a mottled fuzzy ball that hints at resolution. At high magnifications, a 150mm (6-inch) scope will easily resolve many outer stars, the brightest of which shine at magnitude +11. The cluster core appears slightly brighter with subtle changes in brightness visible across the halo. M22 is noticeably elliptical in shape and when viewed through even larger scopes, e.g. a 300mm (12-inch) instrument, it looks absolutely spectacular with hundreds of stars visible across the complete surface. The core is not particularly dense. Due to its southerly declination, M22 never rises high in the sky for Northern Hemisphere observers and therefore appears less impressive than other globulars such asM13 and M5.It's estimated that M22 contains about 80,000 stars that are spread across a spatial diameter of 86 light-years. Of these, 32 variable stars have been identified. The age of the cluster is 12 Billion years and it's unusual in that it is one of only four known globulars (along with M15, NGC 6441 and Palomar 6) known to contain a planetary nebula.

M22 Data Table

Messier	22
NGC	6656
Name	Sagittarius Cluster
Object Type	Globular cluster
Constellation	Sagittarius
Distance (kly)	10.4
Apparent Mag.	5.1
RA (J2000)	18h 36m 24s
DEC (J2000)	-23d 54m 12s

Apparent Size (arcmins)	32 x 32
Radius (light years)	43
Age (years)	12,000M
Number of Stars	80,000
Notable Feature	One of four globulars known to contain a planetary nebula

Messier # 022

Date:	Time:

Site:

Temp:	Wind:	Hum:

Clouds:	Moon:

Scope:

EP:	Mag:
NELM:	See/Trans:
Type:	# Stars:
Mag:	Age:

Const:

Notes:

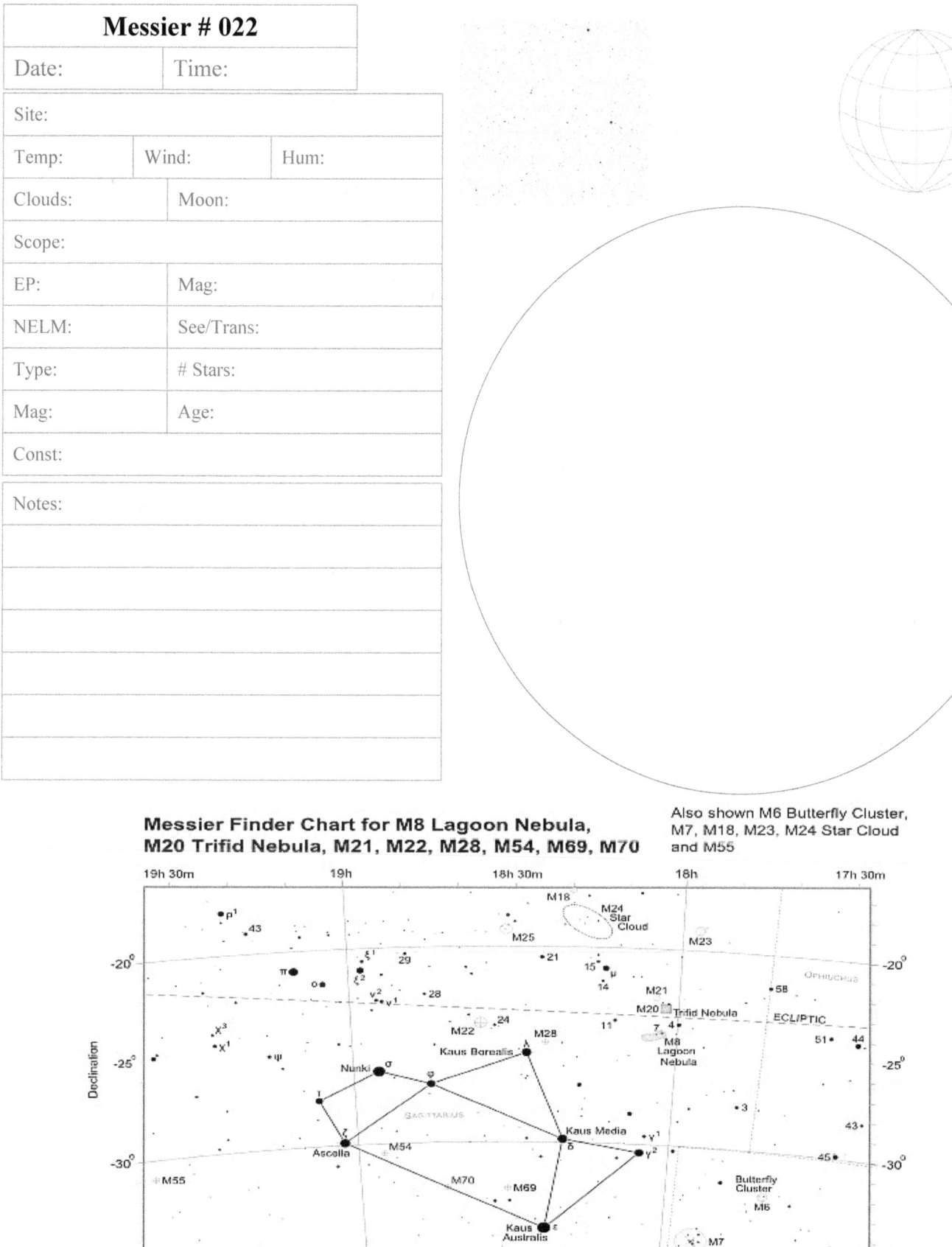

Messier Finder Chart for M8 Lagoon Nebula, M20 Trifid Nebula, M21, M22, M28, M54, M69, M70

Also shown M6 Butterfly Cluster, M7, M18, M23, M24 Star Cloud and M55

Messier 23 - M23 - Open Cluster

M23 is a pretty open cluster that's located in the rich star fields of the Sagittarius Milky Way. With an apparent magnitude of +6.9 it's beyond naked eye visibility but is a superb binocular object and a glorious sight through small telescopes. This vast cloud of about 150 stars is located 2,150 light years from Earth and has an actual diameter of about 20 light years. With an estimated age of at least 220 million years old, it's one of the galaxy's older open clusters.

M23 was discovered by Charles Messier on June 20, 1764. It can be easily found just northwest of the "teapot" asterism of Sagittarius. The three stars that form the top of the teapot are φ Sgr (mag. +3.2), Kaus Borealis (λ Sgr - mag. +2.8) and Kaus Media (δ Sgr - mag. +2.7). Positioned 6 degrees northwest of Kaus Borealis is mag +3.8 star μ Sgr. M23 can be found 4.5 degrees northwest of this star and approximately on a line connecting it with ξ Ser (mag. +3.5). Located 5 degrees east of M23 is M24, the very large Sagittarius Star Cloud.

When viewed through binoculars M23 appears as a grainy smudge that covers an area of 27 arc minutes, about the same as that of the full Moon. The brightest stars are just about resolvable. A small 80mm (3.1-inch) telescope at low magnifications reveals the brightest stars with many fainter background members visible with averted vision. There is a prominent magnitude +6.5 white foreground star positioned at the northwest corner, about one third of a degree from the centre of the cluster.

Medium sized 150mm (6-inch) or 200mm (8-inch) telescopes resolve M23 well with tens of stars visible. The brightest member of M23 is an extremely luminous hot blue star of spectral type B9 that shines at magnitude +9.2. Most stars are between 10th and 13th magnitude with about 100 stars brighter than magnitude +13.5.

M23 Data Table

Messier	23
NGC	6494
Object Type	Open cluster
Constellation	Sagittarius
Distance (kly)	2.15
Apparent Mag.	6.9
RA (J2000)	17h 57m 04s
DEC (J2000)	-18d 59m 07s
Apparent Size (arcmins)	27 x 27
Radius (light years)	10
Age (years)	220M
Number of Stars	150
Other Name	Collinder 356

Messier # 023

Date:	Time:
Site:	
Temp:	Wind:	Hum:
Clouds:	Moon:
Scope:	
EP:	Mag:
NELM:	See/Trans:
Type:	# Stars:
Mag:	Age:
Const:	

Notes:

Messier Finder Chart for M16 Eagle Nebula, M17 Omega Nebula, M18, M23, M24 Star Cloud, M25

Also shown M8 Lagoon Nebula, M9, M20 Trifid Nebula, M21, M22 and M28

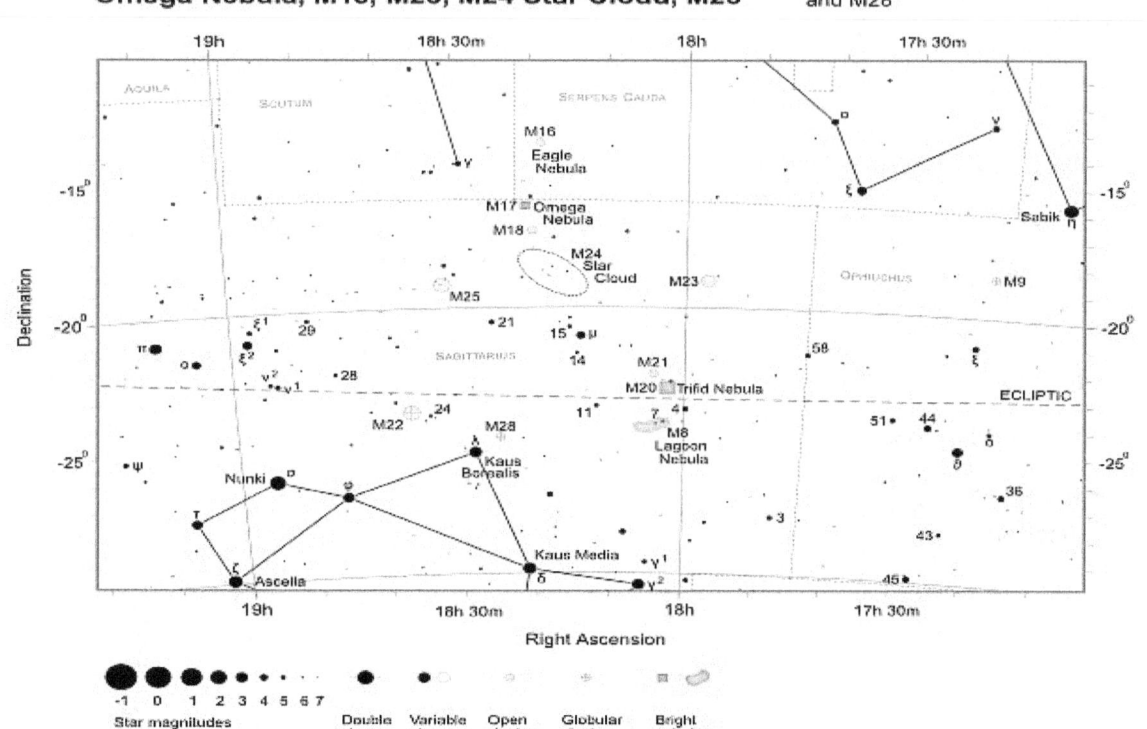

Messier 24 - M24 - Sagittarius Star Cloud

M24 also known as the Sagittarius Star Cloud is a large naked eye expanse of stars, clusters, nebulosity and other objects located in Sagittarius. At apparent magnitude +4.6 and covering 1.5 degrees of sky it's visible to the naked eye as a large detached part of the Milky Way. The object is a fantastic sight in binoculars or small telescopes; it's claimed that M24 has the densest concentration of individual stars visible using binoculars, around 1,000 stars within a single field of view. Spatially, M24 covers a volume of space up to 16,000 light years deep and was discovered by Charles Messier on June 20, 1764. It's best seen during the months of June, July and August from southern or equatorial latitudes.

The Sagittarius Star Cloud is not a "true" deep sky object but results from a chance alignment between the Earth and the centre of our galaxy. We would expect this region to be packed with interstellar dust, however by chance we are looking through a "tunnel" in the interstellar dust, revealing many thousands of distant stars, clusters and nebulae that would otherwise be obscured.

M24 can be found 7 degrees north and a little west of Kaus Borealis (λ Sgr - mag. +2.8) the top star of the bright "teapot" asterism of Sagittarius. Positioned north of M24 is open cluster M18 and the Omega Nebula (M17) with all three objects visible in the same binocular field of view. Open clusters M23 andM25 are located a few degrees west and east of M24 respectively.

Through popular 7x50 or 10x50 binoculars - especially under dark skies - M24 is a wonderful sight with many hundreds of stars visible in striking patterns. It's a showpiece object for binoculars and small telescopes observers. A part of the mystique of M24 is lost in medium and large size telescopes, as they don't show the whole cloud. Nevertheless, through a 200mm (8-inch) scope, it's a fascinating sight with thousands of stars clustered together, intertwined with nebulosity and detailed lines of dust.

Dim open cluster NGC 6603 (mag. +11) lies within M24 and is sometimes incorrectly quoted as being the same object as M24.

M24 Data Table

Messier	24
Object Type	Milky Way star cloud
Constellation	Sagittarius
Distance (kly)	10
Apparent Mag.	4.6
RA (J2000)	18h 17m 00s
DEC (J2000)	-18d 29m 00s
Apparent Size (arcmins)	90 x 40

Radius (light years)	300
Age (years)	220M
Number of Stars	>10,000
Notable Feature	Fills a significant volume of space up to a depth of 16,000 light-years.

Messier # 024

Date:		Time:	
Site:			
Temp:	Wind:		Hum:
Clouds:		Moon:	
Scope:			
EP:		Mag:	
NELM:		See/Trans:	
Type:		# Stars:	
Mag:		Age:	
Const:			

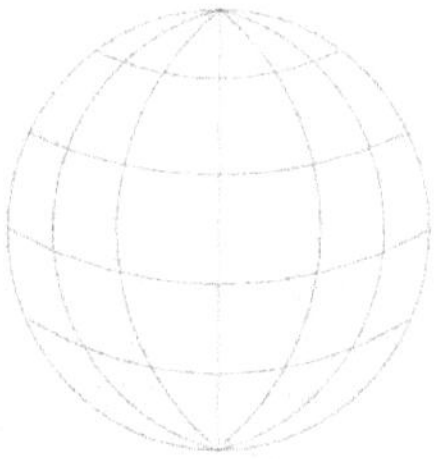

Notes:

Messier Finder Chart for M16 Eagle Nebula, M17 Omega Nebula, M18, M23, M24 Star Cloud, M25

Also shown M8 Lagoon Nebula, M9, M20 Trifid Nebula, M21, M22 and M28

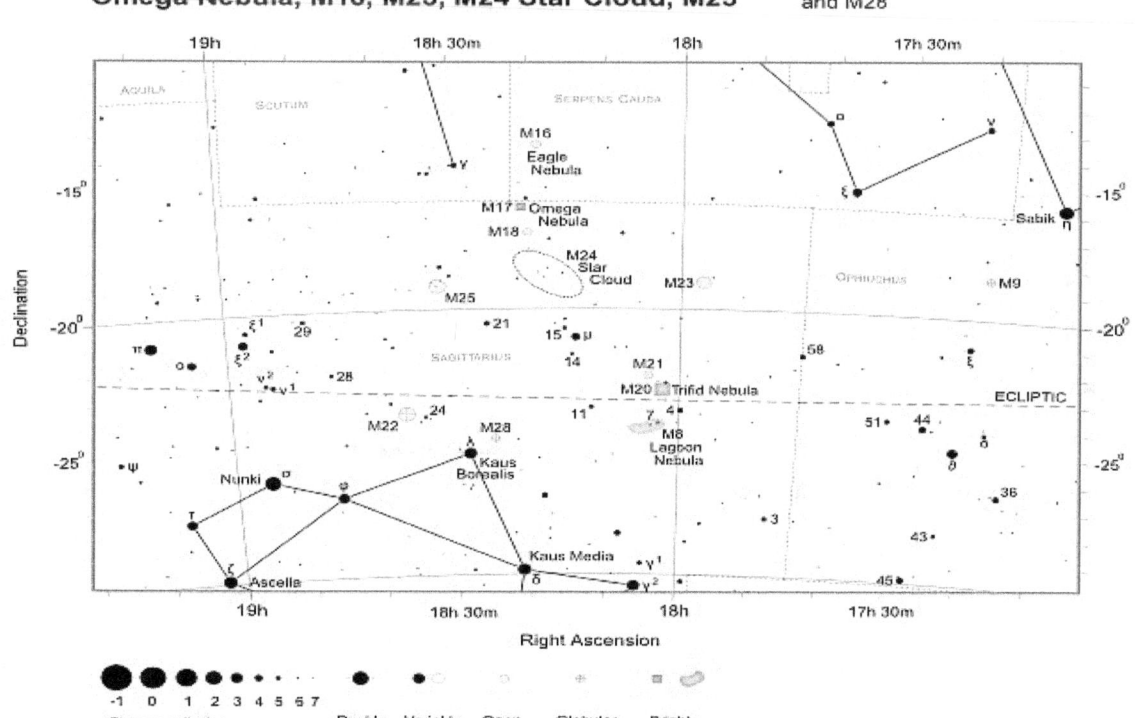

Messier 25 - M25 - Open Cluster

M25 is a bright (mag. +4.6) naked eye open cluster in Sagittarius that's a wonderful sight in binoculars and small telescopes. It was discovered by Philippe Loys de Chéseaux in 1745 and subsequently catalogued by Charles Messier on June 20, 1764. There is however an unusual twist to the history of M25. For such a bright cluster it's reasonable to assume that it would have been included by John Herschel in his comprehensive 19th century General Catalog. For unknown reasons it wasn't despite the cluster been catalogued by Johann Elert Bode in 1777, observed by William Herschel in 1783 and described by Admiral Smyth in 1836. M25 was finally included in the supplementary Index Catalogue (as IC 4725) by J.L.E. Dreyer in 1908.

Finding M25 is relatively easy. It's positioned 6.5 degrees north and a little east of Kaus Borealis (λ Sgr - mag. +2.8) the top star of the bright "Teapot" asterism of Sagittarius. Only 3.5 degrees west of M25 is M24, the very large Sagittarius Star Cloud.

The cluster is best seen from southern and equatorial regions during the months of June, July and August. For mid-latitude northern hemisphere observers, M25 appears low down towards the south during the summer months.

To the naked eye M25 appears as a faint fuzzy patch of light. It's more defined in 10x50 binoculars, covering 32 arc minutes of apparent sky with the brightest stars resolvable. A 80mm (3.1-inch) telescope displays a large loosely defined irregular shaped grouping of about 30 mainly white stars. In larger telescopes it's easily resolvable and more striking as the individual stars colors are pronounced; a 200mm (8-inch) telescope reveals up 60 stars of many colors.

The cluster contains four brighter stars one of which is a Delta Cephei type variable star (U Sagittarii) that varies between magnitudes +6.3 to +7.1 over a period of 6.745 days. It appears yellowish in color and is located towards the centre of the cluster. Since M25 contains many bright comparison stars, it's possible to accurately estimate the brightness changes of U Sagittarii over the full variation period. In total M25 contains at least 80 stars. It's located 2,000 light-years from Earth which corresponds to a spatial diameter of 20 light-years. M25 is estimated to be 90 million years old.

M25 Data Table

Messier	25
IC	4725
Object Type	Open cluster
Constellation	Sagittarius
Distance (kly)	2.0
Apparent Mag.	4.6
RA (J2000)	18h 31m 47s
DEC (J2000)	-19d 06m 54s
Apparent Size	32 x 32

(arcmins)	
Radius (light years)	10
Age (years)	90M
Number of Stars	>80
Other Names	Collinder 382, Melotte 204
Notable Feature	A Delta Cephei type variable star (U Sgr) is a cluster member

Messier # 025

Date:		Time:	
Site:			
Temp:	Wind:		Hum:
Clouds:		Moon:	
Scope:			
EP:		Mag:	
NELM:		See/Trans:	
Type:		# Stars:	
Mag:		Age:	
Const:			

Notes:

Messier Finder Chart for M16 Eagle Nebula, M17 Omega Nebula, M18, M23, M24 Star Cloud, M25

Also shown M8 Lagoon Nebula, M9, M20 Trifid Nebula, M21, M22 and M28

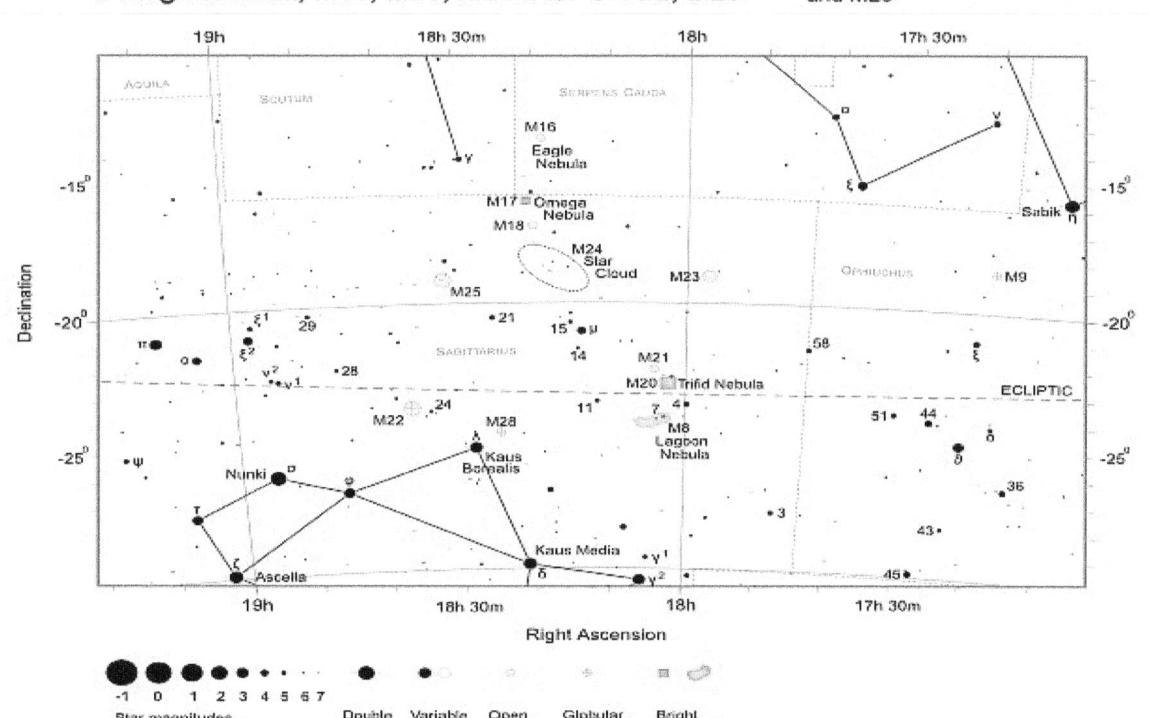

Messier 26 - M26 - Open Cluster

M26 is an 8th magnitude open cluster located in Scutum. It's positioned just a few degrees from the constellations best known deep sky object, M11, the spectacular "Wild Duck open cluster". Despite being not nearly as impressive as it's more illustrious neighbor, M26 is still a nice compact grouping of stars that's best seen through medium size telescopes.

The cluster was discovered by Charles Messier on June 20, 1764. He described it as a cluster that needs a good instrument and contains no nebulosity.

Once one is familiar with the stars that make up faint Scutum, locating M26 is then relatively easy. First find Altair (α Aql - mag. +0.8), the brightest star in Aquila and 12th brightest star in the night sky. Located southwest of Altair is mag. +3.4 star δ Aql. Imagine a line connecting Altair with δ Aql and then curve this line southwards for about the same distance again until you reach a pair of 4th magnitude stars, λ Aql and 12 Aql. To get to M26, image a line connecting these two and then extend the line southwesterly for about 6 degrees to ε Sct (mag. +4.9) and δ Sct (mag. (v) +4.6 -> +4-8). M26 is located one degree southeast of δ Sct.

M26 is not visible to the naked eye but can be spotted with binoculars. A pair of 10x50s depicts the cluster as noticeable slight compression in a rich Milky Way star field. Through a small telescope of 80mm (3.1-inch) aperture, M26 appears as a tight compact misty cluster, hinting on resolution that covers about half the diameter of the full Moon.

The brightest stars in M26 are of 12th magnitude. Of these, about 25 are visible in 150-200mm (6-8 inch) scopes. Four of the brighter stars form a diamond shape near the centre of the cluster with averted vision revealing many fainter stars. Overall there are about 90 members, many of which are resolvable in larger telescopes.

M26 is about 5,000 light years distant and estimated to be 89 million years old. Its apparent size of 15 arc minutes corresponds to a spatial diameter of 22 light years. An interesting feature of M26 is a region of low star density near the nucleus, most likely caused by an obscuring cloud of interstellar matter between us and the cluster.

Overall it's a nice cluster that's visible with binoculars, best viewed at medium magnifications through medium size telescopes. It's best seen from June to September and although often overlooked because of nearby M11, it's well worth a look in its own right.

M26 Data Table

Messier	26	**DEC (J2000)**	-09d 23m 01s
NGC	6694	**Apparent Size (arcmins)**	15 x 15
Object Type	Open cluster	**Radius (light years)**	11
Constellation	Scutum	**Age (years)**	89M
Distance (kly)	5.0	**Number of Stars**	90
Apparent Mag.	8.0	**Other Name**	Collinder 389
RA (J2000)	18h 45m 19s		

Messier # 026

Date:		Time:	
Site:			
Temp:	Wind:		Hum:
Clouds:		Moon:	
Scope:			
EP:		Mag:	
NELM:		See/Trans:	
Type:		# Stars:	
Mag:		Age:	
Const:			

Notes:

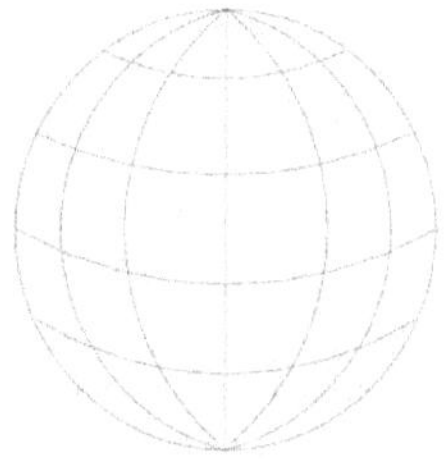

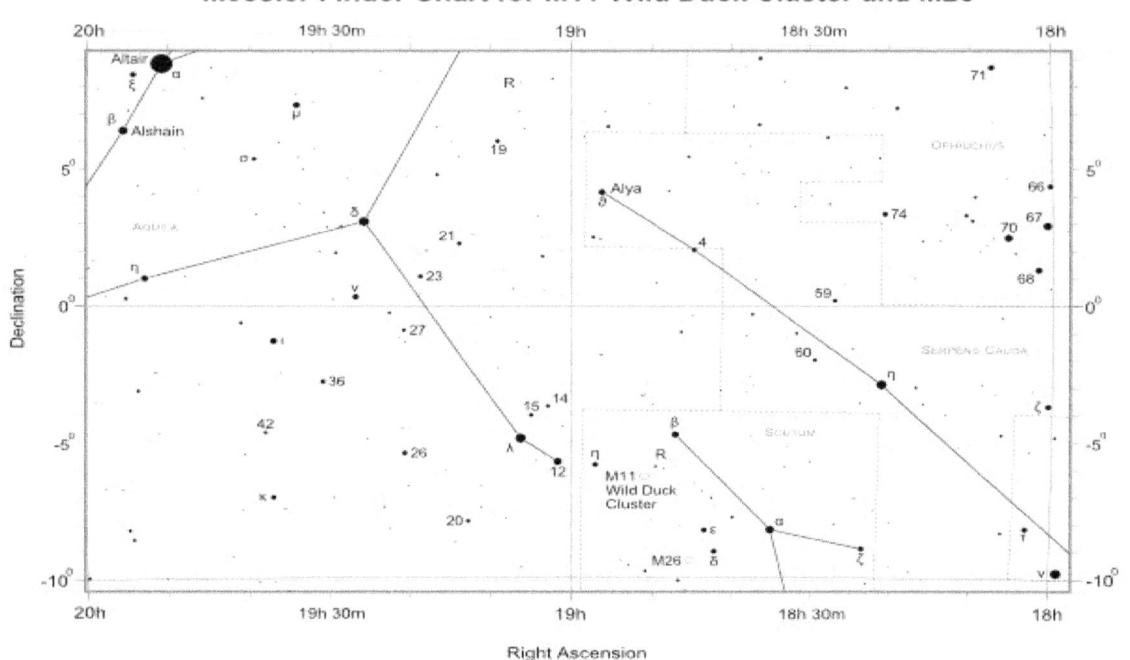

Messier Finder Chart for M11 Wild Duck Cluster and M26

Messier 27 - M27 - The Dumbbell Nebula (Planetary Nebula)

The Dumbbell Nebula or M27 is a showpiece object that is a popular visual and imaging target for amateur astronomers. It is arguably the finest planetary nebula in the night sky and the first of its type to be discovered. The name derives from its resemblance to a dumbbell shape; likewise it has also been compared to an apple core or an hourglass figure. With an apparent mag. of +7.4 it is the second brightest planetary nebula in the sky; only the Helix Nebula (NGC 7293) in Aquarius is marginally brighter. However, the Dumbbell Nebula has a higher surface brightness and therefore the easier target to locate.

M27 is found in the constellation of Vulpecula, easily visible in 10x50 binoculars, appearing as a small oblong shaped patch of light. With 15x70 binoculars it is much larger and brighter with a distinct central region surrounded by fainter outer regions. An 80mm (3.1-inch) telescope will show the famous hourglass shape especially when using averted vision. With larger telescopes M27 displays more intricate surface details. It is a wonderful sight when viewed through a 200mm (8-inch) telescope and as with many objects of this type a nebula filter often enhances the view.

M27 was discovered by Charles Messier on July 12, 1764 and is located 1360 light-years from Earth. It has apparent dimensions of 8.0 x 5.6 arcminutes.

M27 Data Table

Messier	27
NGC	6853
Name	Dumbbell Nebula
Object Type	Planetary Nebula
Constellation	Vulpecula
Distance (kly)	1.36
Apparent Mag.	7.4
RA (J2000)	19h 59m 36s
DEC (J2000)	22d 43m 17s
Apparent Size (arcmins)	8.0 x 5.6
Radius (light years)	1.44

Messier # 027

Date:	Time:
Site:	
Temp:	Wind:	Hum:
Clouds:	Moon:
Scope:	
EP:	Mag:
NELM:	See/Trans:
Type:	# Stars:
Mag:	Age:
Const:	
Notes:	

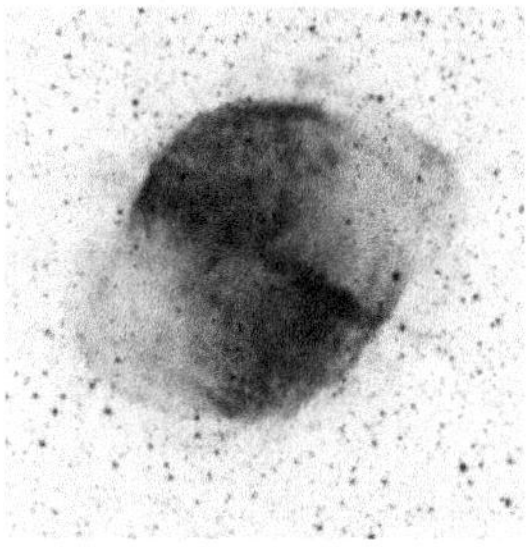

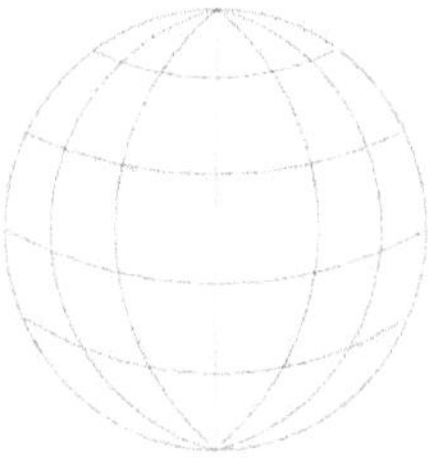

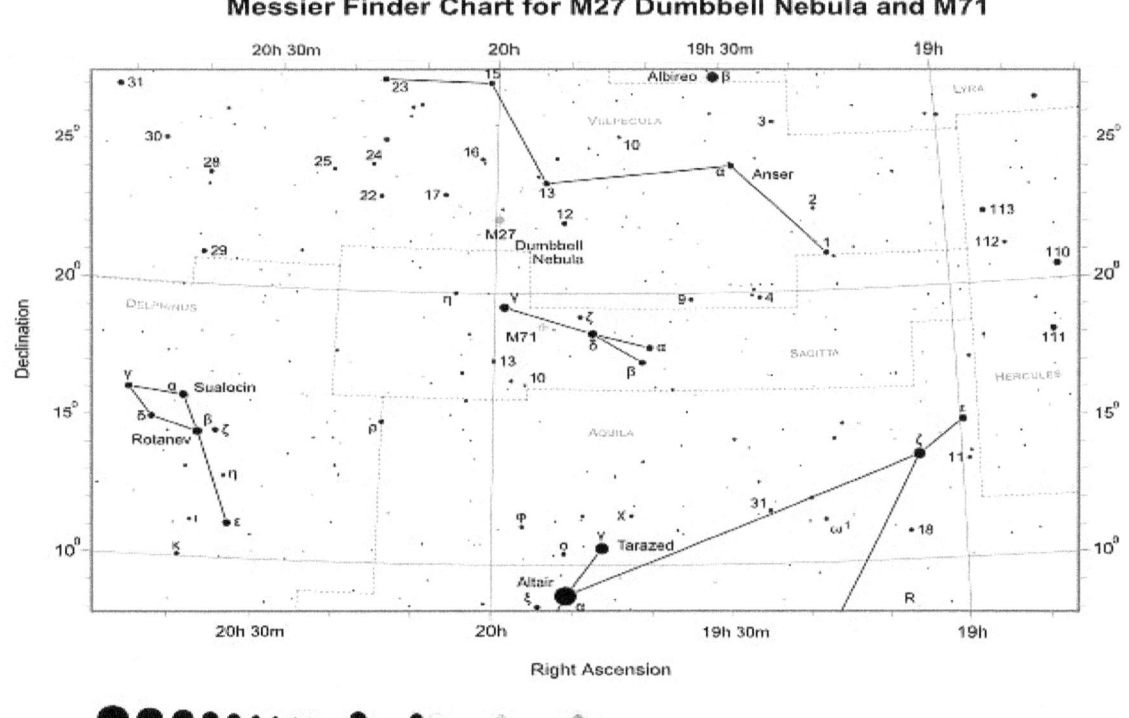

Messier Finder Chart for M27 Dumbbell Nebula and M71

Messier 28 - M28 - Globular Cluster

M28 is a magnitude +7.2 globular cluster located amongst the rich Milky Way star fields of Sagittarius. The cluster is visible with binoculars although unspectacular in appearance. However, telescopes fair better and on nights of good seeing and transparency its possible to resolve a few stars with just a medium size scope.

The cluster was discovered by Charles Messier on July 27, 1764. It's located 18,000 light-years from Earth and spans across 11 arc minutes of apparent sky, corresponding to a relatively small spatial diameter of only 60 light-years. It total M28 contains about 50,000 stars and is best seen from southern and equatorial regions during the months of June, July and August. From northern temperate locations it never rises particularly high above the southern horizon.

M28 is one of the easier globulars to locate as it's positioned less than one-degree northwest of magnitude +2.8 star Kaus Borealis (λ Sag). Visible in the same binocular field of view as M28 is M22, a much larger and brighter globular, which is one of the finest objects of its type in the night sky.

Through a small 80mm (3.1-inch) refractor M28 appears non-stellar without a well-defined core. A 150mm (6-inch) telescope at high power will begin to resolve individual stars although scopes of 250mm (10-inch) aperture or more are essential to fully appreciate this cluster. On closer examination M28 appears slightly elliptical in shape. It's noticeable how considerably smaller and compressed the cluster is when compared to M22.

M28 contains 18 RR Lyrae type variable stars and in 1986 a millisecond pulsar was discovered in the globular. Since then many more pulsars have been detected.

M28 Data Table

Messier	28	**DEC (J2000)**	-24d 52m 12s
NGC	6626	**Apparent Size (arcmins)**	11.2 x 11.2
Object Type	Globular cluster	**Radius (light years)**	30
Constellation	Sagittarius	**Age (years)**	12,000M
Distance (kly)	18	**Number of Stars**	50,000
Apparent Mag.	7.2	**Notable Feature**	Contains millisecond pulsars
RA (J2000)	18h 24m 33s		

Messier # 028

Date:	Time:
Site:	
Temp:	Wind:	Hum:
Clouds:	Moon:
Scope:	
EP:	Mag:
NELM:	See/Trans:
Type:	# Stars:
Mag:	Age:
Const:	
Notes:	

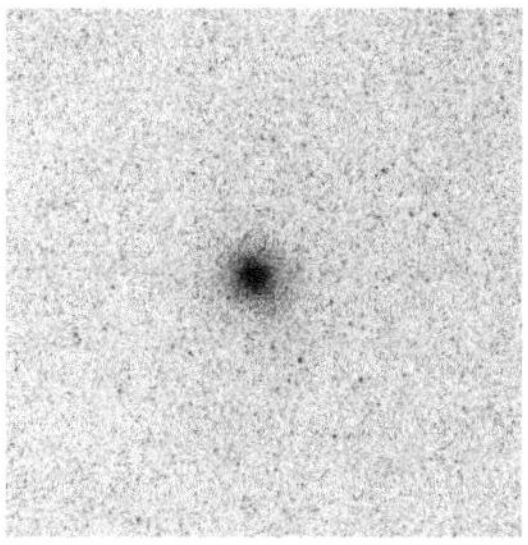

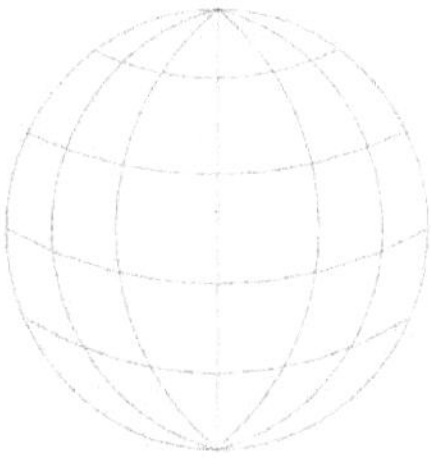

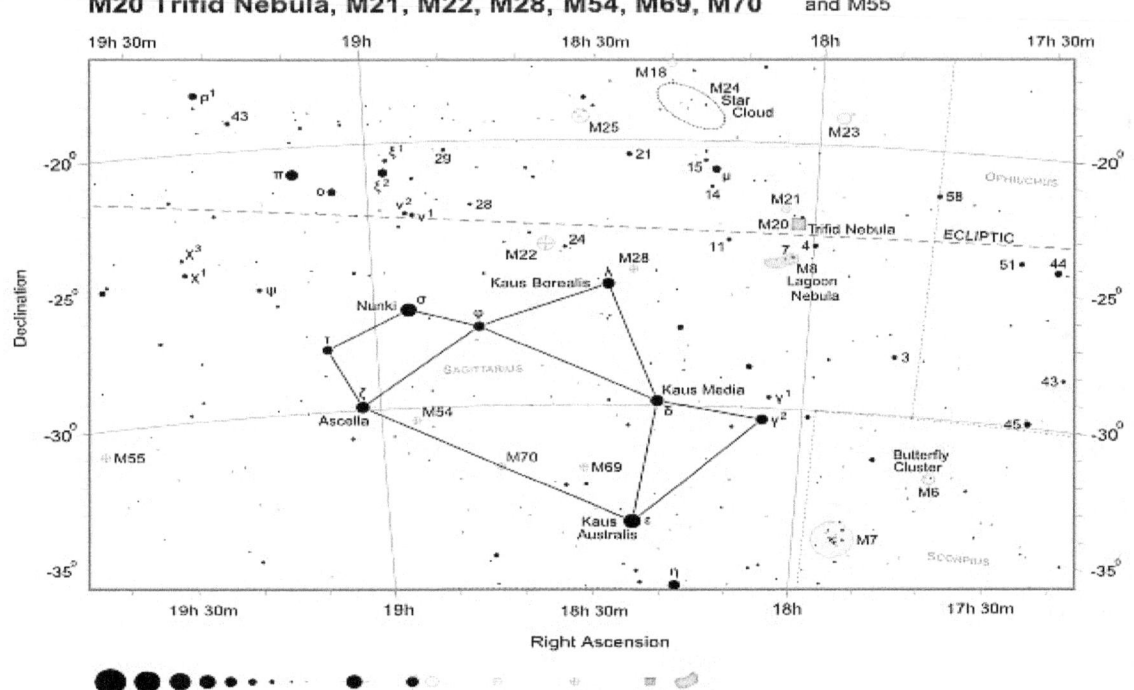

Messier Finder Chart for M8 Lagoon Nebula, M20 Trifid Nebula, M21, M22, M28, M54, M69, M70

Also shown M6 Butterfly Cluster, M7, M18, M23, M24 Star Cloud and M55

Right Ascension

Declination

Star magnitudes   -1 0 1 2 3 4 5 6 7

Double star   Variable stars   Open cluster   Globular cluster   Bright nebulae

Messier 29 – M29 – Open Cluster

M29 is a binocular and telescope open cluster that's situated in the highly crowded Milky Way region of Cygnus. The cluster isn't particularly impressive in terms of brightness, number of stars and compactness but it's certainly worth a look due to its location and unusual shape; a squashed dipper that loosely resembles the main stars of Ursa Major, set against a stunning backdrop of literally thousands of distant Milky Way stars.

M29 was one of Charles Messier original discoveries which he cataloged on July 29, 1764. He described it as "A cluster of 7 or 8 very small stars below Gamma Cygni". What is surprising in such a star rich region of sky is that Messier only catalogued two deep sky objects (in Cygnus), the other being open cluster M39.

Although M29 is unremarkable it's glaringly easy to find; positioned just 1.75 degrees south of Sadr (γ Cyg). At magnitude +2.23, Sadr is the second brightest star in the prominent cross-shaped "Swan" constellation and appears high in the northern hemisphere summer sky.

M29 is visible in binoculars but best seen through telescopes at low to medium powers. When viewed through an 80mm (3.1-inch) scope the brightest stars are visible in a form of a stubby big dipper shape. Remarkably, despite its prominent location M29 does stand out against the background "wall" of stars. However, the cluster is small. It covers just 7 arc minutes of sky - about a quarter of the diameter of the full Moon. An advantage of such a small apparent size is it's possible to push up the magnification, without the cluster over spilling the eyepiece field of view.

Through larger scopes, M29 appears much the same although of course more members are visible along with many non-cluster stars. In the vicinity of M29, there is some diffuse nebulosity which can be detected in images and photographs.

In total M29 contains about 50 member stars of which 6 are brighter than magnitude +9.5. Its distance from Earth is uncertain due to light absorption by surrounding interstellar matter. The current estimate is about 4,000 light years.

M29 Data Table

Messier	29
NGC	6913
Object Type	Open cluster
Constellation	Cygnus
Distance (kly)	4.0
Apparent Mag.	7.1
RA (J2000)	20h 23m 58s
DEC (J2000)	38d 30m 28s
Apparent Size (arcmins)	7.0 x 7.0
Radius (light years)	4.0
Age (years)	10M
Number of Stars	50
Other Name	Collinder 422

Messier # 029

Date:	Time:
Site:	
Temp:	Wind:	Hum:
Clouds:	Moon:
Scope:	
EP:	Mag:
NELM:	See/Trans:
Type:	# Stars:
Mag:	Age:
Const:	
Notes:	

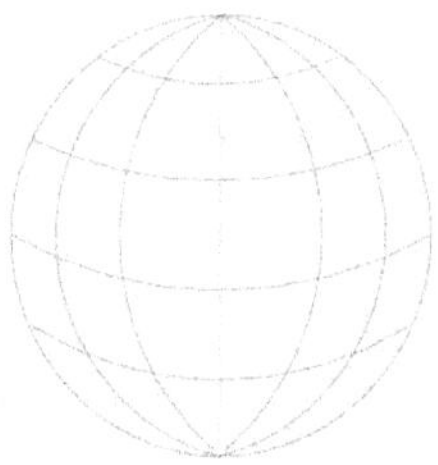

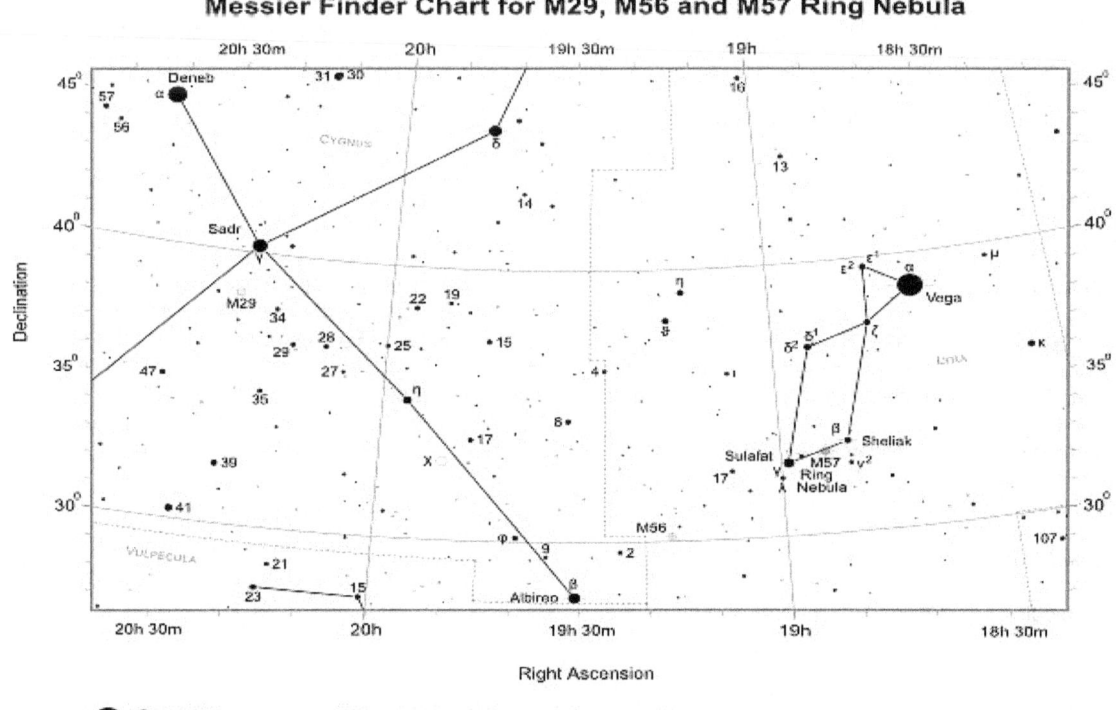

Messier Finder Chart for M29, M56 and M57 Ring Nebula

Right Ascension

Star magnitudes
-1 0 1 2 3 4 5 6 7

Double star   Variable stars   Open cluster   Globular cluster   Planetary nebula

Messier 30 - M30 - Globular Cluster

M30 (mag. +7.4) is a dense globular cluster located in the southern constellation of Capricornus. Located 26,100 light years from Earth, it's visible in binoculars, appearing as a slightly elongated smudge of light. M30 is unusual in that it orbits the galaxy in opposite direction (retrograde) to most other globulars, suggesting that it was acquired from a satellite galaxy rather than forming within the Milky Way. It's best seen during July, August and September from tropical and southern hemisphere locations. The globular was discovered by Charles Messier on August 3, 1764, whom noted it as a round nebula without stars. It was later described in John L. E. Dreyer's New General Catalogue (NGC) as a "remarkable globular, bright, large, slightly oval." Since it's located in the dim zodiac constellation of Capricornus, M30 is one of the more challenging Messier objects to locate, which is especially true for northern hemisphere based observers where the cluster never climbs very high above the southern horizon.

Capricornus is the second faintest zodiac constellation after Cancer. It's positioned to the east of Sagittarius, south of Aquarius and southeast of Aquila. A good starting point to find M30 is the brightest star in the constellation, Deneb Algiedi (δ Cap - mag. +2.9). Once found, move about 7 degrees southwest to arrive at stars 36 Cap (mag. +4.5) and zeta Cap (ζ Cap - mag. +3.8). Located just over 3 degrees southeast of this pair is M30. Right next to M30, on the eastern side, is magnitude +5.2 star 41 Cap, which acts as a good marker.M30 can be easily viewed with a pair of 10x50 binoculars, where it appears non-stellar, a hazy patch of light with a brighter center that's slightly elongated along the east-west axis. Telescopes with apertures starting at 100mm (4-inch) will begin to resolve the outer parts of the cluster, with larger apertures resolving it nicely. Through a medium sized 200mm (8-inch) telescope, M30 appears as bright with a small core and large halo. Many stars are resolved in the halo, which extends up to 12 arc minutes in diameter. The compressed core covers only about 1 arc minute.

M30 is a rewarding globular especially for medium to large size telescopes. It's located at a distance of about 29,400 light-years from Earth and is estimated to be nearly 13 Billion years old. This cluster is similar to at least 20 of the 150 or so globulars in the Milky Way Galaxy, in that it exhibits an extremely dense stellar population, and has undergone a core collapse (like M15, M70 and probably M62).Sadly, M30 is not so loved by Messier Marathoners; astronomers who aim to observe all Messier objects in a single evening. They usually attempt this during a moonless night near the end of March. The reason it's not liked is due to its far southerly declination; a difficult target for the Marathoners, often missed, but otherwise a nice object for amateur astronomers.

M30 Data Table

Messier	30		**DEC (J2000)**	-23d 10m 45s
NGC	7099		**Apparent Size (arcmins)**	12 x 12
Object Type	Globular cluster		**Radius (light years)**	50
Constellation	Capricornus		**Age (years)**	12,930M
Distance (kly)	29.4		**Number of Stars**	150,000
Apparent Mag.	7.4			
RA (J2000)	21h 40m 22s			

Messier # 030

Date:	Time:

Site:

Temp:	Wind:	Hum:

Clouds:	Moon:

Scope:

EP:	Mag:

NELM:	See/Trans:

Type:	# Stars:

Mag:	Age:

Const:

Notes:

Messier Finder Chart for M30, M72 and M73
Also shown M75

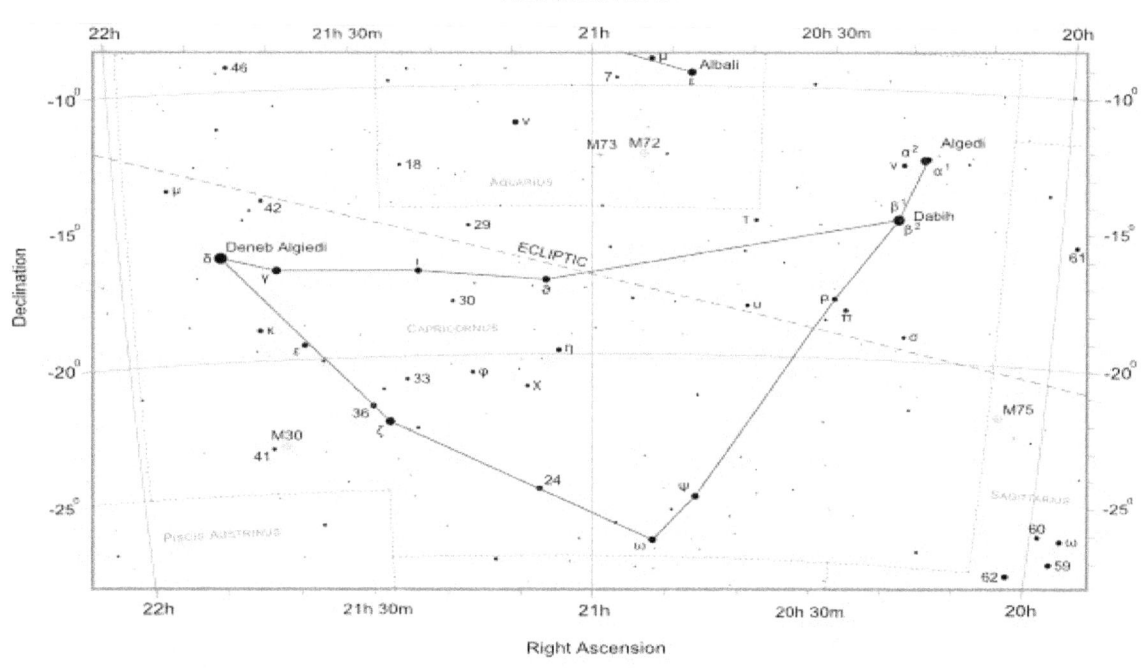

Star magnitudes -1 0 1 2 3 4 5 6 7

Double star   Variable stars   Globular cluster   Asterism

Messier 31 - M31 - Andromeda Galaxy (Spiral Galaxy)

Messier 31 is the famous "Andromeda Galaxy", the largest member of the "Local Group of galaxies" which also includes our Milky Way Galaxy, The Triangulum Galaxy (M33) and about 50 or so other smaller galaxies. With an apparent magnitude of +3.4, it's one of the brightest Messier objects, visible to the naked eye on Moonless nights even in areas with a certain amount of light pollution.

The Andromeda Galaxy has been known for a long time. It was first recorded over 1000 years ago by Isfahan based Persian astronomer Abd-al-Rahman Al-Sufi, who described it in 964 AD as the "little cloud" in his Book of Fixed Stars. This object was almost certainly known to other Persian astronomers at Isfahan for a number of years before this date.

The first person to telescopically observe and describe M31 was German astronomer Simon Marius who did so on December 15, 1612, only a few years after the invention of the telescope. Unaware of both Al Sufi's and Marius' discovery, Giovanni Batista Hodierna independently rediscovered this object sometime before 1654. Then on August 3, 1764, Charles Messier added the great spiral galaxy to his catalog. Incidentally, Messier incorrectly credited Marius as the discoverer, apparently unaware of the earlier work of Al-Sufi.

For all of the above observations, the true nature of M31 was unknown; it was simply regarded as a large nebula located within our galaxy. During this time and up until the 20th century, the object was referred to as the "The Andromeda Nebula" or the "The Great Andromeda Nebula". Sir William Herschel in the 18th century believed, incorrectly, that M31 was one of the nearest nebulae, located at a distance of not more than 17,000 light-years from Earth. However, he was correct in viewing the galaxy as an "island universe" like our Milky Way, although of course at such a small distance it would be much smaller in size than our own galaxy.

In 1887, Isaac Roberts from his private observatory in Sussex, England captured the first photographs of M31. His pictures showed the basic features of its spiral structure for the first time. Clues about the extragalactic nature of the Andromeda galaxy were obtained in 1912 by Vesto Slipher of the Lowell Observatory in Flagstaff, Arizona, who determined its velocity. He found that at about 300 kilometers / second it was the fastest moving astronomical object ever measured.

Around this time there was a "Great debate" surrounding the nature of the Milky Way, spiral nebula like M31 and the dimensions of the universe. One camp believed in "island universes" hypothesis, which held that spiral nebulae were actually independent galaxies, while the other side believed they were located within our galaxy. The debate was settled in 1923 when Edwin Hubble found the first Cepheid variable in the Andromeda galaxy, subsequently measured its variation in brightness and thus established the intergalactic distance and the true nature of M31 as an independent galaxy.

Hubble's original distance measurement of M31 was about 750,000 light-years. At this distance there is no doubt that M31 is a galaxy in its own right but it would be intrinsically much smaller in size than the Milky Way. It was not until 1953 when observations made using the 200-inch Palomar telescope (the then worlds largest) determined that there are two types of Cepheid variables and that the true distance to the Andromeda Galaxy is more than double Hubble's original estimate. Modern measurements place M31 at 2.54 Million light-years distant. This corresponds to an intrinsic diameter of 140,000 light years, which is much larger than the 100,000 light years diameter of our Milky Way Galaxy. It is estimated that M31 may contain up to 1 trillion stars (1×10^{12}). The galaxy is best seen from Northern Hemisphere latitudes during the months of October, November and December. M31 is located in almost the center of the constellation of Andromeda and is positioned to the northeast of the famous "Great Square of Pegasus". Of the four stars of the square, only three of them actually belong to Pegasus. The northeast star and brightest of the four at magnitude 2.1, Alpheratz (α And) is part of neighboring Andromeda. Located 7 degrees to the northeast of Alpheratz is δ And (mag. 3.3) and a further 8 degrees to the northeast of δ And is mag. 2.1, Mirach (β And). The Andromeda galaxy is a further 8 degrees to the northwest of Mirach at the end of a line connecting Mirach with μ And and v And.

To the naked eye, M31 appears as a large easily visible fuzzy patch. Even with a moderate amount of light pollution the galaxy is still visible, although from more densely light polluted areas it is much more difficult to see and may even be invisible. Through 10x50 binoculars, the Andromeda Galaxy appears as an oval shaped small cloud or detached part of the Milky Way that extends around an obvious bright nucleus. A 80mm (3.1-inch) telescope shows subtle details with the centre more pronounced and even at low magnifications the galaxy appears very large and fills most of the eyepiece field of view. With averted vision, some spiral dark lanes are detectable swirling around the central core. In total, M31 covers more than 3.0 x 1.0 degrees of apparent sky, which corresponds to a diameter of over six times that of the full Moon. However, only the brighter central region is visible to the naked eye or when viewed using small telescopes or binoculars. Small telescopes will also show M32 and M110, the two main satellite galaxies of M31.

Through a 200mm (8-inch) telescope, M31 is a spectacular sight. The galaxy easily exceeds the eyepiece field of view at low magnifications with the centre extremely bright, surrounded by a large elongated diffuse shape with prominent dark dust lanes visible. Easily visible, just off the southern edge of the galaxy core is M32 and to the north M110. Even larger telescopes show more intrigue details, although the large apparent size of M31 means that it easily overflows the eyepiece field of view.M31 or the "Great Andromeda Galaxy" is the largest member of the "Local Group of Galaxies" that includes our own Milky Way, the Triangulum Galaxy (M33) and about 50 other smaller galaxies. M31 is located about 2.5 Million light-years from Earth and shines at magnitude 3.4. The Andromeda Galaxy is the most distant object that can easily be seen with the naked eye (M33 is slightly more distant than M31 but requires exceptionally dark skies and good eyesight to be seen with the naked eye).

M31 is a must see on all observing lists and is a superb sight even with just a pair of binoculars or a small telescope. Also visible in small telescopes are M32 and M110, the satellite galaxies of M31. Through larger telescopes it is spectacular, with fine details and dark dust lanes visible. The true beauty of M31 is revealed in images and long exposure photographs where the vast expanse of this magnificent spiral galaxy is apparent.

M31 Data Table

Messier	31
NGC	224
Name	Andromeda Galaxy
Object Type	Spiral galaxy
Classification	SA(s)b
Constellation	Andromeda
Distance (kly)	2,540
Apparent Mag.	3.4

RA (J2000)	00h 42m 44s
DEC (J2000)	41d 16m 06s
Apparent Size (arcmins)	189 x 62
Radius (light years)	70,000
Number of Stars	1 Trillion
Notable Feature	Most distance object that is easily visible to the naked eye

Messier # 031

Date:	Time:

Site:		
Temp:	Wind:	Hum:
Clouds:	Moon:	
Scope:		
EP:	Mag:	
NELM:	See/Trans:	
Type:	# Stars:	
Mag:	Age:	
Const:		

Notes:

Messier Finder Chart for M31 Andromeda Galaxy, M32, M33 Triangulum Galaxy and M110

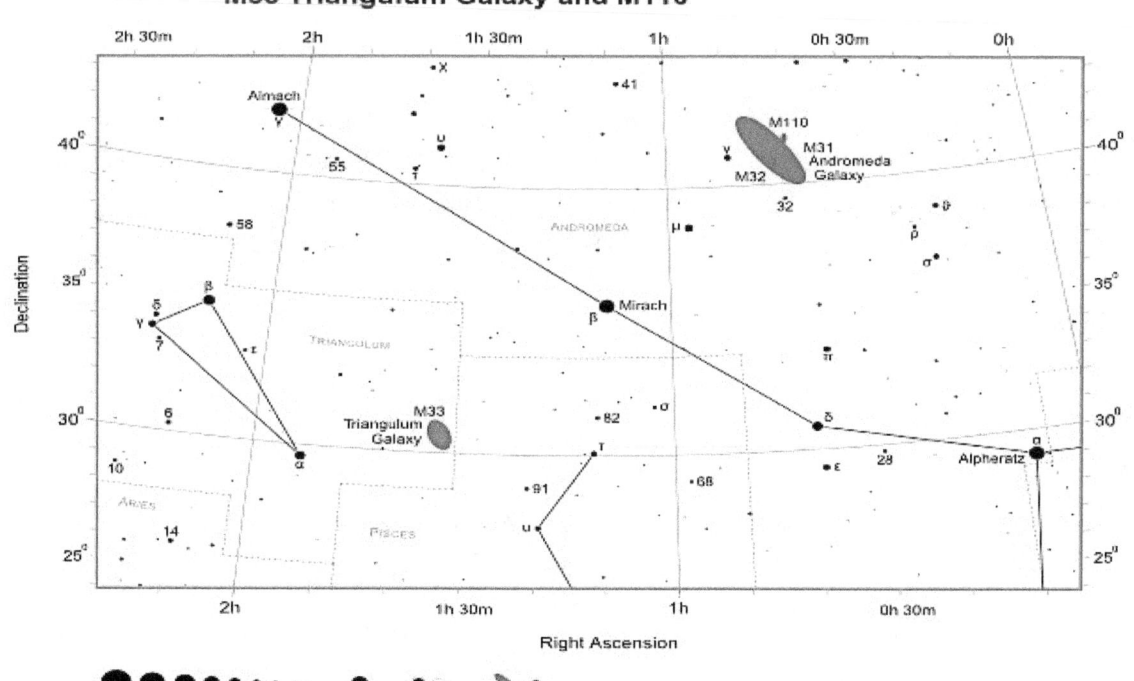

Messier 32 – M32 - Dwarf Elliptical Galaxy

M32 is a dwarf elliptical galaxy located in the constellation of Andromeda. It's a satellite of the famous and much larger Andromeda Galaxy (M31) and the first elliptical galaxy ever observed. French astronomer Guillaume Joseph Hyacinthe Jean-Baptiste Le Gentil de la Galaisière (usually referred to as Guillaume Le Gentil) discovered M32 on October 29, 1749. Le Gentil who was born in Coutances, Normandy also discovered M8 the famous Lagoon Nebula and was the first to catalogue the dark nebula Le Gentil 3 in the constellation Cygnus. He also independently found M36 and M38, objects that were discovered by Giovanni Batista Hodierna around 100 years previously. M32 is best seen from Northern Hemisphere latitudes during the months of October, November and December. At a distance of about 2.65 million light-years, M32 is 110,000 light-years further from us than the Andromeda galaxy. Like most elliptical galaxies, M32 contains mostly older faint red and yellow stars with practically no dust or gas implying no current star formation. However, it has shown hints of star formation in the relatively recent past; the strong tidal forces exerted by nearby massive M31 may have ripped away any spiral arms, resulting in only the central bulge while activating starburst in the core.

Locating M32 is as easy as locating the Andromeda Galaxy, which is positioned to the northeast of the famous "Great Square of Pegasus". Of the four stars of the square, only three of them actually belong to Pegasus. The northeast corner star and brightest of the four at magnitude +2.1, Alpheratz (α And) is part of neighboring Andromeda. Located 7 degrees to the northeast of Alpheratz is δ And (mag. +3.3) and a further 8 degrees to the northeast of δ And is mag. +2.1, Mirach (β And). The Andromeda galaxy is a further 8 degrees to the northwest of Mirach at the end of a line connecting Mirach with μ And ν And. M32 is located 22 arc minutes exactly south of the central region of M31 and appears to be superimposed on one of the outer spiral arms of M31.Despite been small (M32 measures only 8.5 x 6.5 arc minutes) it is relatively bright at magnitude +8.1 and hence is visible in binoculars, appearing non stellar like an out of focus star. The galaxy is an easy target for small telescopes. When viewed through a 80mm (3.1-inch) refractor, it appears as a small round diffuse ball with a small halo and distinctly brighter central core. The galaxy is brighter and more oval shaped when viewed through a 200mm (8-inch) telescope. However for M32, that's about as good as it gets; even larger amateur scopes don't bring out much more detail. Like M110, the other Messier dwarf satellite of M31, M32 is dwarfed in size when compared to the great Andromeda Galaxy.

Intrinsically, M32 has a diameter of 6550 by 5000 light years and contains about 3 billion solar masses. It is the prototype for the relatively rare class of galaxies that are known as compact ellipticals. This galaxy may have been originally a spiral galaxy that at some time in the past has had its arms ripped off by the massive tidal forces from M31. As a result and due to its unusual nature, Halton Arp included M32 as No. 168 in his Catalogue of Peculiar Galaxies.

M32 Data Table

Messier	32		**RA (J2000)**	00h 42m 42s
NGC	221		**DEC (J2000)**	40d 51m 52s
Object Type	Dwarf elliptical galaxy		**Apparent Size (arcmins)**	8.5 x 6.5
Classification	cE2		**Radius (light years)**	3275 x 2500
Constellation	Andromeda		**Notable Feature**	Satellite galaxy of M31
Distance (kly)	2,650		**Other Name**	Arp 168
Apparent Mag.	8.1			

Messier # 032

Date:		Time:	
Site:			
Temp:	Wind:		Hum:
Clouds:		Moon:	
Scope:			
EP:		Mag:	
NELM:		See/Trans:	
Type:		# Stars:	
Mag:		Age:	
Const:			

Notes:

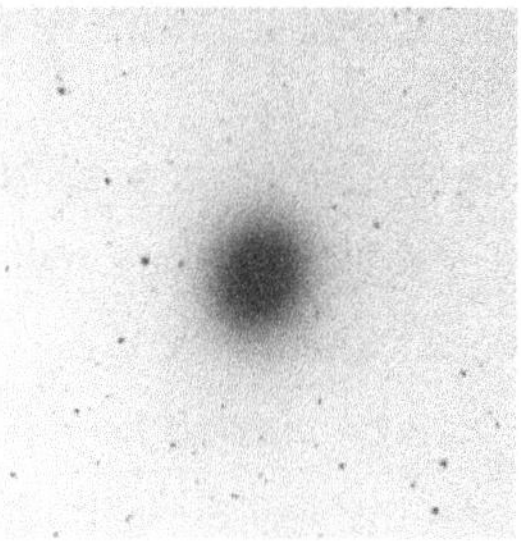

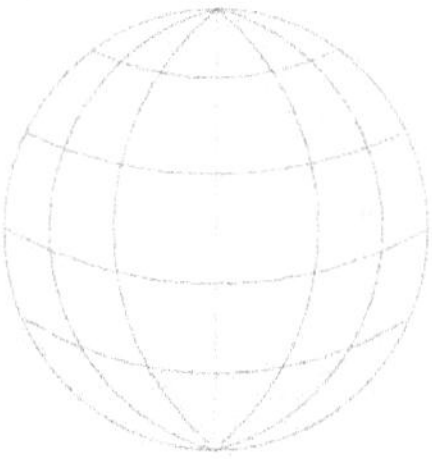

Messier Finder Chart for M31 Andromeda Galaxy, M32, M33 Triangulum Galaxy and M110

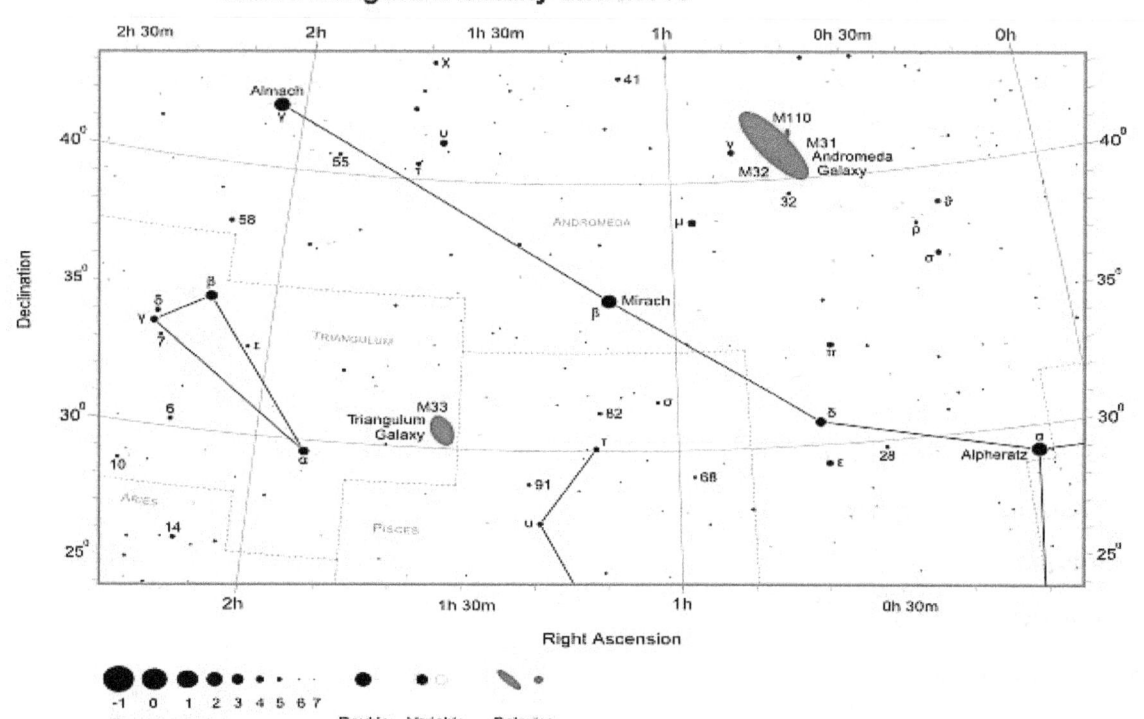

Messier 33 - M33 - Triangulum Galaxy (Spiral Galaxy)

M33, "The Triangulum Galaxy" or "Pinwheel Galaxy" is a spiral galaxy located at a distance of approximately 2.81 million light years in the constellation Triangulum. It appears face-on when viewed from Earth and is the third largest member of the Local Group of galaxies, which also includes our Milky Way Galaxy, M31 "the Andromeda Galaxy" and at least 50 other smaller galaxies. A magnitude +5.7, M33 can be viewed from a very dark site with the naked eye and is widely regarded as the most distant permanent object that can be viewed without a telescope. There are rare reports that some eagle-eyed stargazers have managed, under exceptional conditions, to see the much more distance M81 (11.8 Million light years) but this is incredible viewing by all respects.M33 was probably discovered by Italian astronomer Giovanni Battista Hodierna before 1654. He listed it in his work regarding cometary orbits and admirable objects of the sky. Charles Messier independently re-discovered the galaxy on the night of August 25, 1764.

A good starting point to locate M33 is the "Great Square of Pegasus", focusing on its northeastern star Alpheratz (α And - mag. +2.1). Although it lies at one corner of the square, Alpheratz is officially designated as belonging to Andromeda. Next move 15 degrees northeast to red-giant star Mirach (β And). At magnitude +2.1, Mirach is the same brightness as Alpheratz. Then imagine a line from Mirach towards the southeast for 11 degrees until reaching alpha Trianguli (mag. +3.4) the southernmost star of the Triangulum triangle. Just over halfway along this line is M33.The galaxy is best seen from northern hemisphere latitudes during the months of October, November and December. Despite being visible to the naked eye from a very dark site, the Triangulum Galaxy is often regarded as an elusive object. It appears face on from our perspective and although large and covering more than a degree of sky (71 x 42 arc minutes) it suffers from low surface brightness, hence easily rendering it invisible through binoculars or small telescopes with just a small amount of light pollution. However, it's a different story under dark skies where M33 is relatively easy to find. Binoculars show it as a large mist without detail. An 80mm (3.1-inch) telescope reveals the galaxy as a very large diffuse patch of light covering much of the field of view with a slightly brighter centre that's best viewed using low powers. A medium size 150mm (6-inch) or 200mm (8-inch) scope enhances the core and brings out some mottling but a larger telescope of at least 300mm (12-inch) aperture is required to easily reveal the spiral structure along with dark dust lanes, knotty patches of nebulosity and other subtle features. For the budding astrophotographer, M33 is a rewarding target that's relatively easy to capture the spiral arms and brighter nebulae within.

Overall, M33 has a diameter of about 60,000 light years and is estimated to contain 40 billion stars. For comparison, the Milky Way contains 400 billion stars and M31 about 1 trillion (1,000 billion).

M33 Data Table

Messier	33		**RA (J2000)**	01h 33m 51s
NGC	598		**DEC (J2000)**	30d 39m 37s
Name	Triangulum galaxy		**Apparent Size (arcmins)**	71 x 42
Object Type	Spiral galaxy		**Radius (light years)**	30,000
Classification	SA(s)cd		**Number of Stars**	40 Billion
Constellation	Triangulum		**Other Name**	Pinwheel galaxy
Distance (kly)	2,810			
Apparent Mag.	5.7			

Messier # 033

Date:	Time:
Site:	
Temp:	Wind:	Hum:
Clouds:	Moon:
Scope:	
EP:	Mag:
NELM:	See/Trans:
Type:	# Stars:
Mag:	Age:
Const:	
Notes:	

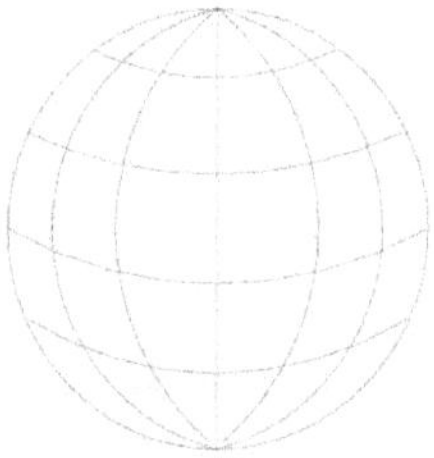

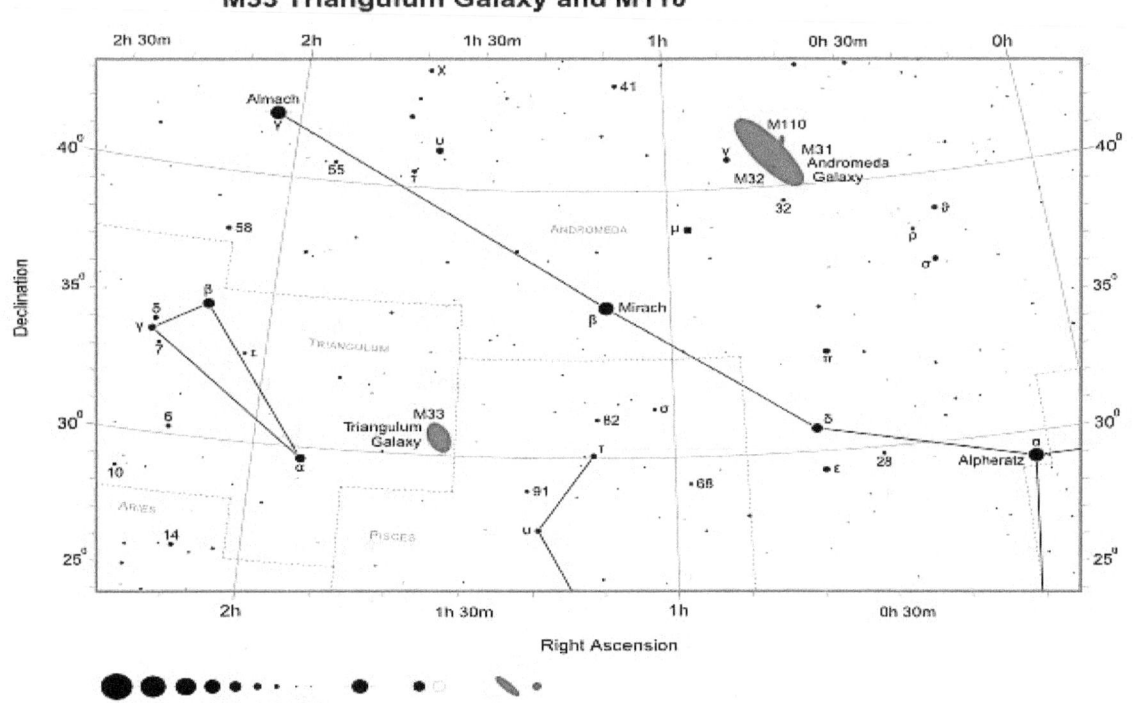

Messier Finder Chart for M31 Andromeda Galaxy, M32, M33 Triangulum Galaxy and M110

Declination

Right Ascension

Star magnitudes
-1 0 1 2 3 4 5 6 7

Double star   Variable stars   Galaxies

Messier 34 - M34 - Open Cluster

M34 is a fine large loose open cluster located in the constellation of Perseus. At magnitude +5.5, it's visible as a faint smudge to the naked eye (from a reasonably dark site) and is easily identifiable with binoculars where the brightest members are resolvable. A small telescope reveals up to 20 bright stars embedded in nebulosity with about 80 members visible in larger amateur scopes.

M34 was probably discovered by Giovanni Batista Hodierna sometime before 1654. Charles Messier rediscovered it on August 25, 1764, then included it in his catalogue describing it as, "A cluster of small stars a little below the parallel of γ (gamma) And. In an ordinary telescope of 3 feet one can distinguish the stars". M34 is one of two Messier objects in Perseus, the other being M76 the Little Dumbbell Nebula. Finding this open cluster is easy; it's located on the western side of Perseus next to the Andromeda boundary and only 5 degrees northwest of famous eclipsing binary star Algol (β Per - mag.(v) +2.1 -> +3.4). It lies just north of a line connecting Algol with beautiful telescopic multiple star Almach(γ And - mag. +2.1).

As previously mentioned, M34 is faintly visible to the naked eye and a fine binocular open cluster. It's a large cluster, covering 35 arc minutes of apparent sky; larger than the diameter of the full Moon. When viewed through a small 80mm (3.1-inch) telescope, the brighter members of the cluster appear prominent with many fainter background members visible, especially when using averted vision. At low powers the cluster nicely fills a good portion of the eyepiece field of view. Towards the center of the cluster a couple of double stars are visible.

Large telescopes show more detail. Through a 150mm (6-inch) or 200mm (8-inch) scope, dozens of stars are revealed and the loose scattered look of the cluster is pronounced. The bright stars towards the centre of the group form three arms of stars radiating outwards in a distinct "Y" or "V" shape.

M34 is about 200 Million years old which is much older than the two components of the Double cluster in Perseus (NGC 869 and NGC 884 - 5.6 and 3.2 million years old respectively), older than the famous M45 Pleiades open cluster (115 million years old) but younger than the Hyades open cluster (625 million years old). The cluster may contain up to 400 stars of which at least 19 members have been identified as white dwarfs.

M34 is a large loose open cluster that's faintly visible to the naked eye, superb through binoculars and a sparkling telescope object. It's best seen from the Northern Hemisphere during October, November and December. Well worth a look!

M34 Data Table

Messier	34
NGC	1039
Object Type	Open cluster
Constellation	Perseus
Distance (kly)	1.5
Apparent Mag.	5.5
RA (J2000)	02h 42m 07s

DEC (J2000)	42d 44m 46s
Apparent Size (arcmins)	35 x 35
Radius (light years)	7.5
Age (years)	200M
Number of Stars	>80
Other Name	Collinder 31

Messier # 034

Date:	Time:

Site:		
Temp:	Wind:	Hum:
Clouds:	Moon:	
Scope:		
EP:	Mag:	
NELM:	See/Trans:	
Type:	# Stars:	
Mag:	Age:	
Const:		
Notes:		

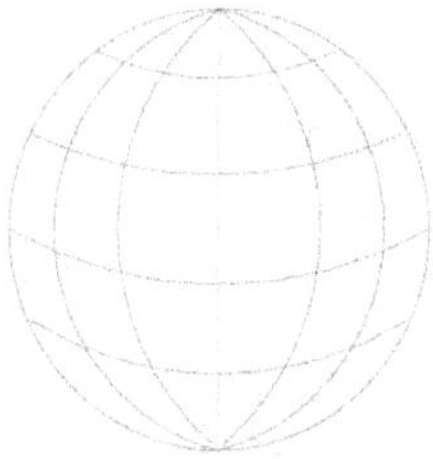

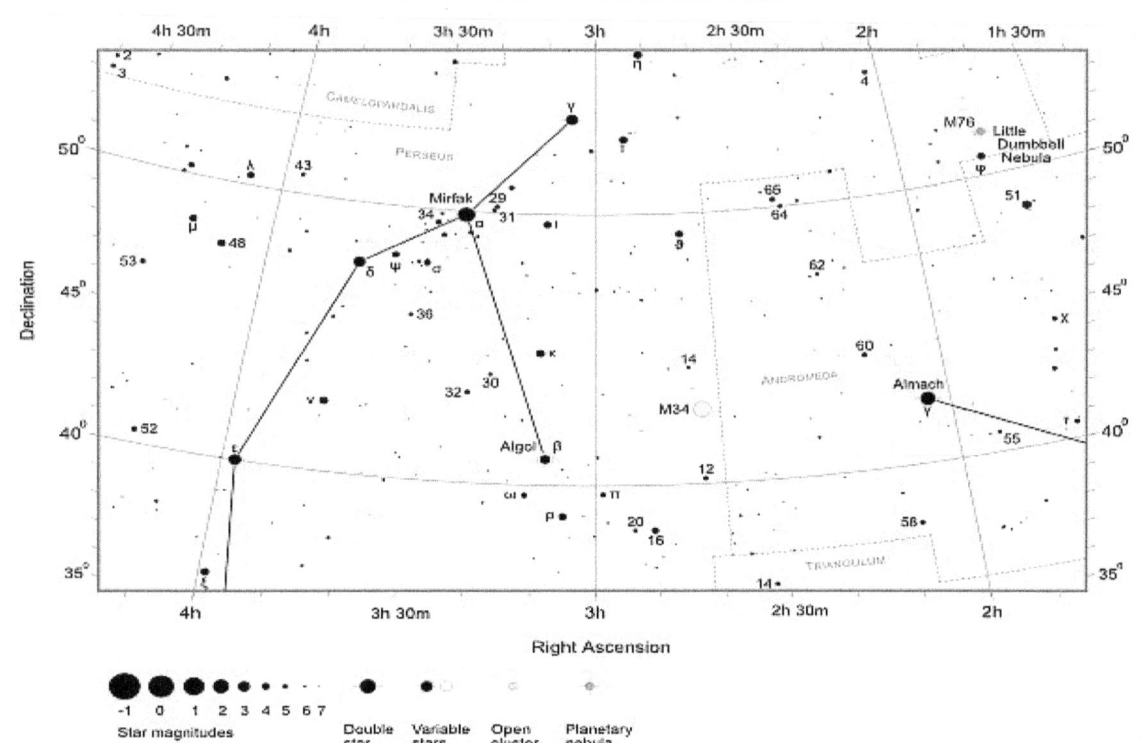

Messier Finder Chart for M34
Also shown M76 Little Dumbbell Nebula

Messier 35 – M35 - Open Cluster

M35 is a fantastic open cluster located in the constellation of Gemini. At magnitude +5.2, it's visible to the naked eye under good conditions. Curiously, despite being a naked eye object it was not discovered until 1745-46 by Philippe Loys de Chéseaux and then independently "re-discovered" by John Bevis sometime before 1750. Since the cluster is reasonable close at 2,800 light-years, it presents a large apparent diameter of some 28 arc minutes - almost exactly the same as that of the full Moon. Without a doubt, M35 is a magnificent sight in all types of optical instruments. The constellation Gemini covers a reasonable 514 sq. degrees and is partly located in the rich star fields of the Milky Way. During the winter months it is positioned high in the sky for Northern Hemisphere observers although less well placed for those located further south. Surprisingly, despite such a prime Milky Way location, Gemini contains only one Messier object. By good chance, it is a superb object: M35. This sole Gemini Messier object is located at the western edge of the constellation in a corner of the sky close to the Taurus, Orion and Auriga borders. Finding M35 is quite easy as it is positioned just over 3.5 degrees to the NW of magnitude 2.9 star Mu (μ) Geminorum.

To the naked eye M35 appears as a misty patch that's easier to spot with averted vision. Through 10x50 binoculars, the brightest dozen or so stars are resolvable. The grouping appears tight, surrounded with haze and shaped like an oblong or even triangular. With larger 20x80 binoculars, M35 appears spectacular with much of the misty view in smaller binoculars easily resolved into stars. A 80mm (3.1 inch) telescope reveals many bright and some faint stars in the cluster with the members of M35 spread evenly about its diameter. Through a 150mm (6-inch) to 200mm (8-inch) telescope at low magnifications, M35 displays a field full of stars with many arranged in chains and lines. A brighter orange star lies near the centre. Larger telescopes do not significantly reveal more stars. In total, M35 contains a couple of hundred stars of which at least 120 are brighter than 13th magnitude.

There is another fainter open cluster that is located only about 15 arc minutes southwest of M35. This cluster is named NGC 2158 and at magnitude +8.6 it is much fainter than M35 although faintly visible in 10x50 binoculars under dark skies. It appears as a more compact cluster than M35 and is a nice sight when viewed through telescopes. M35 and NGC 2158 are intrinsically similar clusters, just that NGC 2158 appears much fainter and smaller as it is five times more distant than M35.

M35 is a wonderful open cluster that can be seen with the naked eye under dark skies. Covering an area of almost half a degree, it is a must see cluster that is a superb site in all types of optical instruments.

M35 Data Table

Messier	35		DEC (J2000)	24d 21m 28s
NGC	2168		Apparent Size (arcmins)	28 x 28
Object Type	Open cluster		Radius (light years)	12
Constellation	Gemini		Age (years)	100M
Distance (kly)	2.8		Number of Stars	200
Apparent Mag.	5.2		Other Name	Collinder 82
RA (J2000)	06h 08m 56s			

Messier # 035

Date:	Time:
Site:	
Temp:	Wind:	Hum:
Clouds:	Moon:
Scope:	
EP:	Mag:
NELM:	See/Trans:
Type:	# Stars:
Mag:	Age:
Const:	

Notes:

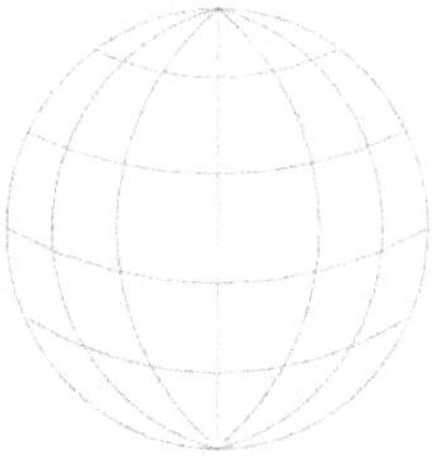

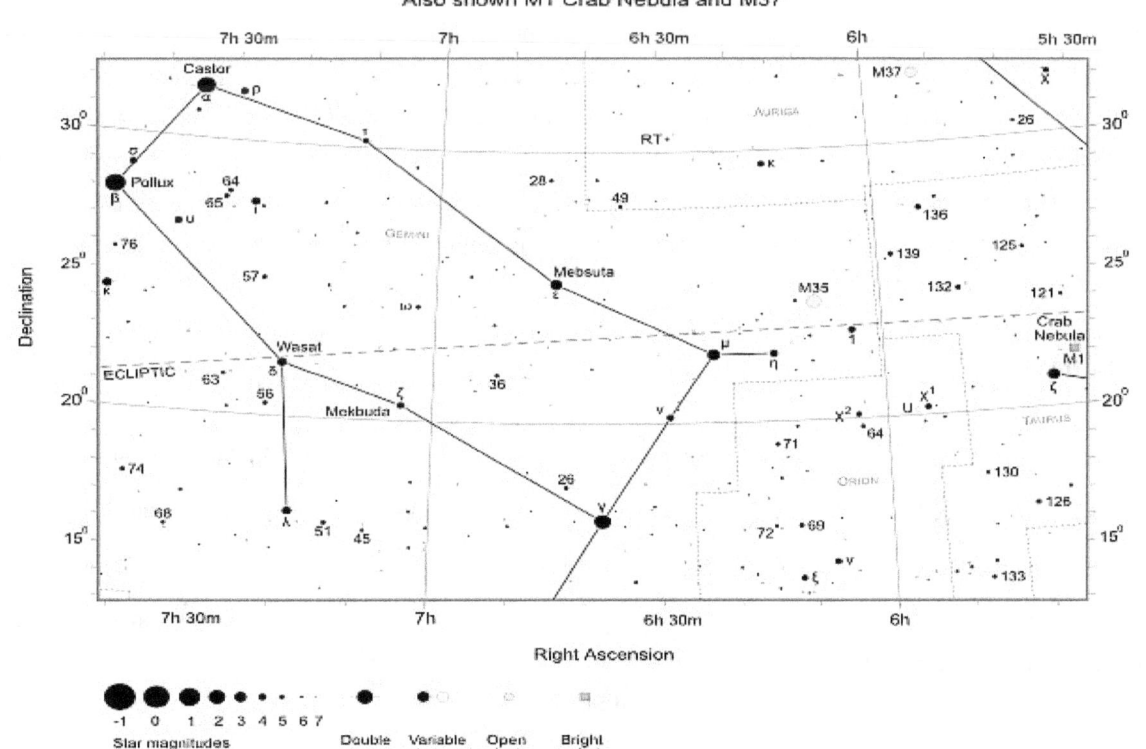

Messier Finder Chart for M35
Also shown M1 Crab Nebula and M37

Right Ascension

Declination

ECLIPTIC

Castor
ρ
α
1
σ
Pollux
β
64
65
ι
υ
76
57
ω
κ
Wasat
63
δ
66
ζ
Mekbuda
74
68
λ
51
45
26
ν
36
Mebsuta
ε
μ
η
ν
GEMINI
AURIGA
M37
X
26
RT
28
49
κ
136
139
125
132
121
M35
1
Crab Nebula
M1
ζ
X²
64
U
X¹
71
ORION
72
69
ν
ξ
130
126
133
TAURUS

Star magnitudes
-1 0 1 2 3 4 5 6 7

Double star   Variable stars   Open cluster   Bright nebula

Messier 36 - M36 - Open Cluster

M36 is the first of three bright Messier open clusters located in the southern part of the constellation of Auriga (the other two been M37 and M38). It was discovered by Italian astronomer Giovanni Batista Hodierna sometime before 1654 and catalogued by Charles Messier on September 2, 1764. At magnitude +6.2, the cluster is at the limit of naked eye visibility and therefore easily visible in binoculars, appearing as a small fuzzy patch of light. M36 is a concentrated cluster that contains at least 60 known members covering 12 arc minutes of apparent sky and is a particularly nice object for owners of small / medium sized telescopes. It's best seen from the Northern Hemisphere during the months of December, January and February. Finding M36 is not difficult once one is familiar with the northern constellation of Auriga, "the Charioteer". Locating Auriga is easy; it's bright with a prominent polygon shape at its core and positioned to the northeast of neighboring Taurus and to the northwest of neighboring Gemini. The northern star of the polygon is brilliant unmistakable magnitude +0.08 yellow star Capella (α Aur).Located 7.5 degrees to the east of Capella is magnitude +1.9 Menkalinan (β Aur) and 8 degrees due south of Menkalinan is θ Aur (mag. +2.65).

From θ Aur draw an imaginary line in the SW direction for 11 degrees until you reach another bright star. This star is El Nath (β Tau – mag. +1.65). El Nath straddles the border between Auriga and Taurus and is easily the brightest star in its region. Located just west of the mid-point of the line connecting θ Aur and El Nath is M36.M36 along with neighbor M37 can be spotted with the naked eye from exceptionally dark sites, both appearing as faint patches of light. M38 shines at magnitude +7.0 and is an extremely difficult naked eye target, even from the darkest locations. All three are easily within the range of good binoculars, with M36 looking like a compact fuzzy patch of light that hints at resolution. Binoculars with a sufficiently wide field of view (at least 6 degrees), allow the observer to view M36, M37 and M38 at the same time. When viewed through larger binoculars or small telescopes, the fuzziness of M36 is transformed into a sprinkling of stars. An 80mm (3.1-inch) telescope at low / medium powers reveals about 15 or so bright members scattered throughout the cluster. Most of the stars appear white or bluish white in color and are arranged in an "X" type shape or a pointed starfish. The brightest stars are of 8th and 9th magnitude and are of the very luminous B type blue stars that are of the order of 350 times more luminous than our Sun. Prominent near the center of the cluster is a double star with 10th magnitude components separated by 10 arc seconds. A large 150mm (6-inch) telescope reveals approximately 25 stars. Switch to averted vision and the view is practically unchanged with hardly any extra stars visible. Progressively larger telescopes will show more and more stars, with a 300mm (12-inch) telescope resolving most of them.M36 is superb open cluster that is about 4,100 light-years distant. It has an apparent diameter of 12 arc minutes, which corresponds to a spatial diameter of 14 light-years. It's the smallest of the three Auriga Messier open clusters and also the least populated. However, M36 does contain a number of individually bright stars compared to the other two clusters. In total, M36 has 60 known members and is easily visible in binoculars. Any telescope will at least partly resolve this fine cluster with larger telescopes completely resolving it. Spatially, M36 is a similar size cluster to that of the famous Pleiades (M45) and would appear at least as bright and spectacular as M45 if it were located as close. Since it's a young cluster of only about 25 million years age, it contains mainly hot white / blue-white stars and no red giants, unlike its neighbors M37 and M38.

M36 Data Table

Messier	36	Apparent Mag.	6.2		Radius (light years)	7
NGC	1960	RA (J2000)	05h 36m 18s		Age (years)	25M
Object Type	Open Cluster				Number of Stars	60
		DEC (J2000)	34d 08m 27s		Other Name	Collinder 71
Constellation	Auriga					
Distance (kly)	4.1	Apparent Size (arcmins)	12 x 12			

Messier # 036

Date:	Time:

Site:	

Temp:	Wind:	Hum:

Clouds:	Moon:

Scope:

EP:	Mag:

NELM:	See/Trans:

Type:	# Stars:

Mag:	Age:

Const:

Notes:

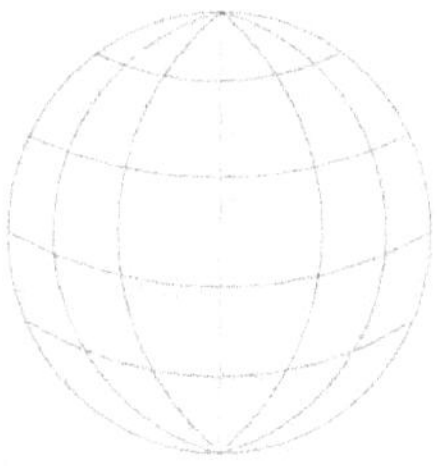

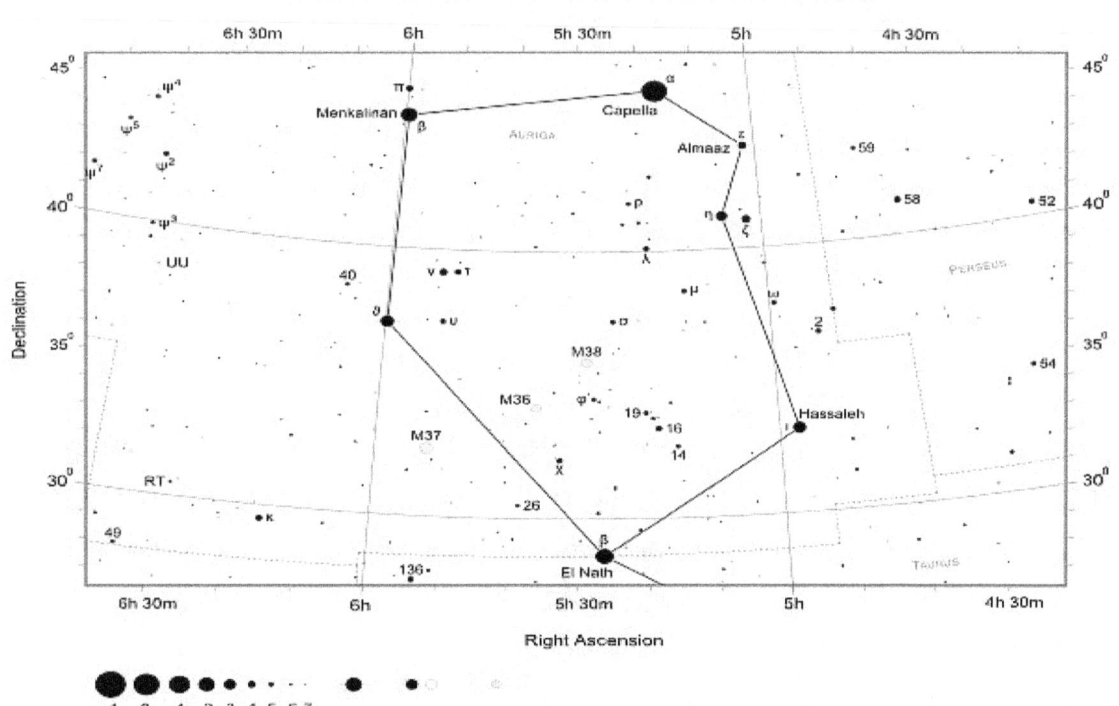

Messier Finder Chart for M36, M37 and M38

Messier 37 - M37 - Open Cluster

M37 is the brightest, largest and richest of the three Messier open clusters located in the constellation of Auriga (the other two been M36 and M38). It's an impressive cluster that shines at magnitude +6.0 and is visible to the naked eye from dark sites. M37 is often referred to as one of the finest open clusters in the northern section of the sky and is best seen from the Northern Hemisphere during the months of December, January and February. The Italian astronomer Giovanni Battista Hodierna discovered M37 sometime before 1654. French astronomer Guillaume Le Gentil rediscovered M36 and M38 in 1749, but surprisingly failed to spot the brighter M37. It was left to Charles Messier to independently rediscover M37, which he did so on September 2, 1764.As with M36 and M38, M37 is not difficult to find once you're familiar with the northern constellation of Auriga, "the Charioteer". Auriga is a medium size constellation with an area of 657 square degrees and ranks as 21st largest of the 88 constellations in the sky. It is positioned to the northeast of neighboring Taurus and to the northwest of neighboring Gemini. As well as been home to a number of bright open clusters it contains the brilliant unmistakable spectroscopic binary star Capella (α Aur) which at combined magnitude +0.08 is the 6th brightest star in the night sky. Auriga is a polygon shaped constellation with Capella located at the northern point. Located 7.5 degrees to the east of Capella is magnitude +1.9 Menkalinan (β Aur) and 8 degrees due south of Menkalinan is θ Aur (mag. +2.65).From θ Aur draw an imaginary line in the SW direction for 11 degrees until you reach another bright star. This star is El Nath (β Tau - mag. +1.65). El Nath straddles the border between Auriga and Taurus and is easily the brightest star in its region. Located just east of the mid-point of the line connecting θ Aur and El Nath is M37. Unlike M36 and M38, M37 is located on the outside of the main Auriga polygon. Both M36 and M38 lie inside the shape. Of the three Messier clusters in Auriga, M37 along with M36 can be spotted with the naked eye from exceptionally dark sites. They both appear as faint patches of light. M38 shines about a magnitude fainter than the other two; hence even from the darkest locations it remains an extremely challenging naked eye target. Together with M36 and M38, all three are easily visible with good binoculars. A pair of 10x50s depicts M37 as a large hazy almost nebula like patch of light. When viewed through larger 20x80 binoculars, M37 appears as a very compact cluster. The brightest stars are just about resolvable, especially when using averted vision. A small 100mm (4-inch) telescope reveals about a dozen or so tenth magnitude stars concentrated towards the centre of the cluster. With averted vision, tens of more stars are visible with the cluster taking on a mottled appearance. The stars appear faint and surrounded by a misty cloak of haze, giving the impression of a sprinkling of diamonds or stardust! A bright orange star towards the center of the cluster is apparent. At about 100x magnification through a 200 mm (8-inch) scope, M37 fills a good proportion of the field of view with hundreds of stars visible. The cluster appears wedged, triangular or even oblong shaped and covers 24 arc minutes of sky. It is believed the cluster contains at least 150 stars brighter than 12th magnitude and probably more than 500 stars in total. Larger telescopes resolve M37 further showing hundreds of mostly white stars, but also bluish-white stars, orange-red stars and even some yellow ones. The cluster is a sensational sight in all types of optical instruments.M37 is a superb open cluster located in the constellation of Auriga. It's brighter and much richer than neighboring open clusters M36 and M38 and is faintly visible to the naked eye at magnitude +6.0. Even a small 80mm (3.1-inch) telescope will start to resolve some of the stars with hundreds visible in larger telescopes. M37 is located 4,400 light years from Earth, which corresponds to a spatial diameter of about 30 light-years. It is estimated to be 300 million years old.

M37 Data Table

Messier	37	Apparent Mag.	6.0	Age (years)	300M
NGC	2099	RA (J2000)	05h 52m 18s	Number of Stars	>150
Object Type	Open Cluster	DEC (J2000)	32d 33m 11s	Other Name	Collinder 75
Constellation	Auriga	Apparent Size (arcmins)	24 x 24		
Distance (kly)	4.4	Radius (light years)	15		

Messier # 037

Date:		Time:	
Site:			
Temp:	Wind:		Hum:
Clouds:		Moon:	
Scope:			
EP:		Mag:	
NELM:		See/Trans:	
Type:		# Stars:	
Mag:		Age:	
Const:			

Notes:

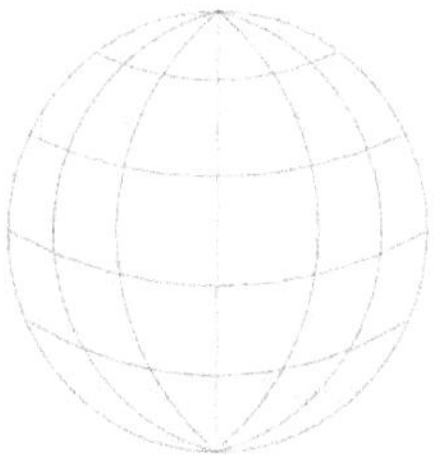

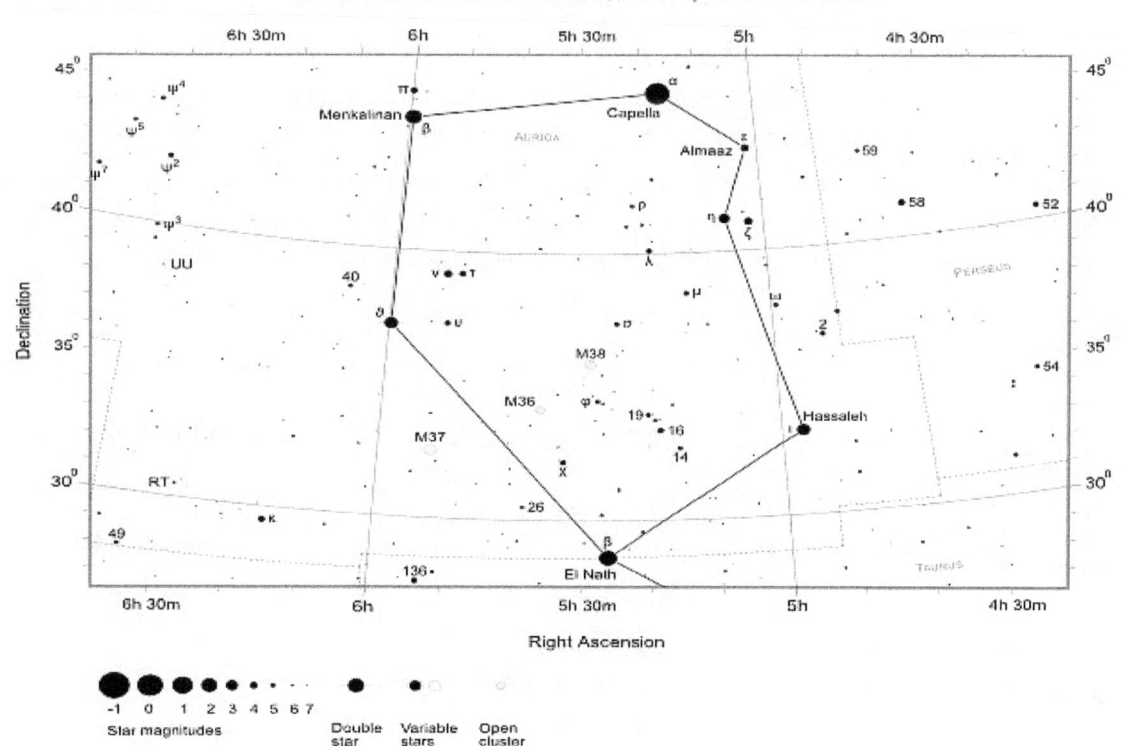

Messier Finder Chart for M36, M37 and M38

Right Ascension

Declination

Star magnitudes
-1 0 1 2 3 4 5 6 7

Double star

Variable stars

Open cluster

Messier 38 – M38 - Open Cluster

M38 is the third and faintest of the three Messier objects located in the constellation of Auriga (the other two been M36 and M37). With an apparent magnitude of +7.0 it's almost a magnitude dimmer than both M36 and M37, but still an easy binocular target. The cluster covers over 20 arc minutes of sky and hence due to its large size is best observed at low magnifications with small to medium size telescopes.

Finding M38 is relatively easy. Start by locating the brightest stars in the constellation of Auriga, "the Charioteer". This relatively large constellation is positioned to the northeast of neighboring Taurus and to the northwest of Gemini. The main stars of Auriga form an easy to find large polygon shape that is marked at the northern point by brilliant Capella (α Aur - mag. +0.08). To trace the polygon start with Capella then move eastwards in a circular like shape to Menkalinan (β Aur - mag. +1.90), followed by θ Aur (mag. +2.65), El Nath (β Tau - mag. +1.68), Hassaleh (ι Aur - mag. +2.69), then η Aur (mag. +3.18), Almaaz (ε Aur. - mag +3.03) and finally back to Capella. Now focus on stars θ Aur and El Nath and imagine a straight line connecting them. Located just east of the mid-point of this line is M37 and to the west M36. Move 2.3 degrees northwest of M36 and you will arrive at M38.As with M37, M38 was discovered by Giovanni Batista Hodierna sometime before 1654. The cluster was then rediscovered by French astronomer Guillaume Le Gentil in 1749 and subsequently added by Charles Messier to his catalog on September 25, 1764. It's best seen from northern latitudes during the months of December, January and February.

M38 is beyond naked eye visibility. When viewed through 10x50 binoculars it appears large and misty with the brightest stars just about resolvable, especially with averted vision. Larger 20x80 binoculars show a round, loose cluster of faint stars. The brightest members are easily visible with direct vision with many fainter stars scattered throughout the cluster. A 100mm (4-inch) telescope reveals many members mainly concentrated towards the center, but with trails of stars streaming outwards. The bright stars are arranged in an irregular pattern that resembles an arrow, "X" shape or as is often referred to, a version of the Greek letter Pi. When viewed through a 200mm (8-inch) scope at low powers, M38 is a nice sight. It almost fills the eyepiece field of view, a loose cluster with many bright stars visible in a circular type pattern. It is easy to make out the "X" shape amongst the stars and separating dark lanes. Quite a few stars are of the order of 9th or 10th magnitude with many fainter stars interspersed amongst them. The brightest member of the cluster is a yellow giant star (type G0) of magnitude +7.9, while many stars appear grouped in pairs. Adjacent to M38 on the southwest side is NGC 1907, a much smaller and fainter open cluster. Although M38 is the faintest of the three Auriga Messier open clusters it is still an excellent object that is best viewed with binoculars or through small / medium sized telescopes at low magnifications. The most prominent feature of the cluster is a grouping of the bright stars that form a noticeable "X", arrow or Pi shaped structure.M38 is located 4,200 light years from Earth and has an apparent diameter of 21 arc minutes, which corresponds to a spatial radius of 25 light-years. In total it contains about 100 stars and has an estimated age of 220 Million years.

M38 Data Table

Messier	38
NGC	1912
Object Type	Open Cluster
Constellation	Auriga
Distance (kly)	4.2

Apparent Mag.	7.0	
RA (J2000)	05h 28m 43s	
DEC (J2000)	35d 51m 18s	
Apparent Size (arcmins)	21 x 21	

Radius (light years)	25
Age (years)	220M
Number of Stars	100
Other Name	Collinder 67

Messier # 038

Date:	Time:
Site:	
Temp:	Wind:	Hum:
Clouds:	Moon:
Scope:	
EP:	Mag:
NELM:	See/Trans:
Type:	# Stars:
Mag:	Age:
Const:	
Notes:	

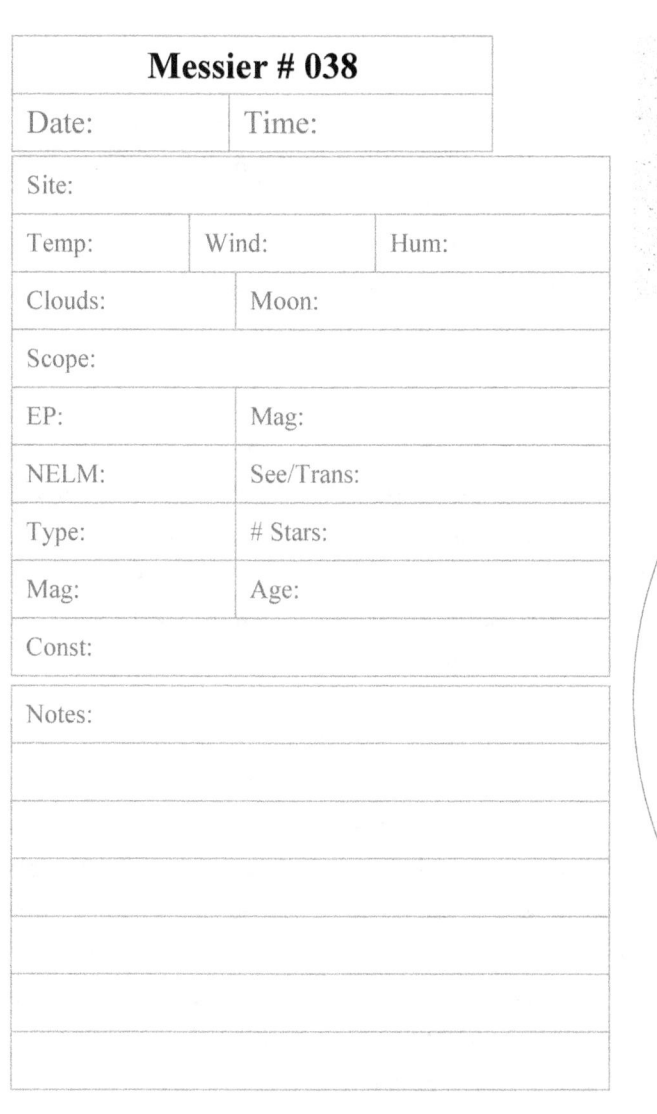

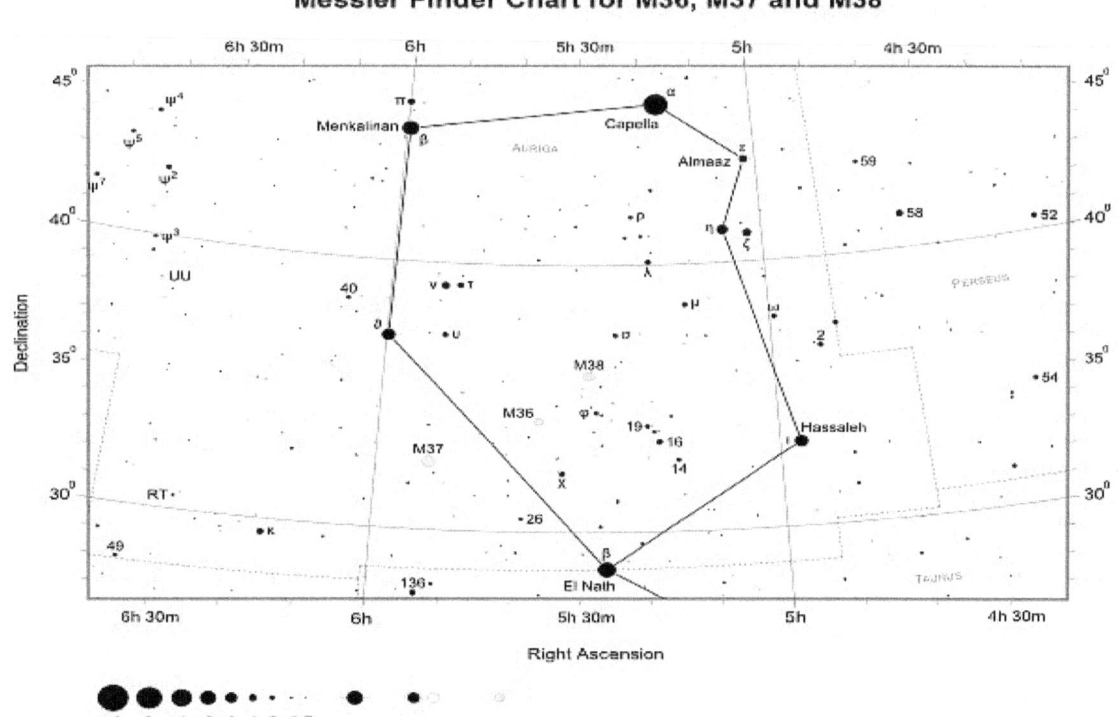

Messier Finder Chart for M36, M37 and M38

Messier 39 - M39 - Open Cluster

M39 is a very loose naked eye open cluster of at least 30 young stars located in a dense portion of the Milky Way in Cygnus "the Swan". The cluster is positioned in a beautiful rich region of the sky, with many young open clusters, thousands of stars, regions of dark nebulae, wispy emission and reflection nebulosity visible nearby.

The discovery of M39 is an interesting story. It's often credited as one of Charles Messier's original discoveries, which he made on October 24, 1764. However, the discovery has also been credited to French astronomer Guillaume Le Gentil in 1750 and even suggested that Aristotle observed it as far back as 325 BC.

The cluster is a relatively easy object to find. It's located in the northeastern part of Cygnus, not far from the constellation's brightest star, Deneb (α Cyg – mag. +1.25). This blue-white luminous super-giant star forms the faintest corner of the well-known and much observed "Summer Triangle". The other two stars that make up the triangle are Vega (α Lyr – mag. +0.03) and Altair (α Aqr – mag. +0.77). Once you have located Deneb, move approx. 4.5 degrees east-southeast to ξ Cyg (mag. +3.72) and then hop another 5.5 degrees northeast to ρ Cyg (mag. +3.98). Positioned 3 degrees directly north of this star is M39. The cluster is best seen from northern latitudes during July, August and September.

Of all of the open clusters in Cygnus, M39 is one of the brightest. With an apparent magnitude of +5.5, it's faintly visible to the naked eye and an easy binocular object. M39 is also one of the constellations largest clusters and is spread over an apparent diameter of 31 minutes, which corresponds to about the same size as that of the full Moon. Through 7x50 or 10x50 binoculars, M39 appears as a large, loose, misty triangular shaped cluster that easily stands out from the background stars. The cluster is probably best seen and at its most impressive when viewed through a small or medium sized telescope at lower powers. For example, a 80mm (3.1 inch) refractor reveals many bright stars in an almost "W" type form, with the brightest member being of magnitude +6.8. Dotted in and around then main shape are numerous fainter stars. Despite its looseness the cluster is impressive when observed at lower powers. Larger scopes reveal at least 30 stars in total, although at higher magnification the cluster can over spill the eyepiece field of view and hence some of its appeal is lost.

M39 is best viewed from northern temperate latitudes and the tropics during the warm summer months where it appears high in the sky. The cluster is only 800 light-years from Earth and has an estimated age of 250 million light-years. What is perhaps surprising is that M39 is one of only two Messier objects located in the rich constellation of Cygnus; the other being M29.

M39 Data Table

Messier	39		**DEC (J2000)**	48d 26m 55s
NGC	7092		**Apparent Size (arcmins)**	31 x 31
Object Type	Open cluster		**Radius (light years)**	3.6
Constellation	Cygnus		**Age (years)**	250M
Distance (kly)	0.8		**Number of Stars**	>30
Apparent Mag.	5.5		**Other Name**	Collinder 438
RA (J2000)	21h 31m 48s			

Messier # 039

Date:	Time:

Site:	

Temp:	Wind:	Hum:

Clouds:	Moon:

Scope:

EP:	Mag:
NELM:	See/Trans:
Type:	# Stars:
Mag:	Age:
Const:	

Notes:

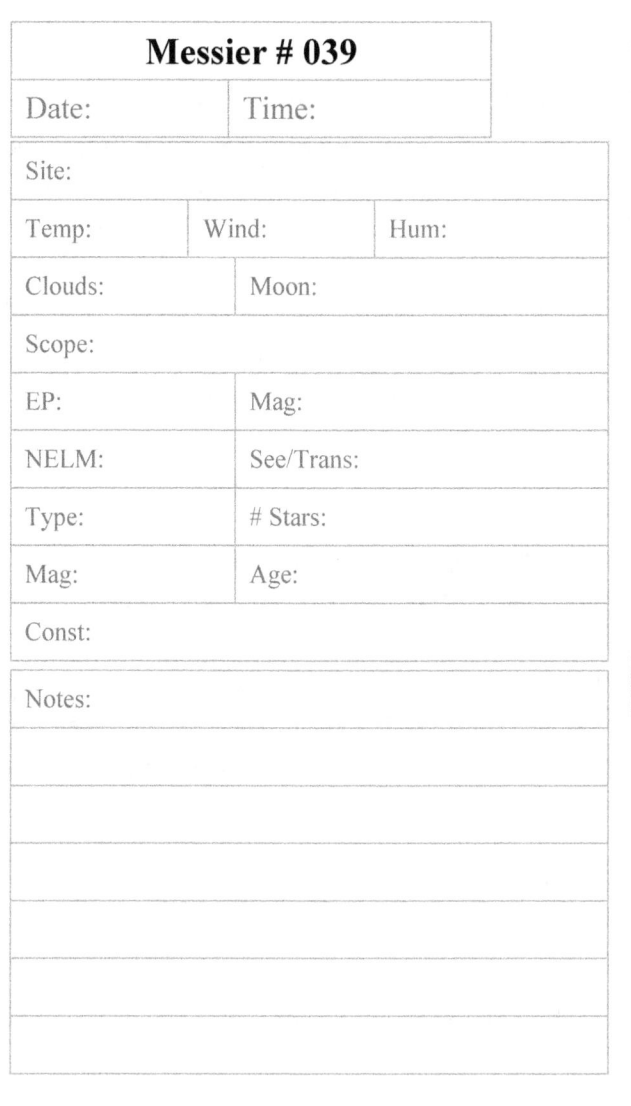

Messier Finder Chart for M39
Also shown M29

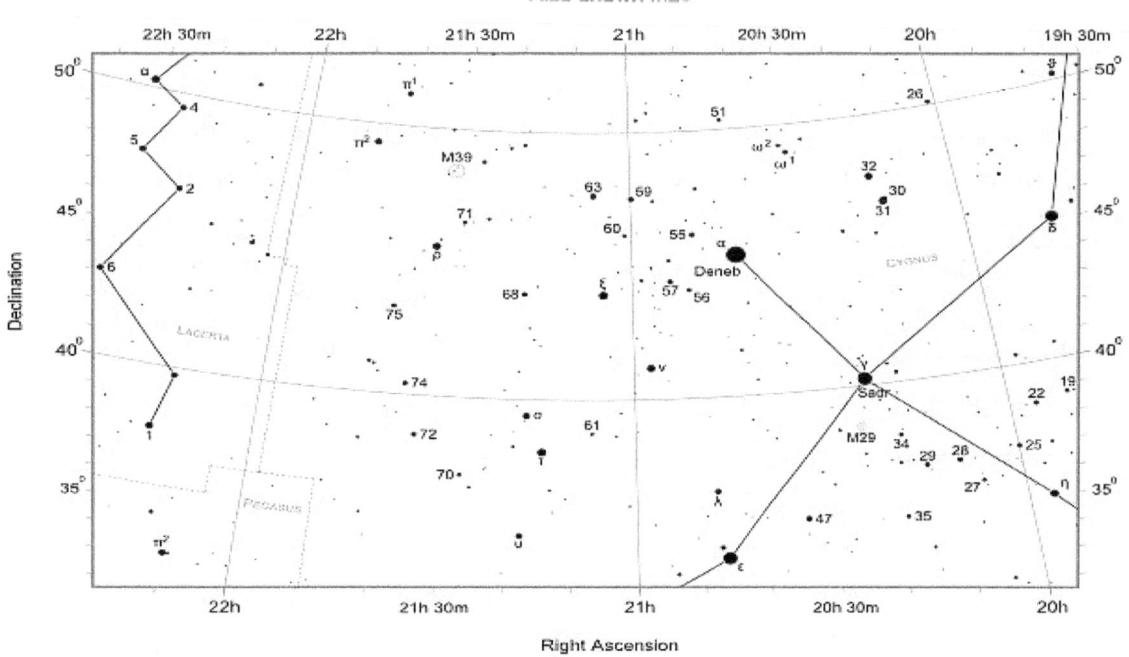

Messier 40 - M40 - Winnecke 4 (Double Star)

M40 is one of three curiosities in the Messier catalog (along with M73 and M102). It's a faint double star in the constellation of Ursa Major that was discovered by Charles Messier on October 24, 1764. Messier was searching for a nebula reported in the area by Johann Hevelius. Although not seeing any nebula, Messier catalogued this double star instead. However, despite no nebulosity existing the double star remained on the list.

American astronomer Robert Burnham called M40 "one of the few real mistakes in the Messier catalogue". He faulted Messier for including it when he found no trace of a nebula and all he saw was a double star. It was rediscovered in 1863 by Friedrich August Theodore Winnecke and hence is sometimes referred to as Winnecke 4 or WNC 4. M40 is best seen from the Northern Hemisphere during the months of February, March and April.

Locating M40 is easy; it lies about 1.5 degrees to the northeast of Megrez (δ UMa). With an apparent magnitude of +3.3, Megrez is the dimmest of the seven stars that make up the famous "Big Dipper" or "Plough" asterism of Ursa Major. Positioned 17 arcminutes southwest of M40, on an imaginary line connecting M40 with Megrez, is star 70 UMa (mag. +5.5).

The two component stars of M40 are of magnitude +9.65 and +10.1 and are currently separated by 51.7 arcseconds. The double star is faintly visible in 10x50 binoculars although much easier through larger 20x80 binoculars, where it's splittable. A small or medium size telescope reveals a pair of widely spaced unimpressive stars. The brighter star is orange-yellow in color, the fainter one white.

The separation between the components of M40 has increased since Messer's days, strongly suggesting that this is merely an optical double star rather than a physically connected system.

The galaxy NGC 4290, a barred spiral of 12th magnitude lies nearby M40. However, this galaxy is faint and could not have been the nebula recorded by Hevelius.

M40 Data Table

Messier	40
Object Type	Double Star
Constellation	Ursa Major
Distance (kly)	0.51
Apparent Mag.	9.0
RA (J2000)	12h 22m 13s
DEC (J2000)	58d 04m 59s
Apparent Separation (arcsecs)	51.7

Number of Stars	2
Star A - Apparent Mag.	9.65
Star A - Spectral Type	G0
Star B - Apparent Mag.	10.1
Star B - Spectral Type	F8
Other Name	Winnecke 4

Messier # 040

Date:	Time:
Site:	
Temp:	Wind:	Hum:
Clouds:	Moon:
Scope:	
EP:	Mag:
NELM:	See/Trans:
Type:	# Stars:
Mag:	Age:
Const:	
Notes:	

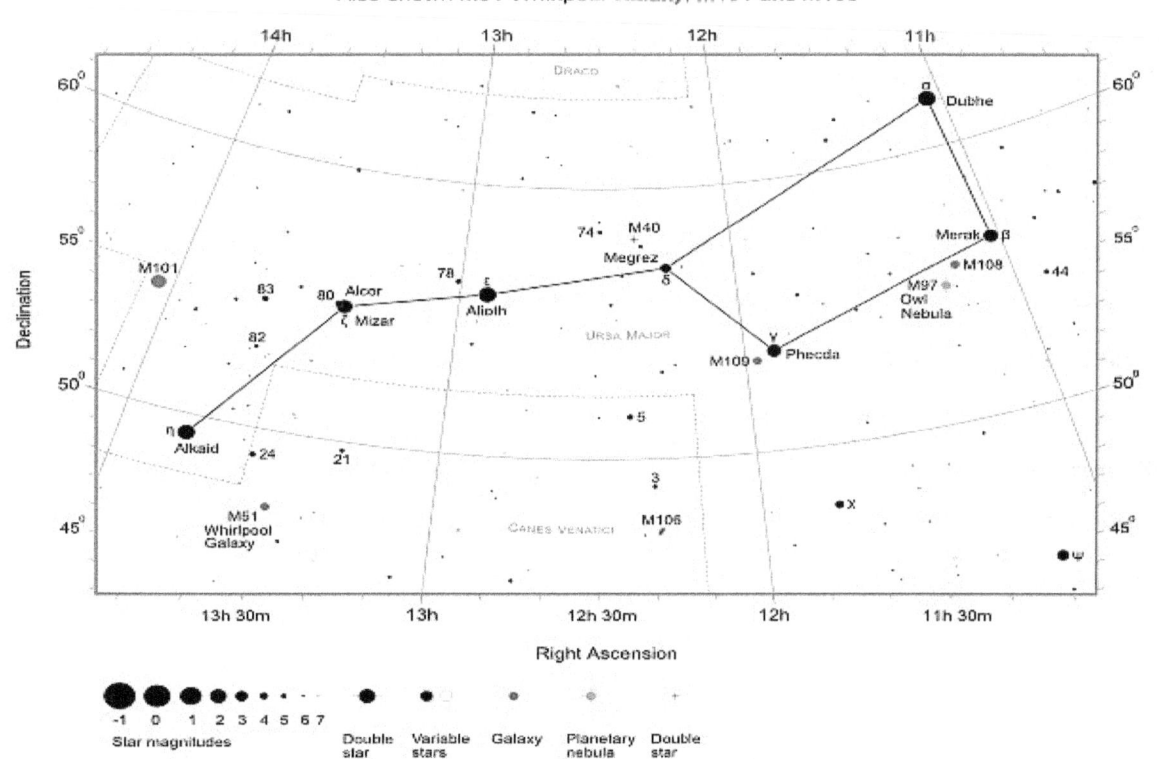

Messier Finder Chart for M40, M97 Owl Nebula, M108 and M109
Also shown M51 Whirlpool Galaxy, M101 and M106

Messier 41 - M41 - Open Cluster

M41 is a stunning, large bright open cluster located in the constellation of Canis Major. Of the many open clusters located in Canis Major, M41 is the stand out cluster. It is the single Messier object in the constellation and a painless item to find; located only four degrees south of Sirius (α CMa) the brightest star in the sky! With an apparent magnitude of +4.5 and diameter of 39 arc minutes, M41 is visible to the naked eye and a rewarding sight when viewed through all types of optical instruments.

The cluster was discovered by Giovanni Batista Hodierna sometime before 1654, although it was perhaps known to Aristotle as far back as 325 BC. If true, this would make it the "faintest object recorded in classical antiquity", but caution prevails as Aristotle may have described part of the nearby Milky Way instead. The cluster was catalogued by Charles Messier on January 16, 1765.

M41 is better seen during the months of December, January and February from the Southern Hemisphere. It's also visible from most parts of the Northern Hemisphere but appears lower down.

Appearing to the naked eye as a cloud like smudge, M41 is easily found with binoculars. In 10x50s, it appears as a large hazy blur with at least half a dozen stars resolvable. Larger 20x80 binoculars reveal many more stars spread across an almost circular shape. Through a small 100mm (4-inch) telescope at low magnifications, M41 is a beautiful sight with 50 or more stars visible against a mottled background. The stars appear to stream outwards from the centre of the cluster. A 150mm (6-inch) to 200mm (8-inch) scope reveals a loosely scattered multi-colored cluster with a multitude of additional fainter stars visible. Most of the stars appear white, but there are a few notable exceptions. These include a pair of bright yellow stars at the heart of the cluster and several red (or orange) giant stars interspersed. At high magnifications, M41 loses its cluster appearance but the individual star colors are much more intense.

M41 is estimated to be about 215 Million years old and is located 2,300 light years from Earth. At the southeast edge of the cluster is 12 Canis Majoris, a blue-white B-type giant star with a magnitude of +6.1. At only 668 light years distant, this star is much closer to Earth and hence an interloper and not a true member of the cluster.

In total, M41 spans an area greater than that of the full Moon and contains about 100 stars with the brightest being of magnitude +6.9. It is superb cluster, visible to the naked eye and a wonderful sight in binoculars and telescopes.

M41 Data Table

Messier	41
NGC	2287
Object Type	Open Cluster
Constellation	Canis Major
Distance (kly)	2.3
Apparent Mag.	4.5
RA (J2000)	06h 46m 00s

DEC (J2000)	-20d 45m 15s
Apparent Size (arcmins)	39 x 39
Radius (light years)	12.5
Age (years)	215M
Number of Stars	100
Other Name	Collinder 118

Messier # 041

Date:	Time:
Site:	
Temp:	Wind:	Hum:
Clouds:	Moon:
Scope:	
EP:	Mag:
NELM:	See/Trans:
Type:	# Stars:
Mag:	Age:
Const:	
Notes:	

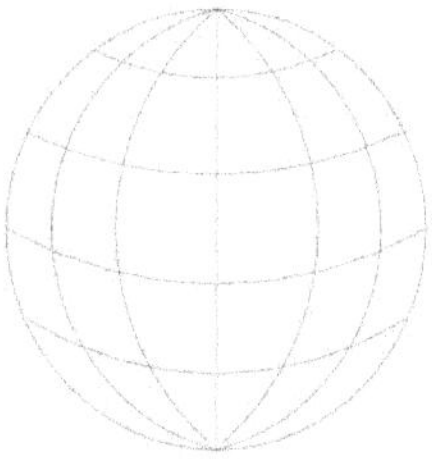

Messier Finder Chart for M41, M46, M47 and M93

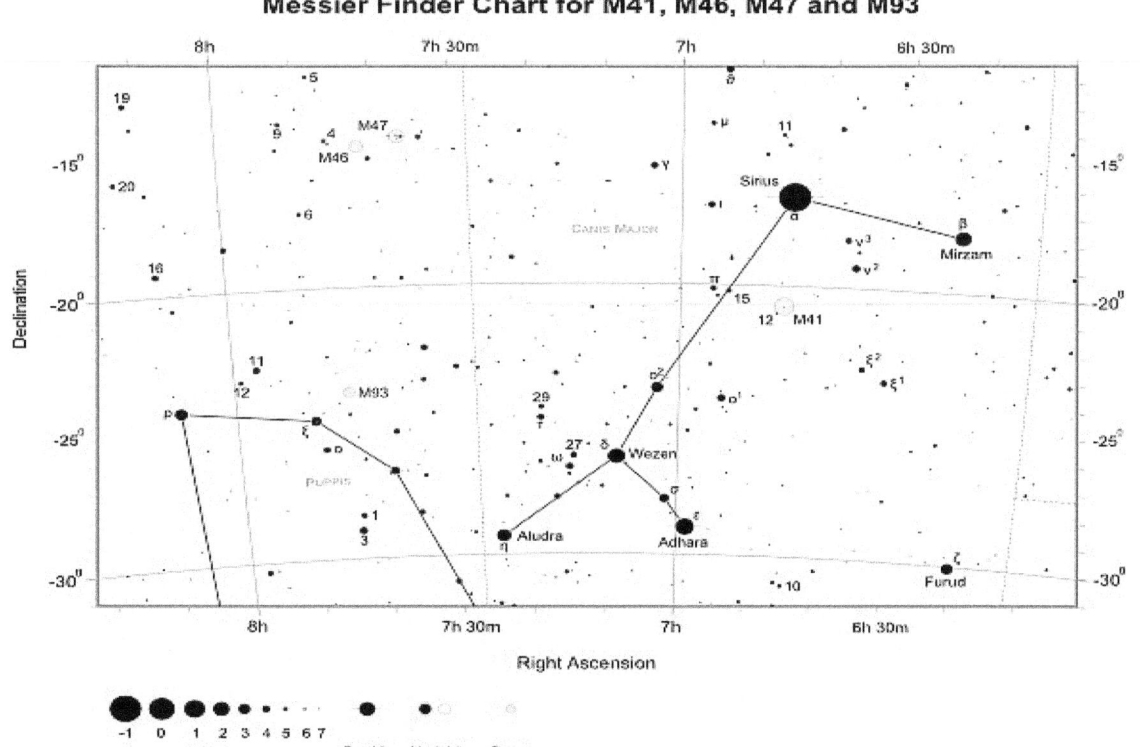

Star magnitudes: -1 0 1 2 3 4 5 6 7
Double star   Variable stars   Open cluster

Messier 42 - M42 - The Great Orion Nebula (Emission/Reflection)

M42 the Great Orion Nebula or Orion Nebula is the prime deep sky attraction in the constellation of Orion and a showpiece object of the night sky. With an apparent magnitude of +4.0, the Orion Nebula is easily visible to the naked eye. Covering more than one degree of apparent sky it's also one of the largest of its type. The nebula is of the emission and reflection variety, a star forming region that is one of the brightest nebulae in the entire sky.

Orion "The Hunter" is a prominent constellation and one of the most recognizable and familiar sights in the night sky. Located on the celestial equator it's visible throughout the world and best seen during the months of December, January and February. The constellation is filled with bright stars including first magnitude stars Rigel and Betelgeuse plus a further five second magnitude stars. Three of those second magnitude stars (Mintaka, Alnilam and Alnitak) form the famous belt of Orion. Positioned just 5 degrees south of the belt is the Orion Nebula itself, which forms part of the Hunters Sword.

To the naked eye M42 appears as a soft diffuse glow that surrounds the stars of the sword of Orion. When viewed through 10x50 binoculars it is a prominent feature, appearing large and bright. The centre region is obvious with parts of fainter nebulosity extending outwards towards the east and west in the shape of two wings. Two prominent bright stars are easily visible at the heart of M42. These stars are embedded within the Orion Nebula and form the famous multiple star Theta1 Orionis, commonly known as the Trapezium. The cluster, which is one of the most observed multiple star systems, consists of four bright members in the shape of a trapezoid plus a few fainter stars. A 80mm (3.1-inch) telescope at low/medium power will easily split the Trapezium into its main components; a fantastic sight combined with the surrounding nebula. The four brightest stars of the Trapezium have magnitudes of +5.1(C), +6.7(D), +6.7->7.7(A) and +8.0->8.7(B) respectively. Unusually, they are lettered in order of right ascension instead of magnitude, as common for other multiple stars. A challenge for observers of the Trapezium is to spot two of the fainter members, 11th magnitude stars E and F. They can be observed with apertures of only 80mm (3.1 inches) on nights of very good seeing, but much easier with telescopes of the order of 150mm (6 inches) or greater.

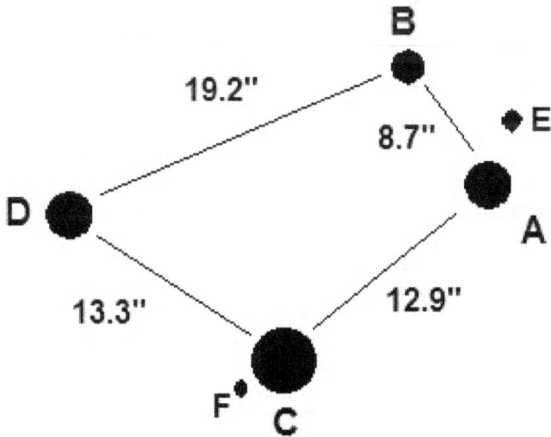

The Trapezium (high magnification view)

Large telescopes reveal more intrigue details in M42. The view through a 200mm (8-inch) telescope is superb. At low magnifications, the Orion Nebula fills the field of view with significant amounts of structural detail visible as twists and wisps of cloud formations. The Trapezium is very evident and bright. Visually the Orion Nebula can often exhibit a green hue but photographically it appears mostly red.

The discovery of the Orion Nebula is generally credited to French astronomer Nicolas-Claude Fabri de Peiresc. He recorded it on November 26, 1610. It was then independently located by Johann Baptist Cysat in 1611. Surprisingly, neither Ptolemy's Almagest nor Al Sufi's Book of Fixed Stars noted this nebula, even though they both listed patches of nebulosity elsewhere in the night sky. Galileo Galilei made telescopic observations of the surrounding region in 1610 and 1617, but also failed to notice the nebula. However, he did discover the Trapezium on February 4, 1617. This has led to some to speculate that a more recent flare-up of the illuminating stars may have increased the brightness of the nebula.

The Orion Nebula is a spectacular deep sky object and one of the most famous of all. It's visible to the naked eye as a faint haze, is a wonderful sight in binoculars and spectacular in telescopes. You are looking at a stellar nursery where stars are been born. At the heart of the cluster and illuminating the surrounding area is a group of stars known as the Trapezium; a multiple star system consisting of four main bright members that are easily resolvable in small telescopes.

M42 is located 1340 light years from Earth and has a diameter of 24 light years. Located only 8 arc minutes north of M42 is M43, which is also part of the Orion Nebula. It is separated from the main nebula by a dark dust lane.

M42 Data Table

Messier	42
NGC	1976
Name	Orion Nebula
Object Type	Emission and Reflection Nebula
Constellation	Orion
Distance (kly)	1.34
Apparent Mag.	4.0
RA (J2000)	05h 35m 17s
DEC (J2000)	-05d 23m 27s
Apparent Size (arcmins)	65 x 60
Radius (light years)	12
Other Name	Sharpless 281
Notable Feature	Trapezium Cluster

Messier # 042

Date:	Time:

Site:

Temp:	Wind:	Hum:

Clouds:	Moon:

Scope:

EP:	Mag:

NELM:	See/Trans:

Type:	# Stars:

Mag:	Age:

Const:

Notes:

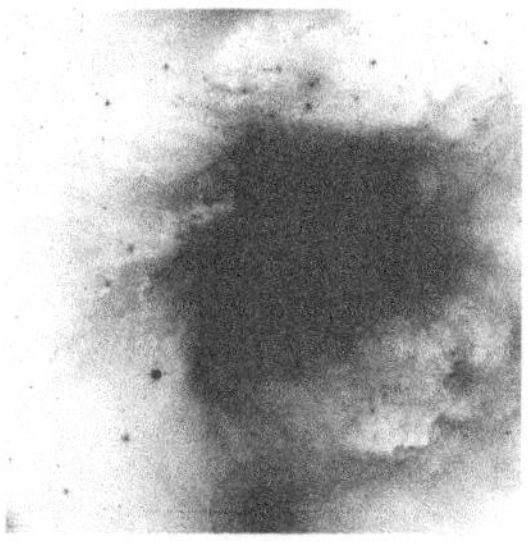

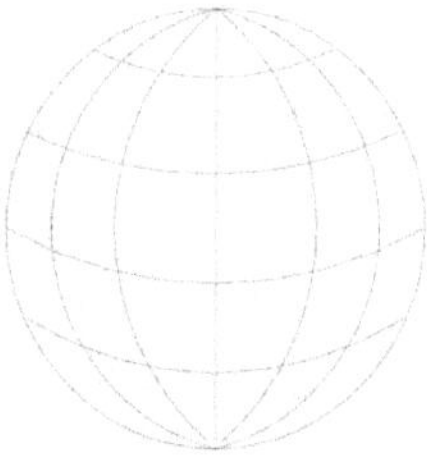

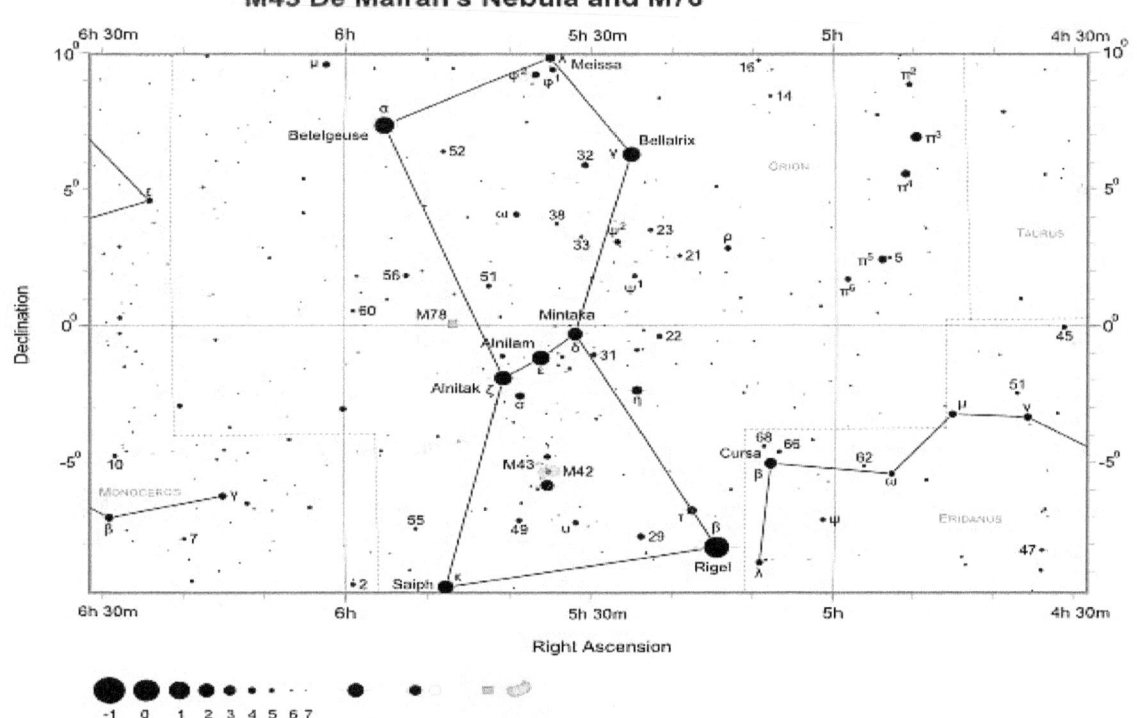

**Messier Finder Chart for M42 Great Orion Nebula,
M43 De Mairan's Nebula and M78**

Messier 43 - M43 - De Mairan's Nebula (Emission/Reflection)

M43 is an H II region located in the constellation of Orion that was discovered by Jean-Jacques Dortous de Mairan sometime before 1731. As part of the famous Orion Nebula (M42), it's positioned just north of the main nebula and separated from it by a narrow dust lane. With an apparent magnitude of +9.0, the nebula is about 100 times fainter than M42, but still bright enough to be seen with binoculars.

Occasionally, 9th magnitude nebulae like M43 can be difficult to find - especially if located in barren parts of the sky - but not M43. Firstly, it's located in majestic Orion, perhaps the most recognizable constellation of all, secondly it's part of the Orion Nebula and hence positioned right next to the great showpiece object and finally it has a relatively high surface brightness. Of course, finding M42 is easy, it's positioned just 5 degrees south of the three bright stars that form Orion's belt (Mintaka, Alnilam and Alnitak). M43 is located just 8 arc minutes north of M42 surrounding a 7th magnitude star. M43 (and M42) are best seen during the months of December, January and February.

In 10x50 binoculars, M43 appears as a small, faint fuzzy elongated patch with a noticeably brighter centre. NU Orionis the irregular young star at the heart of M43 that excites the gas causing the nebula to shine is easily visible in binoculars. It varies in brightness between magnitudes +6.5 and +7.6.

M43 is a nice sight in larger 20x80 binoculars, with the nebula appearing defined and looking somewhat comma shaped. With a 100mm (4-inch) telescope, the centre star is bright. M43 doesn't show as much range in brightness or detail as M42 but it has a bright ring around the star which in turn is surrounded by the fainter comma shaped nebulosity. The dark lane separating M43 from the gigantic Orion Nebula is easily visible. Larger telescopes of the order 200mm (8-inch) reveal subtle details such as tendrils, smoky wisps and dark and light areas across the surface of M43, particularly along the eastern side.

M43 is 1600 light-years from Earth and has an apparent size of 20 x 15 arc minutes, which corresponds to a spatial diameter of 9 light-years. It's dwarfed by the much brighter M42 Orion Nebula which is separated from M43 by a dark dust lane. Physically M43 is a part of the Orion Nebula and both M42 and M43 are part of the much larger Orion Molecular Cloud Complex.

Historically, when the first photograph of the Orion Nebula was made by Henry Draper on March 14, 1882, M43 was also visible.

M43 Data Table

Messier	43
NGC	1982
Name	De Mairan's Nebula
Object Type	H II region
Constellation	Orion
Distance (kly)	1.6
Apparent Mag.	9.0

RA (J2000)	05h 35m 31s
DEC (J2000)	-05d 16m 03s
Apparent Size (arcmins)	20 x 15
Radius (light years)	4.5
Notable Feature	Part of the Orion Nebula

Messier # 043

Date:	Time:

Site:

Temp:	Wind:	Hum:

Clouds:	Moon:

Scope:

EP:	Mag:
NELM:	See/Trans:
Type:	# Stars:
Mag:	Age:

Const:

Notes:

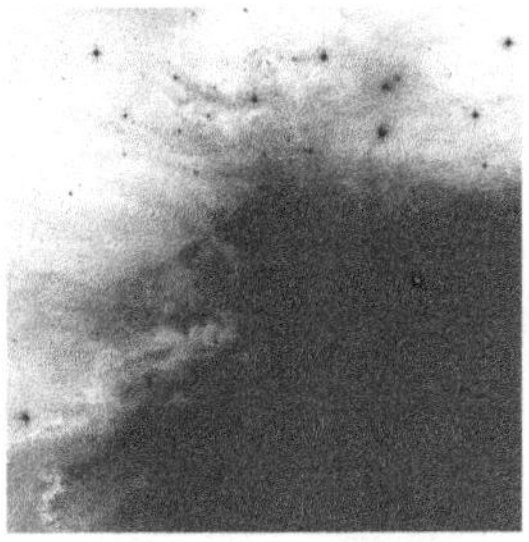

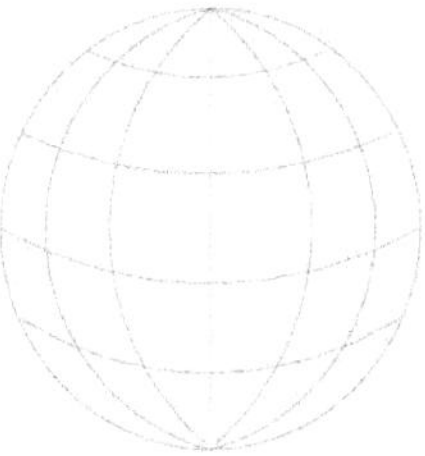

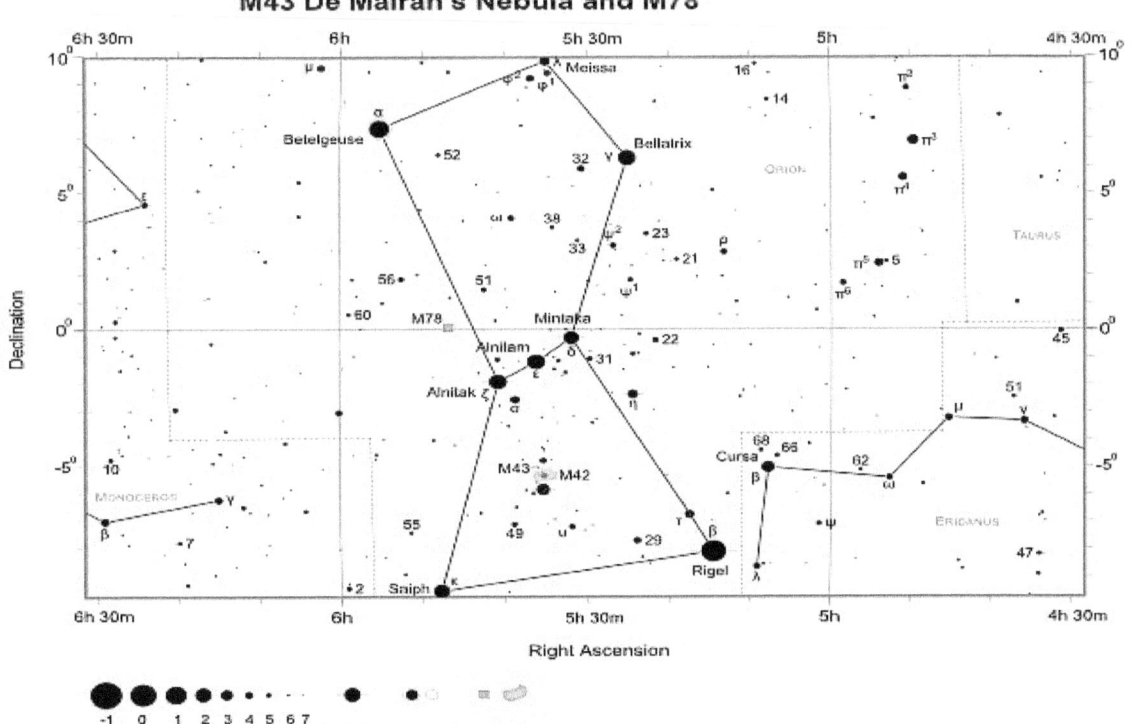

Messier Finder Chart for M42 Great Orion Nebula, M43 De Mairan's Nebula and M78

Messier 44 - M44 - The Praesepe (Open Cluster)

M44 is a sprawling open cluster that is also known as the Praesepe or Beehive cluster. It's the brightest and most prominent deep sky object in the constellation of Cancer. Visible to the naked eye under dark skies, the cluster appears like a large misty cloud covering over 1.5 degrees of sky. The brightness and size of M44 results from its close proximity to Earth; it lies a mere 577 light-years distant. Only the Hyades (at 153 light-years), Coma cluster (280 light-years), M45 Pleiades (425 light-years), Southern Pleiades (480 light-years) and IC 2391 (500 light-years) are nearer. Consequently, M44 is one of the brightest and largest objects of its type in the night sky.

The constellation of Cancer "the Crab" is a faint zodiac constellation that is bordered by much brighter Leo to the east and Gemini to the west. To the north is faint Lynx, with Canis Minor and Hydra located on the southern side. At the heart of Cancer is a grouping of four faint stars. They are Asellus Australis (δ Cnc - mag. +3.9), Asellus Borealis (γ Cnc - mag. +4.7), η Cnc (mag. +5.3) and θ Cnc (mag. +5.3). Of these, the brighter two stars are relatively easy naked eye objects, the fainter ones more difficult. Positioned at the centre of this grouping is M44. An alternative method to locate M44 is to imagine a line extending in a southeastern direction from Pollux (β Gem - mag. +1.1) for 37 degrees to Regulus (α Leo - mag. +1.4). M44 is positioned approximately at the mid-point of this line.

Nearly 2000 years ago, the astronomer Ptolemy described M44 as "the nebulous mass in the breast of Cancer", but it was Galileo who first telescopically observed the Praesepe. He did this in 1609 and was able to resolve 40 stars. Charles Messier added the item to his catalog on March 4, 1769. The cluster is best seen from the Northern Hemisphere during February, March and April.

With an apparent magnitude of +3.7, the Praesepe is a relatively easy naked eye object. It can be detected with even moderate amounts of light pollution. Under good conditions it's clearly visible as a large patch of nebulosity covering 3x the diameter of the full Moon. Although better seen with averted vision, no stars are resolvable. Due to its large size, M44 is a superb binocular and small telescope object. When viewed through 10x50 binoculars it bursts into life, neatly filling a good proportion of the field of view with dozens of stars sprinkled throughout a hazy background.

The extra aperture and magnification range afforded by small or medium size telescopes bodes well when observing M44. With a 80mm (3.1-inch) telescope at about 30 to 40x magnification, the Praesepe is wonderful with many stars arranged in pairs and triplets in equatorial shape triangle groupings. Most stars appear bluish-white, but there are at least four orange stars visible. Higher magnification reveals fainter background stars. In medium aperture telescopes the Praesepe is still a superb cluster. The star colors are prominent but the cluster appears loose and tends to lose some of its awe. With a 200mm (8-inch) telescope, it doesn't take much magnification before the cluster overflows the eyepiece field of view.

In total, M44 contains at least 350 members. The brightest member is ε Cnc, a magnitude +6.3 hot blue-white A type star. There are at least another twenty stars brighter than magnitude +8.0. Many different types of stars exist including red giants and white dwarfs along with main sequence stars of spectral classes A, F, G, K, and M. Since M44 is located just north of the ecliptic plane, you often find the Moon or a planet weaving their way through the cluster.

M44 is a fantastic open cluster and one of the nearest open clusters to Earth. It's a large sprawling naked eye object that covers 3x the size of the full Moon and as a result is best seen with binoculars or a small telescope. Any optical aid will start to resolve the brightest stars with dozens visible in telescopes, although through larger instruments it appears loose and consequently some of its star appeal is lost.

M44 Data Table

Messier	44
NGC	2632
Name	Praesepe
Object Type	Open cluster
Constellation	Cancer
Distance (kly)	0.577
Apparent Mag.	3.7
RA (J2000)	08h 40m 22s
DEC (J2000)	19d 40m 19s
Apparent Size (arcmins)	95 x 95
Radius (light years)	8
Age (years)	650M
Number of Stars	350
Other Name	Beehive Cluster

Messier # 044

Date:		Time:	
Site:			
Temp:	Wind:		Hum:
Clouds:	Moon:		
Scope:			
EP:		Mag:	
NELM:		See/Trans:	
Type:		# Stars:	
Mag:		Age:	
Const:			

Notes:

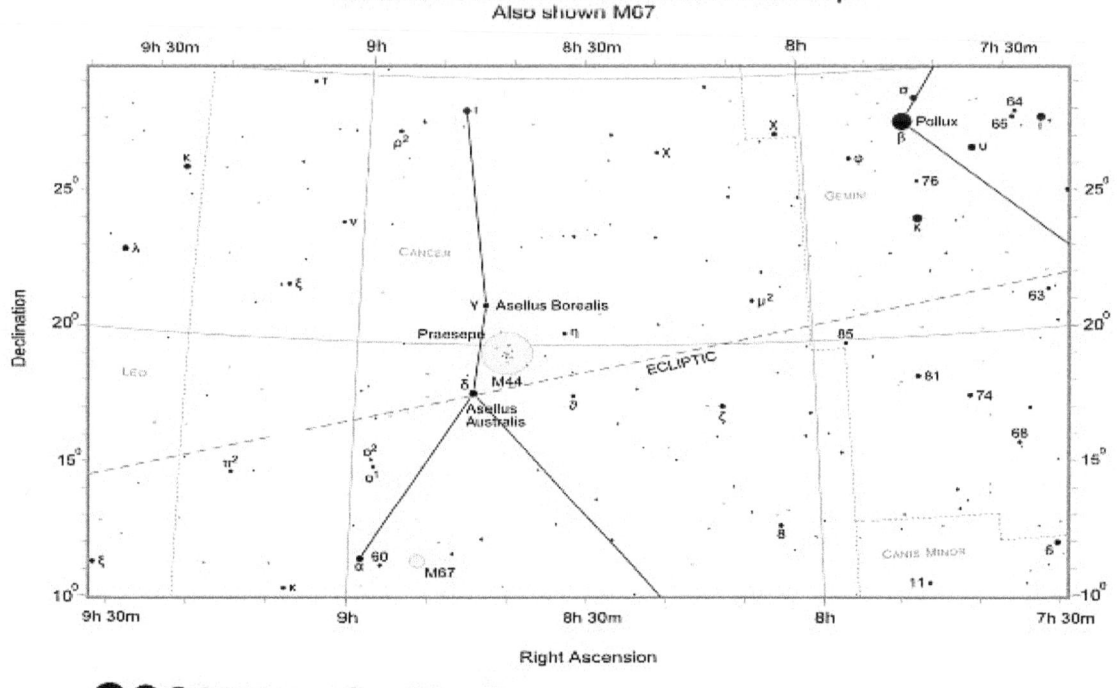

Messier Finder Chart for M44 Praesepe
Also shown M67

Star magnitudes

Double star Variable stars Open cluster

Messier 45 - M45 - The Pleiades (Open Cluster)

M45, commonly known as the Pleiades or Seven Sisters is the finest open cluster in the sky. It's a breathtaking site, known since antiquity and easily visible to the naked eye. Located in the large and prominent zodiac constellation of Taurus "the Bull", it's best seen during the Northern Hemisphere winter and the Southern Hemisphere summer.

Finding M45 is easy. The cluster is positioned about 14 degrees to the northwest of orange giant star Aldebaran (mag. +0.9), the brightest star in Taurus. At first glance with the naked eye it's obvious that something is special about this small section of sky. On closer inspection, M45 reveals itself to be a beautiful cluster, looking like a hazy grouping of about half a dozen white stars covering an area much greater than that of the full Moon. At the heart of the cluster is a set of stars that form a small dipper shape, similar to the brightest stars of Ursa Major. Even under moderately light polluted skies the dipper shape is readily visible. From dark sites, the Pleiades is an outstanding naked eye object. The main stars appear bright and striking with up to 10 or more visible under ideal conditions.

In total, the Pleiades cover nearly two degrees of sky. Due to its large apparent size, it's best seen either with binoculars, a small or wide field telescope or even a finder scope. The view through a pair of 10x50 binoculars is fantastic. The main stars are bright and distinct with dozens more fainter stars visible. A trail of 7th magnitude stars extends to the southeast from one corner of the dipper. When viewed through a small telescope of around 80mm (3.1-inch) or 100mm (4-inch) aperture the cluster is breathtaking. It neatly fills the field of view at low magnifications with the bright bluish-white stars of the cluster standing out brilliantly against the dark background. Many fainter background stars are also visible especially when using averted vision. Some of the awe of the Pleiades is lost in larger scopes due to its large size. A 200mm (8-inch) telescope reveals dozens of additional stars. However, the observer is limited to eyeing only part of the cluster and then scanning around to see the remainder.

M45 is a young cluster of hot blue and extremely luminous stars that formed about 115 million years ago. The brightest stars are all of class B:

Star	Designation	Magnitude
Alcyone	Eta Tauri	2.86
Atlas	27 Tauri	3.62
Electra	17 Tauri	3.70
Maia	20 Tauri	3.86
Merope	23 Tauri	4.17
Taygeta	19 Tauri	4.29
Pleione	28 Tauri	5.09v
Celaeno	16 Tauri	5.44

There is a faint reflection nebulosity surrounding the brightest stars that is easy to image or photograph, although more difficult to observe visually. Under excellent seeing conditions some faint nebulosity may be seen in small/medium sized telescopes especially around the star Merope (23 Tauri). Originally it was thought at first that the nebulosity was left over from the formation of the cluster, but it's now believed to be an unrelated dust cloud in the interstellar medium that the stars are currently passing through.

The Pleiades are located 425 light-years from Earth. In total it is estimated that there are around 500 members belonging to the cluster.

M45 Data Table

Messier	45
Name	Pleiades
Object Type	Open cluster
Constellation	Taurus
Distance (kly)	0.425
Apparent Mag.	1.6
RA (J2000)	03h 47m 06s
DEC (J2000)	24d 07m 00s
Apparent Size (arcmins)	110 x 110
Radius (light years)	45
Age (years)	115M
Number of Stars	500
Other Name	Collinder 42, Melotte 22

Messier # 045

Date:	Time:

Site:		
Temp:	Wind:	Hum:
Clouds:	Moon:	
Scope:		
EP:	Mag:	
NELM:	See/Trans:	
Type:	# Stars:	
Mag:	Age:	
Const:		

Notes:

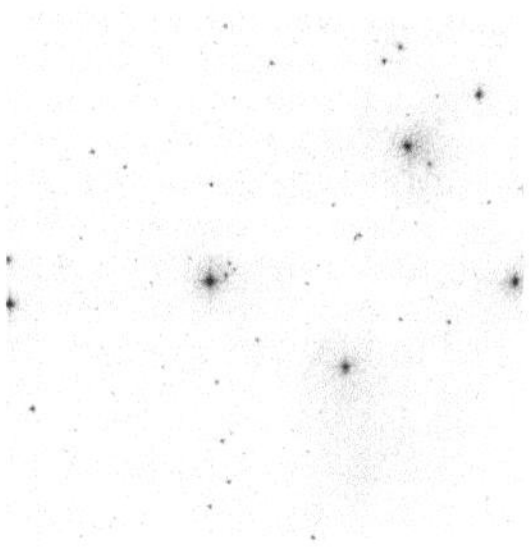

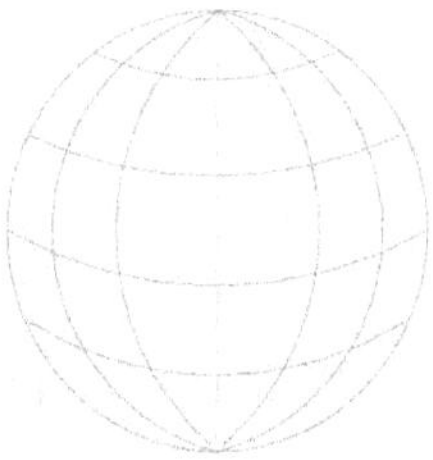

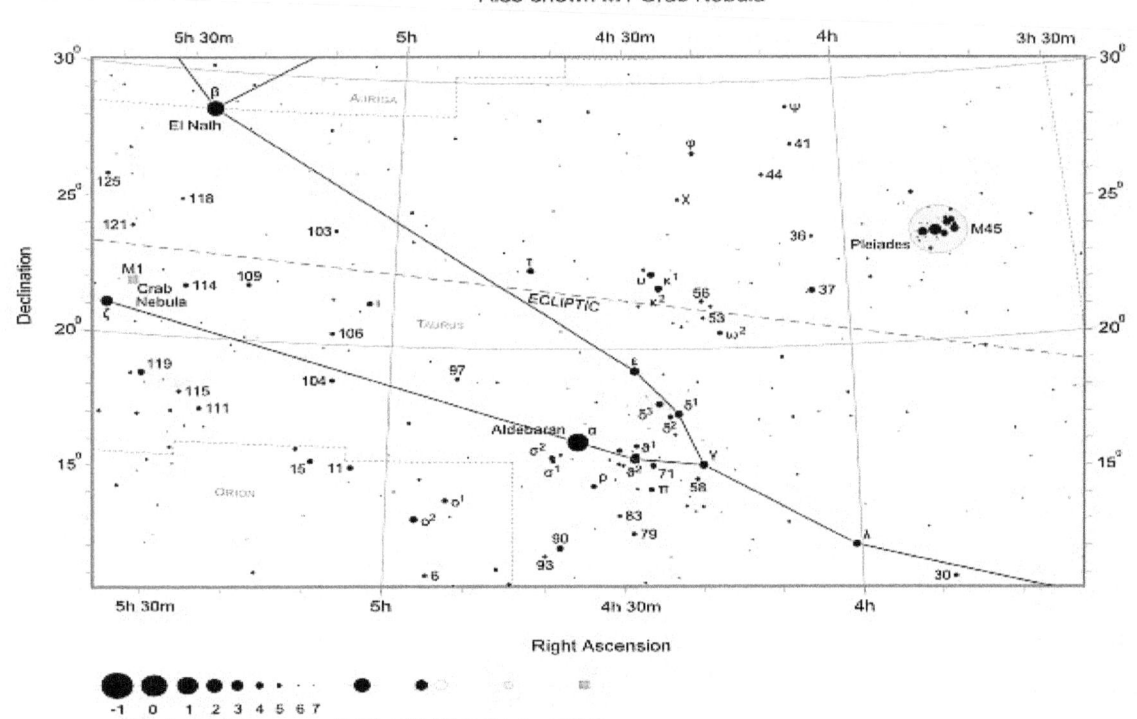

Messier Finder Chart for M45 Pleiades
Also shown M1 Crab Nebula

Declination

Right Ascension

Star magnitudes -1 0 1 2 3 4 5 6 7

Double star   Variable stars   Open cluster   Bright nebula

Messier 46 - M46 - Open Cluster

M46 is a rich open cluster, easily visible with binoculars, that's located in the rich Milky Way star fields of the constellation of Puppis. The cluster is a relatively large object with an apparent diameter almost equal to that of the full Moon. Charles Messier discovered it on February 19, 1771. It's best seen from southern and equatorial latitudes during the months of December, January and February.

Puppis borders many constellations including Canis Major, the home of the brightest star in the sky, Sirius (α CMa - mag. - 1.46). The brilliant Sirius is in the same general region of the sky as M46 and therefore acts as a good starting point; the cluster is located just 14 degrees to the east and 2 degrees north of the star. An imaginary line connecting Mirzam (β CMa - mag. +1.98) with Sirius and then extending for about 14 degrees in the northeast direction leads to the area of sky where M46 is located. In addition to M46, there are two further Messier objects in Puppis, M47 and M93. All three are superb open clusters with M47 positioned just over a degree to the northwest of M46. The third cluster M93 is located about 9 degrees south of the M46, M47 pair.

With a magnitude of +6.1, M46 is on the fringe of naked eye visibility. When viewed through 10x50 binoculars, it appears of mottled appearance, more like a nebula than an open cluster. Also visible in the same field of view is the much brighter, similar sized but less rich cluster M47. About 0.5 degrees to the north of M47 is another open cluster NGC 2423 (7th magnitude). In 20x80 binoculars, the view of M46 is enhanced. The cluster appears large and dense with at least 50 faint members visible. Small telescopes show M46 as a sprinkling of stardust. A 80mm (3.1-inch) scope at high powers reveals many fainter stars of remarkably uniform brightness, especially when using averted vision. M46 is a superb sight through a 200mm (8-inch) telescope with dozens of white, blue-white stars visible. At a magnification of about 100x, the field of view is completely filled with stars. In total, there are up to 150 stars brighter than magnitude +13 with the brightest stars being of magnitude +9. It's believed that there are at least 250 stars that are members of M46.

One notable feature of M46 is a planetary nebula (NGC 2438) located at the northern edge of the cluster. It is visible in small telescopes of the order of 100mm (4-inch). The nebula appears as a diffuse circular patch that seems to be a part of the open cluster, but is believed to be unrelated and purely a line of site phenomena. It is estimated that M46 is 5400 light-years distant, with the planetary nebula closer at 3000 light-years. M46 has an estimated age of about 300 million years and a linear diameter of 30 light-years. It's a compact cluster with an apparent diameter of 27 arc minutes, almost the same as that of the full Moon. The cluster is a nice sight when viewed through all types of optical instruments. In the same wide field of view as M46 is M47, a much brighter but less compact cluster. Also visible, is an unrelated planetary nebula (NGC 2438) towards the northern edge of M46.

M46 Data Table

Messier	46
NGC	2437
Object Type	Open Cluster
Constellation	Puppis
Distance (kly)	5.4
Apparent Mag.	6.1

RA (J2000)	07h 41m 47s
DEC (J2000)	-14d 48m 36s
Apparent Size (arcmins)	27 x 27
Radius (light years)	15

Age (years)	300M
Number of Stars	250
Other Name	Collinder 159

Messier # 046

Date:	Time:	
Site:		
Temp:	Wind:	Hum:
Clouds:	Moon:	
Scope:		
EP:	Mag:	
NELM:	See/Trans:	
Type:	# Stars:	
Mag:	Age:	
Const:		
Notes:		

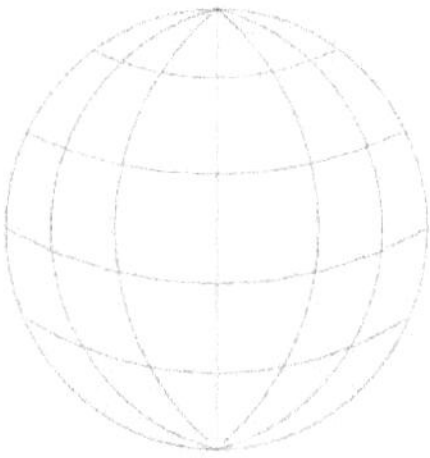

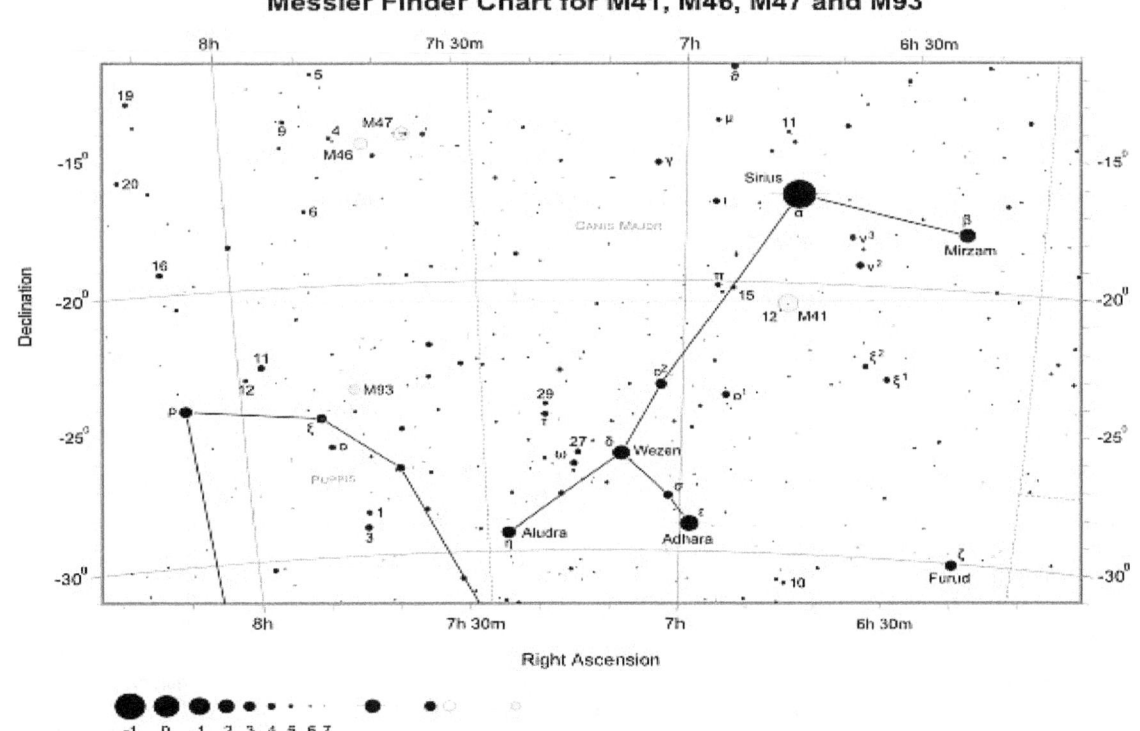

Messier Finder Chart for M41, M46, M47 and M93

Star magnitudes: -1 0 1 2 3 4 5 6 7

Double star Variable stars Open cluster

Messier 47 - M47 - Open Cluster

M47 is a loose, bright naked eye open cluster located in the Milky Way star fields of the constellation of Puppis that was discovered by Giovanni Batista Hodierna before 1654 and independently discovered by Charles Messier on February 19, 1771. It's best seen from southern and equatorial latitudes during the months of December, January and February.

Finding M47 is quite easy as the brightest star in the sky Sirius (α CMa - mag. -1.46) can be used as the perfect starting point. The cluster is located 12 degrees east and 2 degrees north of the star. An imaginary line connecting Mirzam (β CMa - mag. +1.98) with Sirius and then extending for just over 12 degrees in a northeasterly direction also leads to the area of sky where M47 is located. Just over a degree to the southeast of M47 and in the same binocular field of view is another Messier open cluster, M46. Both clusters have about the same apparent size (approx. 0.5 degrees) although M47 is noticeably brighter - magnitude +4.3 compared with magnitude +6.1 for M46. However although fainter, M46 is the more compact and richer cluster. M47 appears as a large fuzzy patch to the naked eye with the brightest stars just about resolvable under dark skies. With 10x50 binoculars it is an easy object. There are about 10-15 stars visible including at least half a dozen bright stars, spread over a diameter of 30 arc minutes. Larger 20x80 binoculars or small telescopes reveal a myriad of fainter stars of varying magnitudes, interspersed between the brighter members of this loose cluster. Also visible in the same wide field of view is M46 and another open cluster NGC 2423 (7th magnitude), which is located only 0.5 degrees to the north of M47. In stark contrast to M47, M46 is a very compact cluster with many more stars visible. The two clusters form a fine double act in binoculars and small telescopes.

A 150mm (6-inch) telescope at low magnifications reveals M47 as a large loose cluster, with at least 10 or so brighter members and a few dozen fainter stars visible. A very close double star of equal magnitude components lies near the center of the cluster (Sigma 1121). These stars are of magnitude 7.9 and separated by 7.4 arc seconds. In total, M47 contains about 50 stars. M47 has an interesting history in that although Charles Messier discovered the cluster on the same night he discovered M46 (February 19, 1771), much later it emerged that Giovanni Batista Hodierna had observed M47 sometime before 1654. Credit is therefore usually given to both men regarding the discovery. In another twist, Messier described M47 as a cluster of stars brighter than those of apparently neighbored M46, but made a sign error when recording its position. Subsequently M47 was regarded as a missing object until T.F. Morris identified it (together with the also missing M48) in 1959.

M47 is a nice bright loose open cluster located in the constellation of Puppis. It is 1600 light-years from Earth, easily visible to the naked eye and a superb site in binoculars or small telescopes. The cluster contains a handful of bright stars and many dozen faint ones. In the same wide field of view is the fainter but much more compact cluster M46.

M47 Data Table

Messier	47		**DEC (J2000)**	-14d 28m 57s
NGC	2422		**Apparent Size (arcmins)**	30 x 30
Object Type	Open Cluster		**Radius (light years)**	6
Constellation	Puppis		**Age (years)**	78M
Distance (kly)	1.6		**Number of Stars**	50
Apparent Mag.	4.3		**Other Name**	Collinder 152
RA (J2000)	07h 36m 35s			

Messier # 047

Date:	Time:	
Site:		
Temp:	Wind:	Hum:
Clouds:	Moon:	
Scope:		
EP:	Mag:	
NELM:	See/Trans:	
Type:	# Stars:	
Mag:	Age:	
Const:		

Notes:

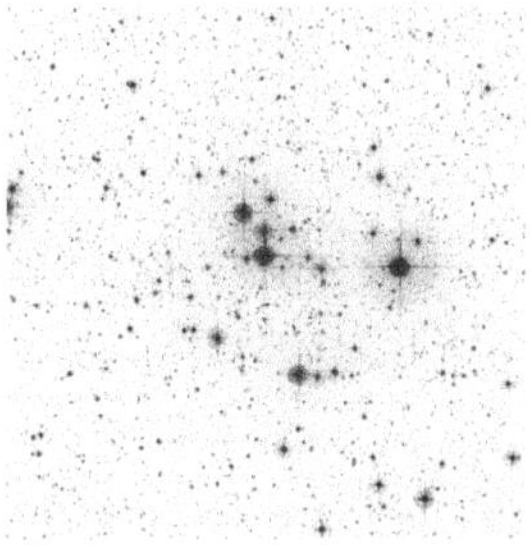

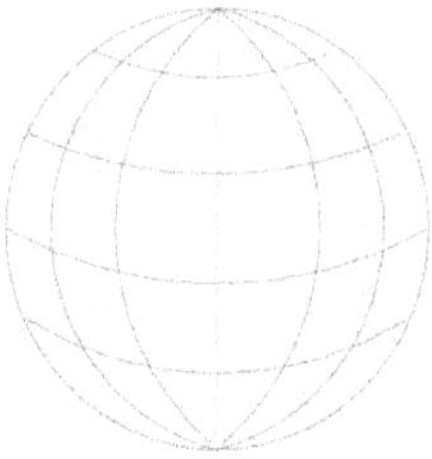

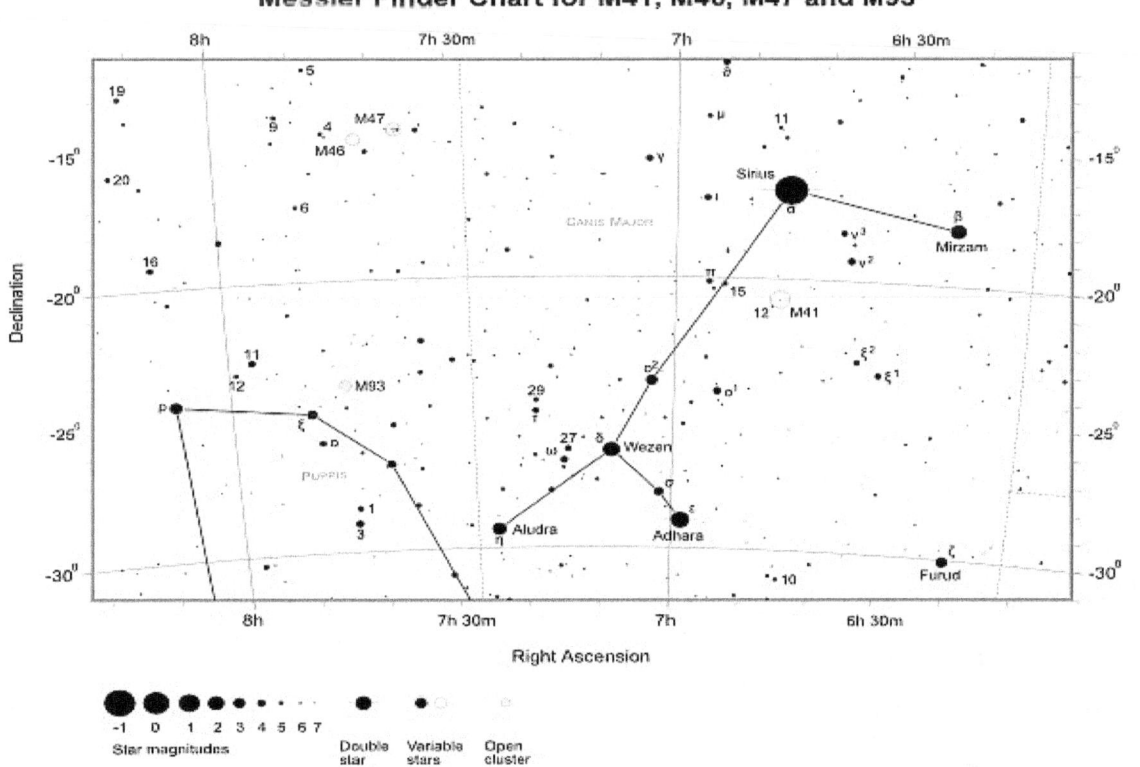

Messier Finder Chart for M41, M46, M47 and M93

8h
7h 30m
7h
6h 30m
-15°
-20°
-25°
-30°
Declination
Right Ascension
M47
M46
M93
M41
Sirius
Mirzam
Wezen
Aludra
Adhara
Furud
Canis Major
Puppis
Star magnitudes
-1 0 1 2 3 4 5 6 7
Double star
Variable stars
Open cluster

Messier 48 - M48 - Open Cluster

M48 is a large conspicuous open cluster covering almost a degree of sky in the constellation of Hydra. It's located close to the border with Monoceros. At magnitude +5.5, the cluster is visible to the naked eye under reasonably dark skies and is a superb binocular or small telescope object. It's best seen during the months of December, January and February.

Charles Messier originally discovered and cataloged M48 on February 19, 1771. However, he made a mistake in the positioning and hence the object was missing for over 150 years until it was identified by Oswald Thomas in 1934, and independently by T.F. Morris in 1959. Since M48 was lost, two subsequent independent rediscoveries occurred. The first was by Johann Elert Bode who found it before 1782 and then Caroline Herschel located it on March 8, 1783.

Hydra is the largest constellation in the sky but finding M48 is relatively easy as it's positioned just 14 degrees southeast of Procyon (α CMi) in Canis Minor. At magnitude +0.34, Procyon is the seventh brightest star in the night sky. Also visible to the naked eye but better seen with binoculars is a triangle of 4th and 5th magnitude stars located about 5 degrees northwest of M48. The stars in question are ζ Mon (zeta Mon - mag. +4.4), 28 Mon (mag. + 4.7) and 27 Mon (mag. +4.9).

M48 appears as a large faint misty patch of light to the naked eye. Through 7x50 or 10x50 binoculars it looks somewhat like a fainter and smaller version of M44 The Praesepe open cluster. With a small telescope of 80mm (3.1-inch) aperture, the cluster looks superb with many stars visible arranged in a triangle shape. The stars are concentrated more towards the centre part of the cluster, which in total spans 54 arc minutes of apparent sky. Since it covers nearly twice the size of the full Moon, M48 is best viewed at low magnifications. The brightest stars in M48 are of 8th magnitude.

Larger telescopes of 150mm (6-inch) aperture reveal about 50 stars brighter than magnitude 13, with the total number of stars estimated to be about 80. M48 is 1,500 light years distant which corresponds to a spatial diameter of 24 light years and is about 300 million years old.

M48 Data Table

Messier	48
NGC	2548
Object Type	Open cluster
Constellation	Hydra
Distance (kly)	1.5
Apparent Mag.	5.5
RA (J2000)	08h 13m 43s
DEC (J2000)	-05d 45m 02s
Apparent Size (arcmins)	54 x 54
Radius (light years)	12
Age (years)	300M
Number of Stars	80
Other Name	Collinder 179

Messier # 048

Date:	Time:

| Site: | |

Temp:	Wind:	Hum:

Clouds:	Moon:

| Scope: | |

EP:	Mag:

NELM:	See/Trans:

Type:	# Stars:

Mag:	Age:

| Const: | |

| Notes: | |

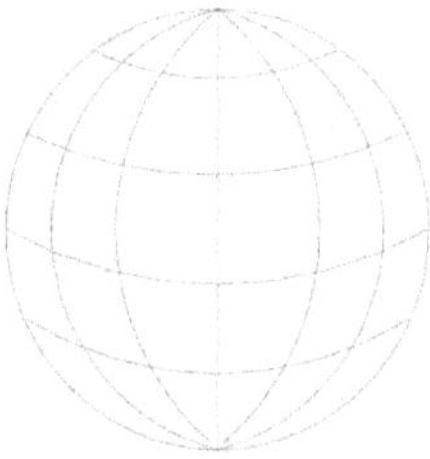

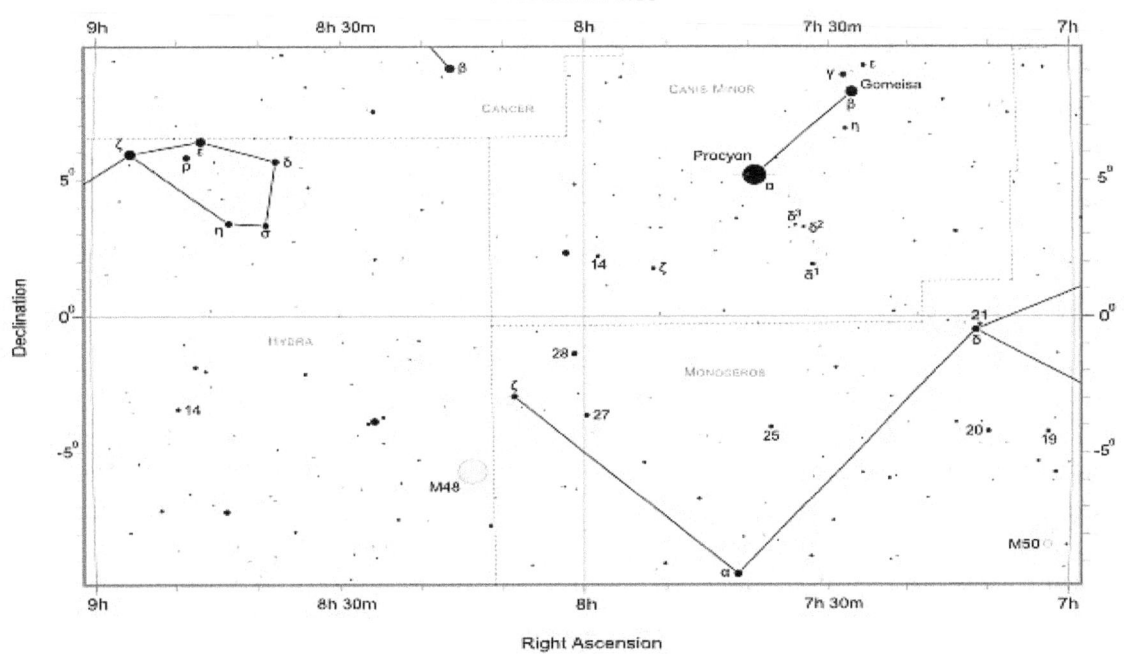

Messier Finder Chart for M48
Also shown M50

Star magnitudes -1 0 1 2 3 4 5 6 7

Double star   Variable stars   Open cluster

Messier 49 - M49 - Elliptical Galaxy

M49 is an elliptical galaxy located in the constellation of Virgo. It belongs to the Virgo cluster of galaxies, a large group of galaxies centered about 54 Million light-years from Earth. With an apparent magnitude of +8.4, M49 is the brightest member of this famous cluster. It's best seen during the months of March, April and May.

The galaxy was discovered by Charles Messier on February 19, 1771. He also discovered M46 andM48 on the same night and re-discovered M47, which was originally found by Giovanni Batista Hodierna sometime before 1654. After M83, M49 was the second galaxy to be discovered that's located beyond the Local Group and the first member of the Virgo cluster to be found.

To locate M49, first imagine a line connecting Denebola (β Leo - mag. +2.1) to Vindemiatrix (ε Vir - mag. +2.8). The centre of the Virgo cluster is positioned about halfway along this line and M49 is located about 5 degrees due south of this point.

M49 a fine example of an elliptical galaxy and along with M60 and M87, it's one of the great giant elliptical galaxies of the Virgo cluster. These galaxies are all very large and massive. M49 has an apparent size of 10 x 8 arc minutes, which at a distance of 58 million light years corresponds to an actual diameter of 170,000 light-years. For comparison, M31 has a diameter of 140,000 light years and our Milky Way galaxy 100,000 light years. It's estimated that M49 contains at least 200 billion stars.

Large 15x70 or 20x80 binoculars show M49 as a small faint fuzzy patch of light that's non-stellar in appearance. Through a 100mm (4-inch) telescope, the galaxy appears small and diffuse but with a bright compressed central core. Larger telescopes do better, a 200mm (8-inch) aperture scope shows the bright centre with the large but featureless halo surrounding it.

Only one supernova has been observed within M49, SN 1969Q in June 1969. Slightly unusual is that the galaxy has the physical form of a radio galaxy, but its radio emissions are only of the levels of a normal galaxy. M49 also has a large collection of globular clusters, estimated at about 5,900. The Milky Way has only about 150 but even the number for M49 is dwarfed by M87, which has in excess of 13,400.

M49 Data Table

Messier	49
NGC	4472
Object Type	Elliptical galaxy
Classification	E4
Constellation	Virgo
Distance (kly)	58,000
Apparent Mag.	8.4
RA (J2000)	12h 29m 47s

DEC (J2000)	08d 00m 00s
Apparent Size (arcmins)	10.2 x 8.3
Radius (light years)	85,000
Number of Stars	>200 Billion
Notable Feature	Only Supernova observed within this galaxy is SN 1969Q

Messier # 049

Date:	Time:

Site:

Temp:	Wind:	Hum:

Clouds:	Moon:

Scope:

EP:	Mag:
NELM:	See/Trans:
Type:	# Stars:
Mag:	Age:

Const:

Notes:

Messier Finder Chart for M49, M58, M59, M60, M84, M85, M86, M87, M88, M89, M90, M91, M98, M99 and M100   Also shown M53, M64 Black Eye Galaxy, M65 and M66

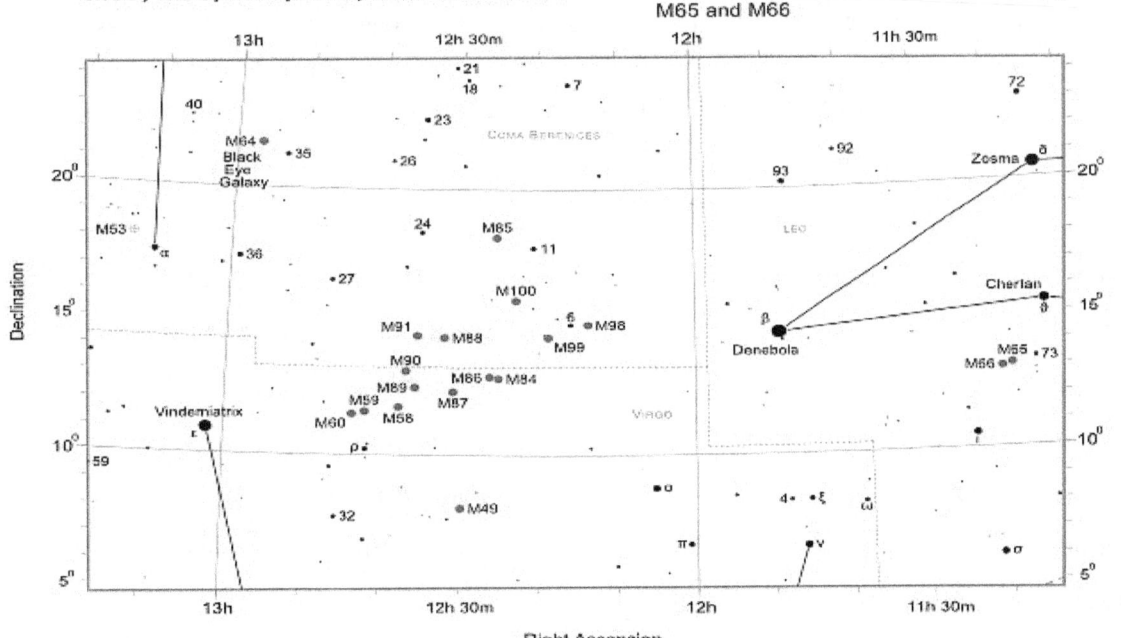

Messier 50 - M50 - Open Cluster

M50 is an appealing and relatively bright open cluster located in the rich star fields of Monoceros, close to the Canis Major border. With an apparent magnitude of +5.9, it's just about visible to the naked eye as a faint patch of nebulosity. Telescopically the cluster is moderately dense, contains a number of bright stars and spans about half the diameter of the full Moon; a nice sized object for owners of small to medium sized scopes. It's best seen during the months of December, January and February.

The constellation of Monoceros straddles the equatorial equator, east of majestic Orion and to the north of bright Canis Major. Despite lying in rich Milky Way star fields, Monoceros is a faint constellation containing no stars brighter than 4th magnitude. Consequently, tracing the outline of this dim grouping can require some patience. However, although devoid of bright stars Monoceros does contain a number of bright and interesting deep sky objects. In addition to M50, the constellation is home to the Rosette Nebula (NGC 2237, 2238, 2239, and 2246), the Christmas Tree Cluster (NGC 2264) and Hubble's Variable Nebula (NGC 2261).Although Monoceros is faint, finding M50 is relatively easy. Start by aiming your sights on the brightest star in the sky, Sirius (α CMa - mag. -1.46). Sirius can be found by extending an imaginary line from the three bright stars of Orion's belt southwards for about 20 degrees. Located 9.5 degrees NNE of Sirius is M50. Sandwiched halfway between Sirius and M50 is θ CMa (mag. 4.1), a well-placed stepping-stone star.

The cluster was one of Charles Messier's discoveries. He found it on April 5, 1772, although possibly G.D. Cassini had already discovered it before 1711, according to a report by his son, Jacques Cassini, in his book "Elements of Astronomy". With binoculars, M50 appears as a relatively large bright patch of nebulosity with at least 2 or 3 stars resolvable. An 80mm (3.1-inch) scope reveals a heart-shaped figure containing a sprinkling of mostly 7th and 8th magnitude blue-white stars set against a background haze. Towards the southern edge of the cluster appears a distinct orange star.

The cluster is as an attractive grouping of stars when viewed at low to medium powers (30 to 50x) through medium sized 150mm (6-inch) or 200mm (8-inch) telescopes. At least 40 mainly blue-white stars but also an orange and some yellow stars can be resolved in a somewhat irregular shaped grouping. At about 100x magnification the stars begin to overflow the eyepiece field of view with many of the fainter background members now resolvable, bringing the total visible to more than 80.M50 covers 16 arc-minutes of apparent sky. At a distance of 3200 light-years from Earth, this corresponds to a spatial diameter of 20 light-years. The cluster is a nice sight in small / medium size telescopes with many blue-white stars visible but some colored ones also. In total it contains over 100 stars and has an estimated age of 78 Million years.

M50 Data Table

Messier	50
NGC	2323
Object Type	Open cluster
Constellation	Monoceros
Distance (kly)	3.2
Apparent Mag.	5.9
RA (J2000)	07h 02m 42s

DEC (J2000)	-08d 23m 26s
Apparent Size (arcmins)	16 x 16
Radius (light years)	10
Age (years)	78M
Number of Stars	>100
Other Name	Collinder 124

Messier # 050

Date:	Time:
Site:	
Temp:	Wind:	Hum:
Clouds:	Moon:
Scope:	
EP:	Mag:
NELM:	See/Trans:
Type:	# Stars:
Mag:	Age:
Const:	

Notes:

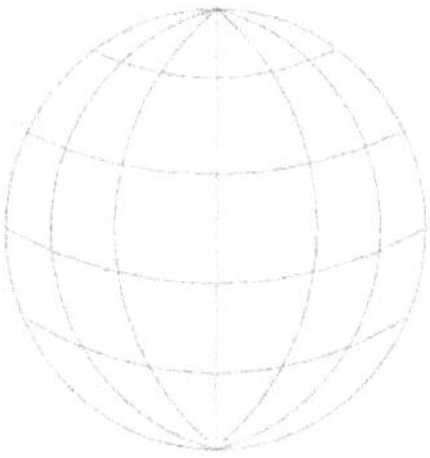

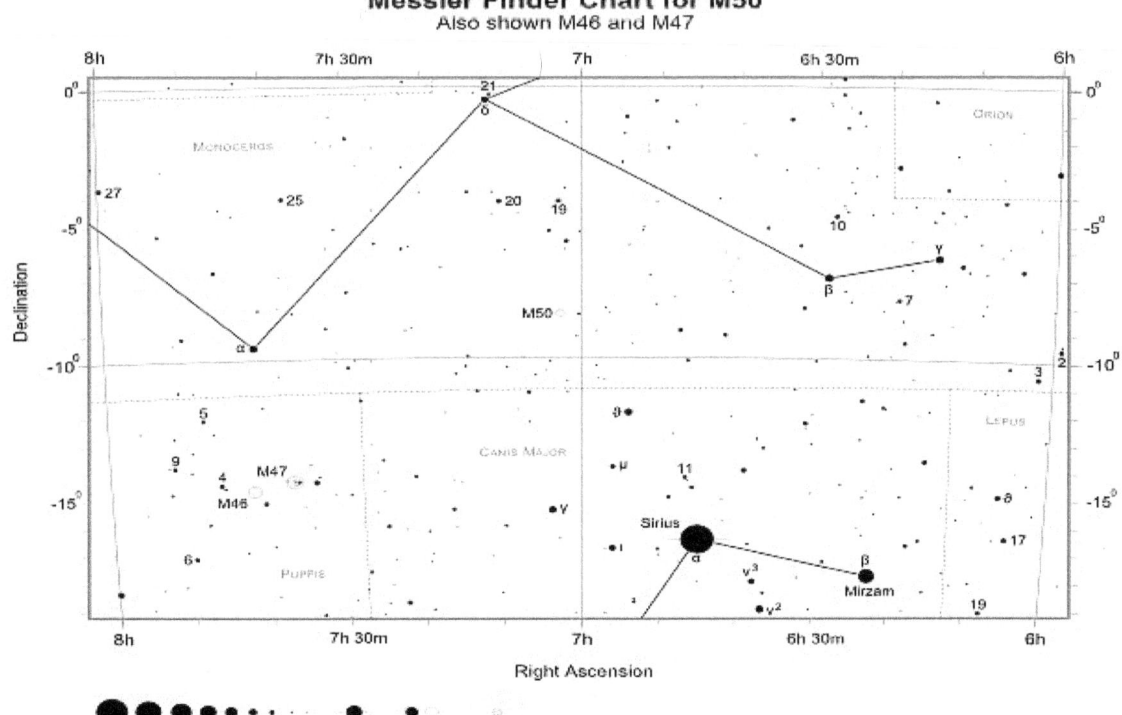

Messier Finder Chart for M50
Also shown M46 and M47

8h   7h 30m   7h   6h 30m   6h

0°   21   ORION   0°

MONOCEROS

•27   •25   •20   19   10   γ   -5°
-5°

β   •7

α   2   -10°
-10°   3

5   LEPUS

9   M47   CANIS MAJOR   θ   μ   11   •δ   -15°
-15°   4   •V   •17
M46   Sirius
α   ν³
6   •ι   β   19
PUPPIS   Mirzam
ν²

8h   7h 30m   7h   6h 30m   6h

Declination

Right Ascension

-1 0 1 2 3 4 5 6 7
Star magnitudes   Double star   Variable stars   Open cluster

Messier 51 - M51 - The Whirlpool Galaxy (Spiral Galaxy)

M51 (NGC 5194) The Whirlpool Galaxy is a grand design spiral galaxy located in the constellation of Canes Venatici, "the Hunting Dogs". It's one of the most famous galaxies in the sky, appearing face-on when viewed from Earth. At magnitude +8.4, it's relatively bright and visible in binoculars especially from a dark site. M51 has a much smaller dwarf companion NGC 5195 and together they are well-known as the finest and most studied example of an interacting galaxy pair. They are best seen from the Northern Hemisphere during the months of March, April or May. From latitudes greater than 42N the galaxies are circumpolar and therefore never set.

M51 was one of Charles Messier original discoveries on October 13, 1773 while his friend Pierre Méchain discovered NGC 5195 on March 20, 1781. Messier described M51 as a faint nebula without stars that was difficult to see. In his catalogue of 1781, Messier describes both M51 and NGC 5195 in the same note and hence there is some confusion over the exact designation of M51. Is he referring to M51 as just the larger galaxy or does he actually mean the pair? If it's the pair then NGC 5194 is sometimes referred to as "M51A", with NGC 5195 separately known as "M51B".

Canes Venatici is a small northern constellation of faint stars that was created by Johannes Hevelius in the 17th century. Apart from its brightest star Cor Caroli (α CVn - mag. +2.9), the constellation contains no stars brighter than 4th magnitude. However, locating M51 isn't difficult as it positioned towards the northeast border of Canes Venatici and only a few degrees from the handle of the seven stars that form the famous "Plough" or "Big Dipper" asterism of Ursa Major.

To locate M51, first identify Alkaid (η UMa - mag +1.9) the end star of the handle of the bowl of the "Plough". Located 3 degrees directly west of Alkaid is magnitude +4.5 star 24 UMa. Positioned a degree to the northeast of 24 UMa is a magnitude +6.5 star. Now imagine a line from this star to 24 UMa and then continue it southwards for a further two degrees. This leads to a triangle of stars of magnitudes +7.1, +7.1 and +7.5. All three stars are easily visible in binoculars with M51 located just west of the southernmost star.

The Whirlpool galaxy is visible through 10x50 binoculars and finder scopes as a smudge or patch of light, not unlike an out of focus star. An 80mm (3.1-inch) telescope reveals M51 as a diffuse patch of light with a brighter core although not much more detail. When viewed through a 200mm (8-inch) scope, M51 has a well-defined bright core surrounded by a large, fainter diffuse halo.

The galaxy's surface appears mottled with some evidence of faint dark dust lanes and the spiral structure. Satellite galaxy NGC 5195 is also visible although it's difficult to detect evidence of the connection between the two. When viewed through larger instruments (at least 300mm – 12-inch), numerous spiral bands are apparent in M51 with large HII regions of gas also visible. The attaching band of light between M51 and NGC 5195 can also be seen. The true beauty of this galaxy pair is apparent in long exposure images and for this reason M51 is a popular and sought after amateur astrophotography target.

Despite been one of the most studied galaxies of all the distance to M51 is uncertain. Recent estimates placed it at 23 million light-years but other values between 15 and 35 million light-years have also been quoted. The figure of 23 million light-years is based on a magnitude 13.5 type II supernova (SN 2011dh) that was detected in M51 on May 31, 2011. To date, three supernovae have been observed in M51 (SN 1994I in April 1994, SN 2005cs in June 2005 and SN 2011dh in May 2011).

It's believed that a black hole, surrounded by a ring of dust, exists at the heart of M51 with the centre part currently undergoing a period of enhanced star formation.

The Whirlpool Galaxy has the distinction of the first to have its spiral nature recognized. This was achieved in 1845 by the 3rd Earl of Rosse, William Parsons, using his 72-inch (1.83 m) reflecting telescope at Birr Castle in Ireland; the largest optical telescope in the world at the time. Overall, M51 is a showpiece object of the night sky. It's the brightest and the most spectacular example of an interacting spiral galaxy and is a much sought after object for visual observers and imagers alike.

M51 Data Table

Messier	51
NGC	5194
Name	Whirlpool Galaxy
Object Type	Spiral galaxy
Classification	SA(S) bc
Constellation	Canes Venatici
Distance (kly)	23,000
Apparent Mag.	8.4
RA (J2000)	13h 29m 52s
DEC (J2000)	47d 11m 43s
Apparent Size (arcmins)	11.2 x 6.9
Radius (light years)	37,500
Number of Stars	>100 Billion
Notable Feature	Interacting with dwarf galaxy NGC 5195

Messier # 051

Date:		Time:	
Site:			
Temp:	Wind:		Hum:
Clouds:		Moon:	
Scope:			
EP:		Mag:	
NELM:		See/Trans:	
Type:		# Stars:	
Mag:		Age:	
Const:			

Notes:

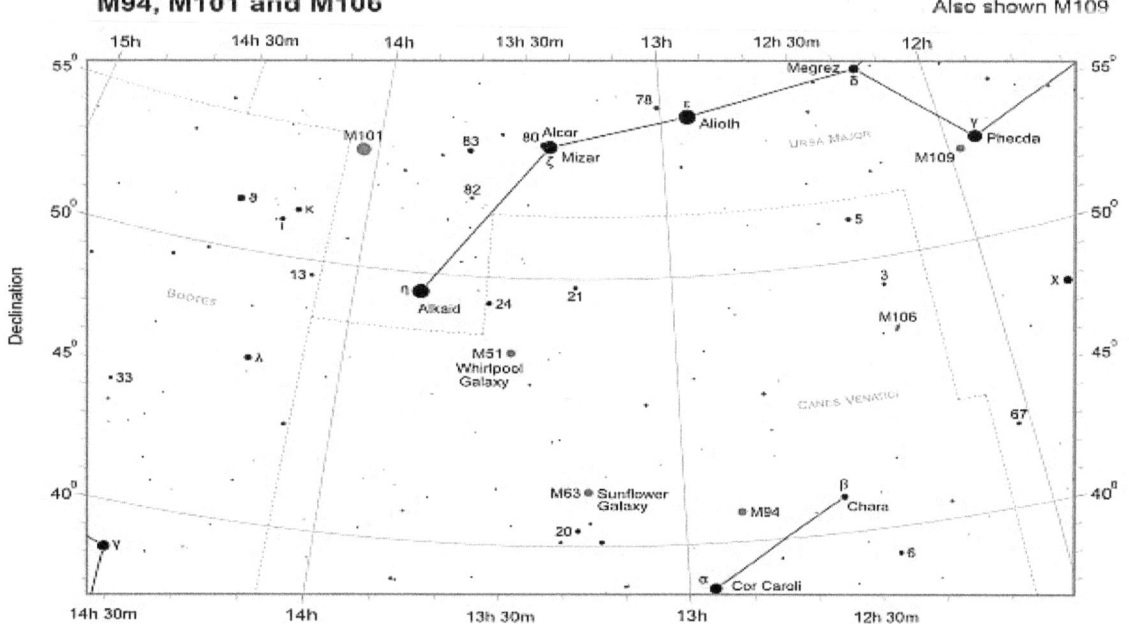

Messier Finder Chart for M51 Whirlpool Galaxy, M63 Sunflower Galaxy, M94, M101 and M106

Also shown M109

Messier 52 - M52 - Open Cluster

M52 is a magnitude +7.2 open cluster located in western Cassiopeia. This excellent grouping of stars is one of the best clusters in the northern Milky Way; containing some 200 members spread across 13 arc minutes of sky. It's visible in binoculars and a wonderful sight through telescopes, appearing as a rich, bright mass of dozens of white stars.

M52 is one of Charles Messier's original discoveries, who cataloged it on September 7, 1774. He noticed it after a comet he was observing passed close by. The cluster is easy to find as it's located one degree south of star 4 Cas (mag. +5.0), which can be found by drawing an imaginary line from Schedar (α Cas - mag. +2.2) to Caph (β Cas - mag. +2.3) and then extending it for about the same distance again.

Through 10x50 binoculars, M52 is a somewhat fan, kidney or "V" shaped patch of light with a prominent 8th magnitude yellow star at the southwest corner. An 80mm (3.1-inch) telescope reveals M52 as a concentrated cluster with mostly faint members that are best seen using averted vision. It appears compressed, somewhat like a sprinkling of "salt and pepper" spread over an area just less than half the size of the full Moon. A small aperture increase enhances the fan shape and overall appearance of the cluster. Through a 150mm (6-inch) scope, M52 appears as a grouping of at least a dozen stars surrounded by a round diffuse haze when using direct vision. The bright yellow star is prominent. Averted vision resolves the background haze into about 50 or so faint stars. In even larger telescopes M52 is brilliant, with many more stars visible in this gem of a cluster.

In total M52 contains about 200 stars. The cluster is well suited to light polluted skies and hence visible under moonlit or hazy conditions. Due to interstellar absorption of light, the distance to M52 is difficult to accurately calculate. Estimates range between 3,000 and 7,000 light-years. If we assume a value of 7,000 light-years then M52 has a spatial diameter of 26.5 light-years. It's estimated to be 35 million years old.

M52 is best seen from the Northern Hemisphere during August, September and October when it appears high in the sky and even overhead from some locations. From latitudes greater than 28N, the cluster is circumpolar. Along with NGC 457 and NGC 7789, it's among the finest clusters in the northern Milky Way.

The unusual emission nebula NGC 7635 (the Bubble Nebula) is located one degree SW of M52.

M52 Data Table

Messier	52		**DEC (J2000)**	61d 36m 24s
NGC	7654		**Apparent Size (arcmins)**	13 x 13
Object Type	Open Cluster		**Radius (light years)**	13.25
Constellation	Cassiopeia		**Age (years)**	35M
Distance (kly)	7.0		**Number of Stars**	200
Apparent Mag.	7.2		**Other Name**	Collinder 455
RA (J2000)	23h 24m 50s			

Messier # 052

Date:	Time:
Site:	
Temp:	Wind:	Hum:
Clouds:	Moon:
Scope:	
EP:	Mag:
NELM:	See/Trans:
Type:	# Stars:
Mag:	Age:
Const:	
Notes:	

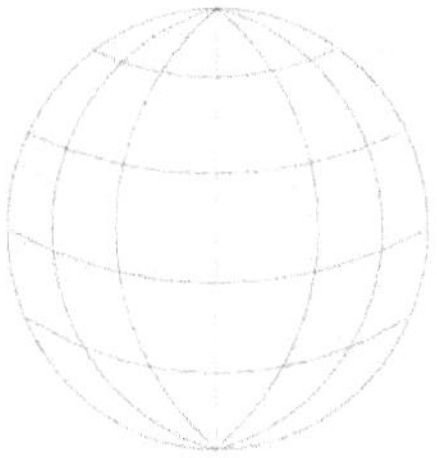

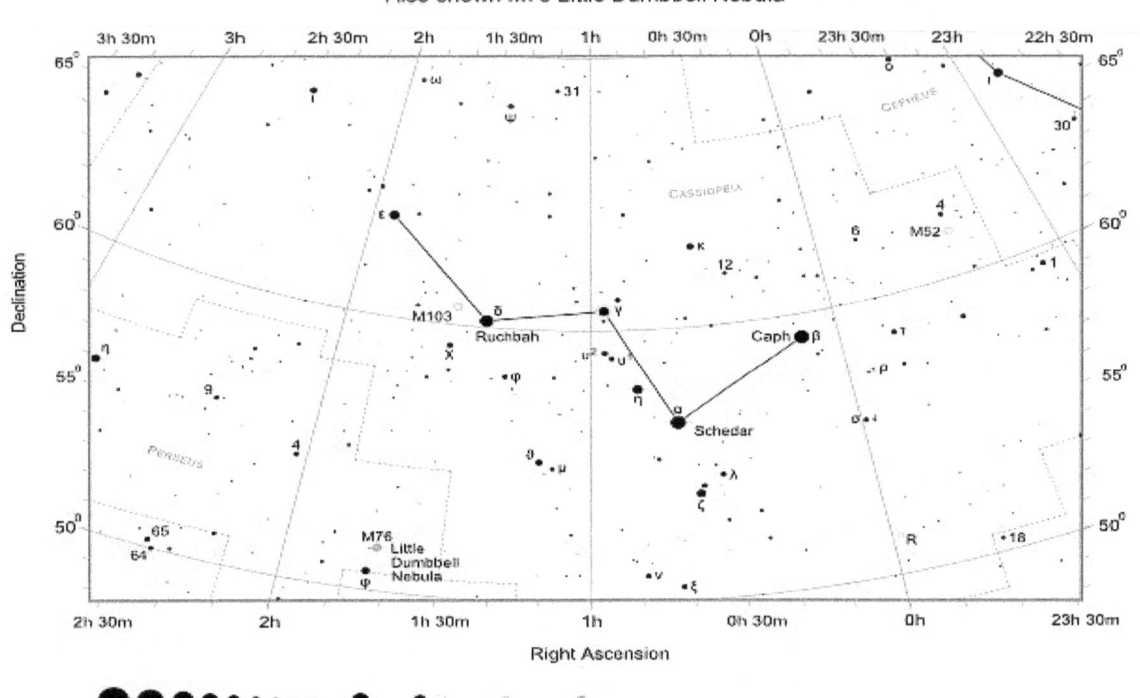

Messier Finder Chart for M52 and M103
Also shown M76 Little Dumbbell Nebula

Messier 53 - M53 - Globular Cluster

M53, mag. +8.0, is a distant globular cluster that's positioned in the eastern part of the constellation of Coma Berenices. Located about 60,000 light years from the galactic center and 58,000 light years from Earth, M53 is one of the Milky Way's more outlying globulars. For comparison, M13, the Great Hercules globular cluster is a mere 25,100 light years distant.

M53 was discovered by German astronomer Johann Elert Bode on February 3, 1775. He described it as a "rather vivid and of round shape" nebula. Charles Messier independently rediscovered M53 on February 26, 1777. When observing it later in 1781 he compared it to another recently discovered distant globular M79. As with most globulars it was William Herschel who first resolved M53 into stars, finding it similar in appearance to M10.

The cluster is quite easy to find, lying just 1 degree northeast of mag. +4.3 star Diadem (α Com). Located 15 degrees directly east of M53 is orange giant star Arcturus (α Boo - mag. -0.05) the fourth brightest star in the night sky. M53 is best seen from northern latitudes during the months of March, April and May.

Through 10x50 binoculars, M53 appears as a very faint point of light that appears slightly fuzzy. A 80mm (3.1-inch) telescope reveals an oval shaped object with a bright centre and extended halo. To resolve stars a medium size scope is required; scopes of aperture 150mm (6-inch) to 200mm (8-inch) resolve some of the outer stars, especially under dark skies and using averted vision. The central part appears bright and crisp but unresolved. In total, the large outer halo extends for 13 arc minutes, although only about half of this is visible through medium size scopes. Large instruments of aperture 300mm (12-inch) or more resolve the globular well with a reasonably concentrated nucleus and stars spread across its full diameter.

Located a degree southeast of M53 is the peculiar globular cluster NGC 5053, a sparse magnitude 10 system containing only 3,500 stars. It appears as a grainy patch of weak light through an 8-inch scope.

Considering its vast distance from us the brightness and apparent size of M53 is impressive; it's an intrinsically large globular with a linear diameter of 220 light-years. The cluster contains at least 500,000 stars of which at least 67 are variable stars. It's estimated to be 12.67 billion years old.

M53 Data Table

Messier	53
NGC	5024
Object Type	Globular cluster
Constellation	Coma Berenices
Distance (kly)	58.0
Apparent Mag.	8.0
RA (J2000)	13h 12m 55s
DEC (J2000)	18d 10m 08s

Apparent Size (arcmins)	13 x 13
Radius (light years)	110
Age (years)	12,670M
Number of Stars	>500,000
Notable Feature	A number of blue stragglers have been identified in the cluster

Messier # 053

Date:	Time:

Site:

Temp:	Wind:	Hum:

Clouds:	Moon:

Scope:

EP:	Mag:
NELM:	See/Trans:
Type:	# Stars:
Mag:	Age:
Const:	

Notes:

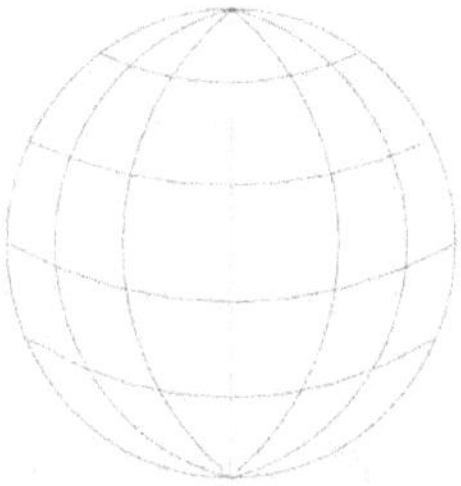

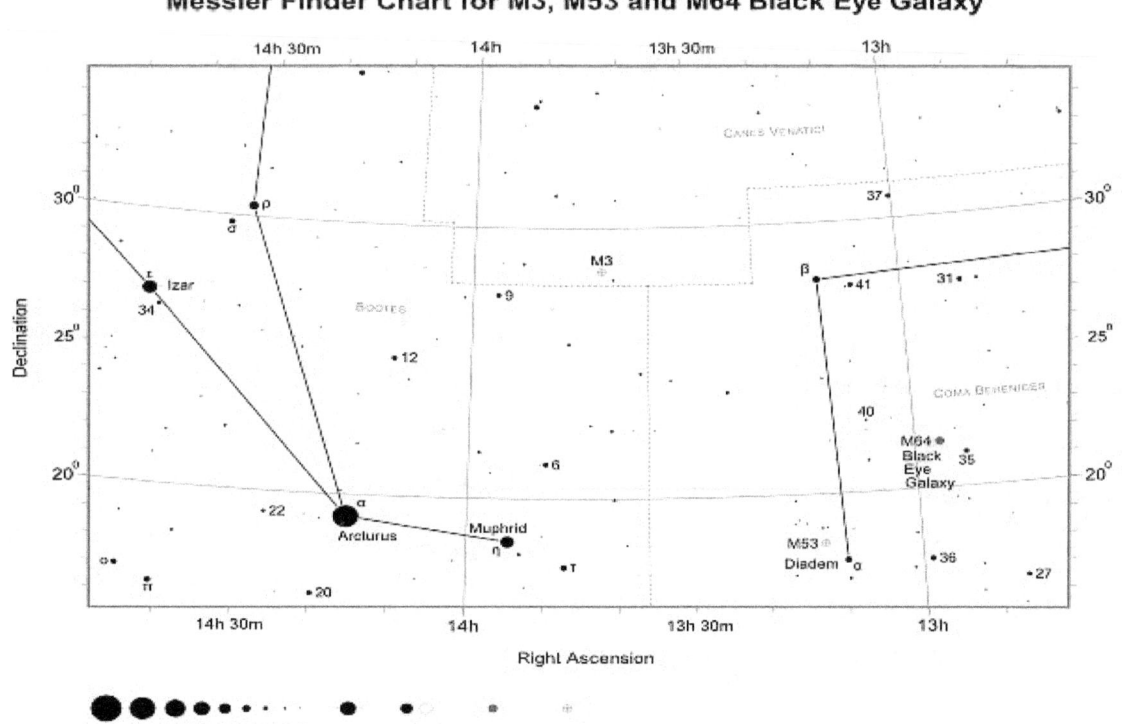

Messier Finder Chart for M3, M53 and M64 Black Eye Galaxy

Messier 54 - M54 - Globular cluster

M54 is a globular cluster located in Sagittarius that's a staggering 87,400 light years from Earth. It was discovered by Charles Messier on July 24, 1778 and was for many years thought to be part of the Milky Way but is now believed to belong to the nearby Sagittarius Dwarf Elliptical Galaxy. It owns the distinction of being the first extragalactic globular cluster ever discovered, even though it wasn't recognized as such for over 200 years. Despite its vast distance, M54 is visible in binoculars albeit faintly (mag. +7.9). The fact that it can be seen in binoculars at all from such a distance is incredible and is due to its large intrinsic size and high absolute brightness. With a diameter of over 300 light-years diameter, this globular is enormous and one of the largest known.

Finding M54 is easy as it lies within the teapot asterism of Sagittarius. The starting point is to focus on the base of the teapot and image a line connecting Ascella (ζ Sgr - mag. +2.6) with Kaus Australis (ε Sgr - mag. +1.8). Positioned about 1.75 degrees along this line and slightly north is M54. With a declination of -30 degrees, the globular is best seen from the Southern Hemisphere during the months of June, July and August. From northern temperate latitudes, it's a much more difficult target as it never rises very high above the southern horizon.

Binocular observers may at first overlook this fascinating globular cluster. Although the core is bright, the globular looks small and may appear starlight on initial observation. However, further inspection shows a small degree of nebulosity - like an out of focus star - hinting at the true nature of this massive object. A small 80mm (3.1-inch) telescope under good seeing reveals a bright point like core surrounding by nebulosity that tails off gradually from the centre. Larger scopes of the order of 150mm (6-inches) or 200mm (8-inches) may reveal twists and knots especially under good seeing conditions. However, due to its considerable distance from Earth even the largest amateur telescopes fail to resolve M54.

M54 has a luminosity of 850,000 Sun's and an absolute magnitude of -10. At such a large distance from Earth, the brilliance of M54 is not obvious and this is also reflected in the apparent magnitude of its brightest stars; a mere +15.5. In reality, it does contains over 1 million stars of which at least 82 variable stars, mostly of the RR Lyrae type and there are also two semi-regular red variable stars with periods of 77 and 101 days respectively.

M79 in Lepus is the only other extragalactic globular cluster in the Messier catalogue and is part of the tiny Canis Major Dwarf galaxy.

M54 Data Table

Messier	54
NGC	6715
Object Type	Globular cluster
Constellation	Sagittarius
Distance (kly)	87.4
Apparent Mag.	7.9
RA (J2000)	18h 55m 03s
DEC (J2000)	-30d 28m 42s

Apparent Size (arcmins)	12 x 12
Radius (light years)	153
Age (years)	13,000M
Number of Stars	>1 Million
Notable Feature	Likely belongs to the Sagittarius Dwarf Elliptical Galaxy

Messier # 054

Date:	Time:
Site:	
Temp:	Wind:	Hum:
Clouds:	Moon:
Scope:	
EP:	Mag:
NELM:	See/Trans:
Type:	# Stars:
Mag:	Age:
Const:	
Notes:	

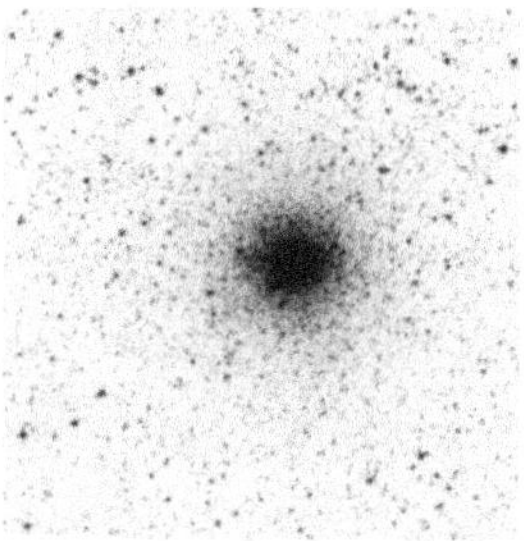

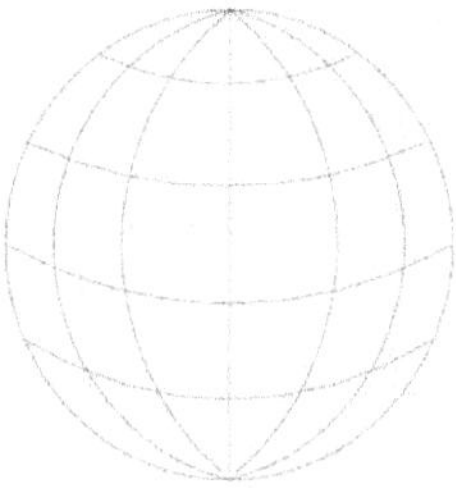

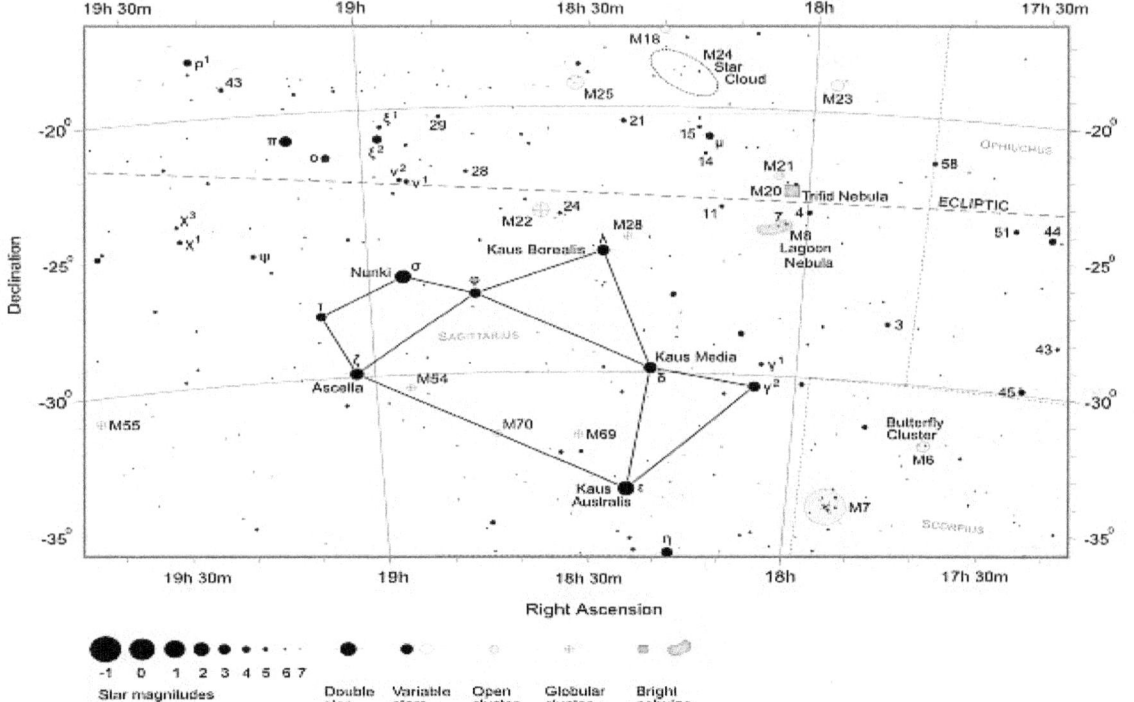

Messier Finder Chart for M8 Lagoon Nebula, M20 Trifid Nebula, M21, M22, M28, M54, M69, M70

Also shown M6 Butterfly Cluster, M7, M18, M23, M24 Star Cloud and M55

Star magnitudes -1 0 1 2 3 4 5 6 7

Double star · Variable stars · Open cluster · Globular cluster · Bright nebulae

Messier 55 – M55 – Globular Cluster

M55 is a globular cluster located in eastern Sagittarius towards its border with Capricornus and Microscopium. At magnitude +6.7, it's beyond naked eye visibility but bright enough to be seen with binoculars. However, it's not an easy globular to locate since there aren't any particular bright stars nearby. With a declination of -30 degrees, M55 is one of the more southerly objects in Messier's catalogue and therefore especially difficult for observers based at northern temperate latitudes. It's best seen from southern or equatorial latitudes during the months of June, July and August.

M55 was discovered by Nicholas Louis de Lacaille on June 16, 1752 while observing from South Africa. Charles Messier then catalogued it on July 24, 1778. From Paris, Messier had difficulty finding M55, it took him 14 years to spot it!

Finding M55 can be challenging. One method is to begin with the "teapot" asterism of Sagittarius. Start by locating stars Kaus Media (δ Sgr - mag. +2.7) and Ascella (ζ Sgr - mag. +2.6). Then imagine a line from Kaus Media moving eastwards towards and passing through Ascella. Curve this line for another 17 degrees to arrive at M55.

When viewed through 7x50 or 10x50 binoculars, M55 appears as a diffuse ball of light, non-stellar without a bright core. Under dark sky conditions, it appears like a round hazy comet. However, M55 has a low surface brightness meaning that only a small amount of light pollution renders it practically invisible to binocular observers. Through a 100mm (4-inch) scope, it appears loose and hints at resolution. M55 in total covers 19 arc minutes of apparent sky but appears much smaller through the eyepiece. A 200mm (8-inch) instrument will resolve many stars with the cluster bursting into life; a magnificent swarm of thousands of pinpoints of light spread across the complete surface.

M55 is located 17,600 light-years distant and has a spatial diameter of about 94 light years. Only about 55 variable stars have been discovered in the cluster that's estimated to contain 100,000 stars. It's about 12,700 billion years old.

M55 Data Table

Messier	55
NGC	6809
Object Type	Globular cluster
Constellation	Sagittarius
Distance (kly)	17.6
Apparent Mag.	6.7
RA (J2000)	19h 39m 59s
DEC (J2000)	-30d 57m 44s
Apparent Size (arcmins)	19 x 19
Radius (light years)	47
Age (years)	12,300M
Number of Stars	100,000

Messier # 055

Date:	Time:
Site:	
Temp:	Wind:	Hum:
Clouds:	Moon:
Scope:	
EP:	Mag:
NELM:	See/Trans:
Type:	# Stars:
Mag:	Age:
Const:	
Notes:	

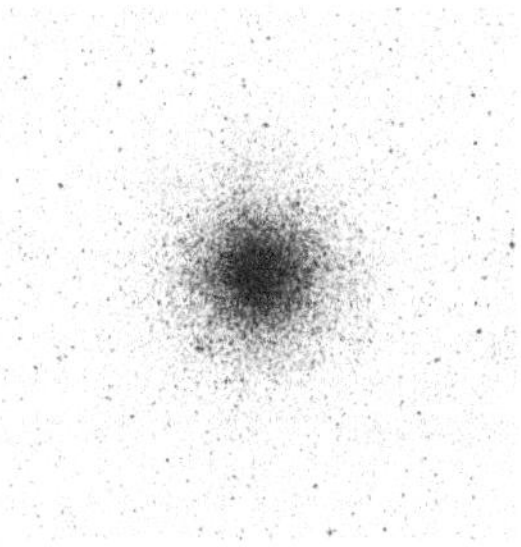

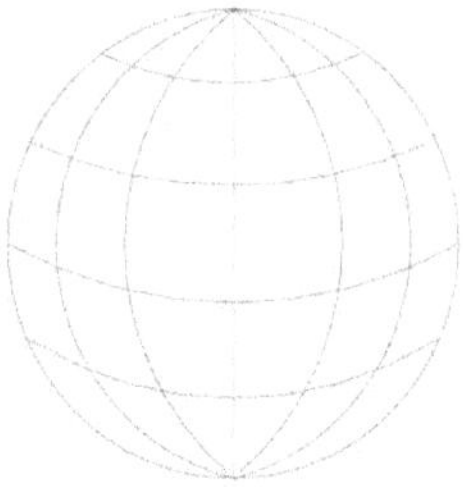

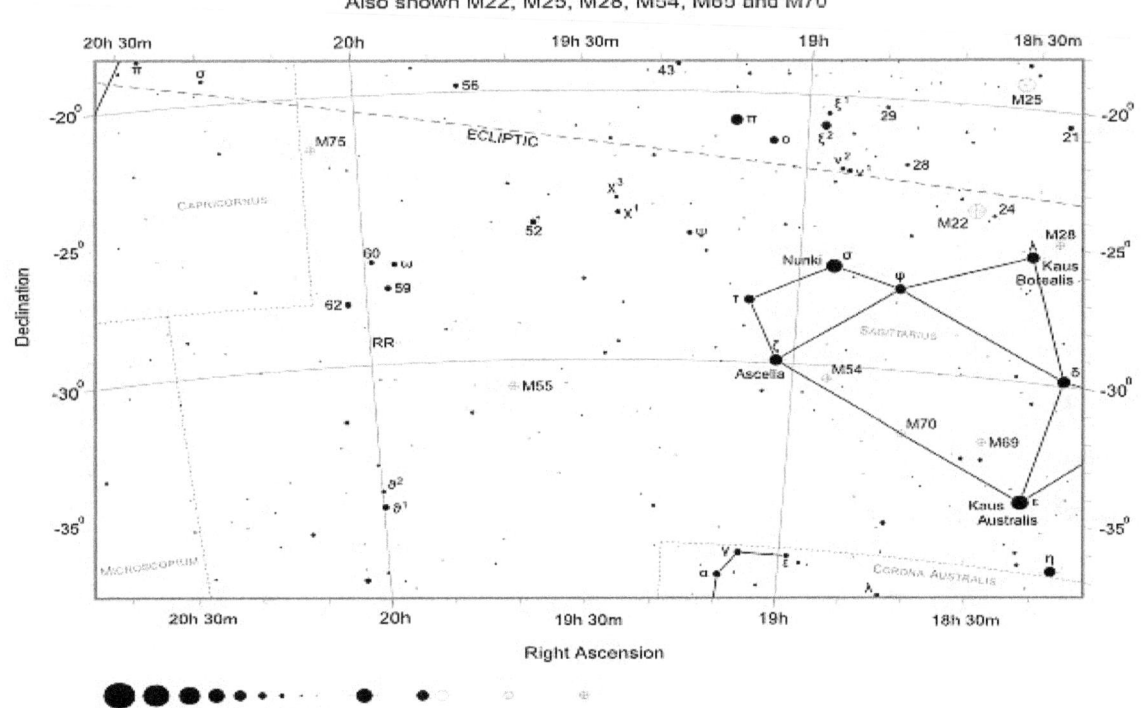

Messier Finder Chart for M55 and M75
Also shown M22, M25, M28, M54, M69 and M70

Star magnitudes -1 0 1 2 3 4 5 6 7

Double star   Variable stars   Open cluster   Globular cluster

Messier 56 - M56 - Globular Cluster

M56 is a faint distant globular cluster in Lyra positioned close to its border with Cygnus. At apparent magnitude +8.3, it's one of the dimmer Messier globulars and unlike most objects of its type lacks a bright core, resulting in it being a challenging binocular object. Nevertheless, the cluster is visible in small telescopes and amateur scopes of the order of 250mm (10-inch) will resolve some stars, despite its relatively large distance.

M56 was discovered by Charles Messier discovered on January 23, 1779. He described it as a "nebula without stars" and like many globular clusters was first resolved into stars by William Herschel five years later. Another unusual feature about this object is that it follows a retrograde orbit through the Milky Way. It has been suggested that M56 may have been acquired during the merger of a dwarf galaxy, of which Omega Centauri forms the surviving nucleus.

The globular is located almost halfway along an imaginary line connecting beautiful double star Albireo (β Cyg - mag. +3.1) with Sulafat (γ Lyr - mag. +3.3). However, since it's located in a dense part of the Milky Way it's easy to miss especially with small telescopes. Not far from M56 is the only other Messier object in Lyra, M57 the "Ring Nebula".

At best M56 appears as a faint slightly fuzzy star through 7x50 or 10x50 binoculars. It's obviously non-stellar when seen with larger 70 or 80mm models. A 100mm (4-inch) telescope shows the cluster as a faint, round, diffuse ball of light with very little or no detail discernible. A noticeable 5th magnitude star is located less than a degree northwest of the cluster. It's possible to resolve some of the outer stars using a 250mm (10-inch) telescope with the brightest members being of 13th magnitude. M56 displays a gradual, soft brightening from the outer regions towards the core. In total, the cluster measures 8.8 arc minutes across although visually it appears less than half this size. However, despite being overshadowed by M13 in Hercules, M56 is a fine globular in its own right.

M56 is 32,000 light-years distant, which corresponds to a spatial diameter of 84 light-years. It contains only a dozen or so variable stars and is estimated to be 13.7 billion years old. The globular is best seen from the Northern Hemisphere during the months of June, July and August.

M56 Data Table

Messier	56		**Apparent Size (arcmins)**	8.8 x 8.8
NGC	6779		**Radius (light years)**	42
Object Type	Globular cluster		**Age (years)**	13,700M
Constellation	Lyra		**Number of Stars**	80,000
Distance (kly)	32.9		**Notable Feature**	Moving in a retrograde orbit through the Milky Way
Apparent Mag.	8.3			
RA (J2000)	19h 16m 35s			
DEC (J2000)	30d 11m 05s			

Messier # 056

Date:	Time:

Site:		

Temp:	Wind:		Hum:

Clouds:	Moon:	

Scope:	

EP:	Mag:
NELM:	See/Trans:
Type:	# Stars:
Mag:	Age:
Const:	

Notes:

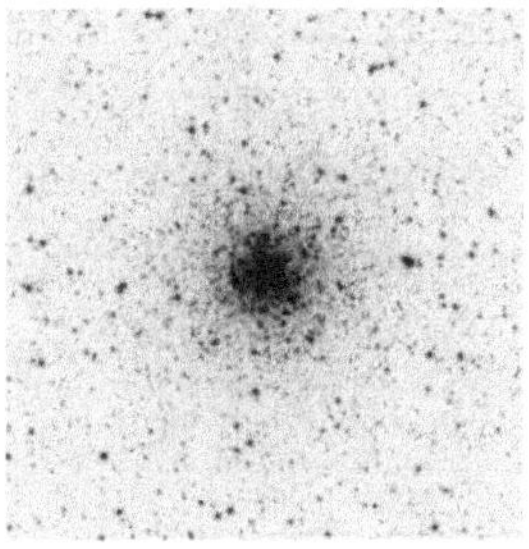

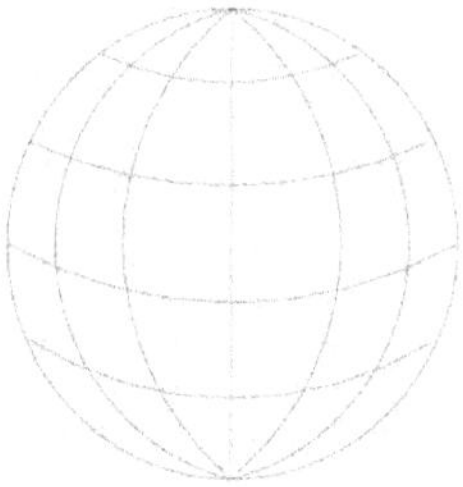

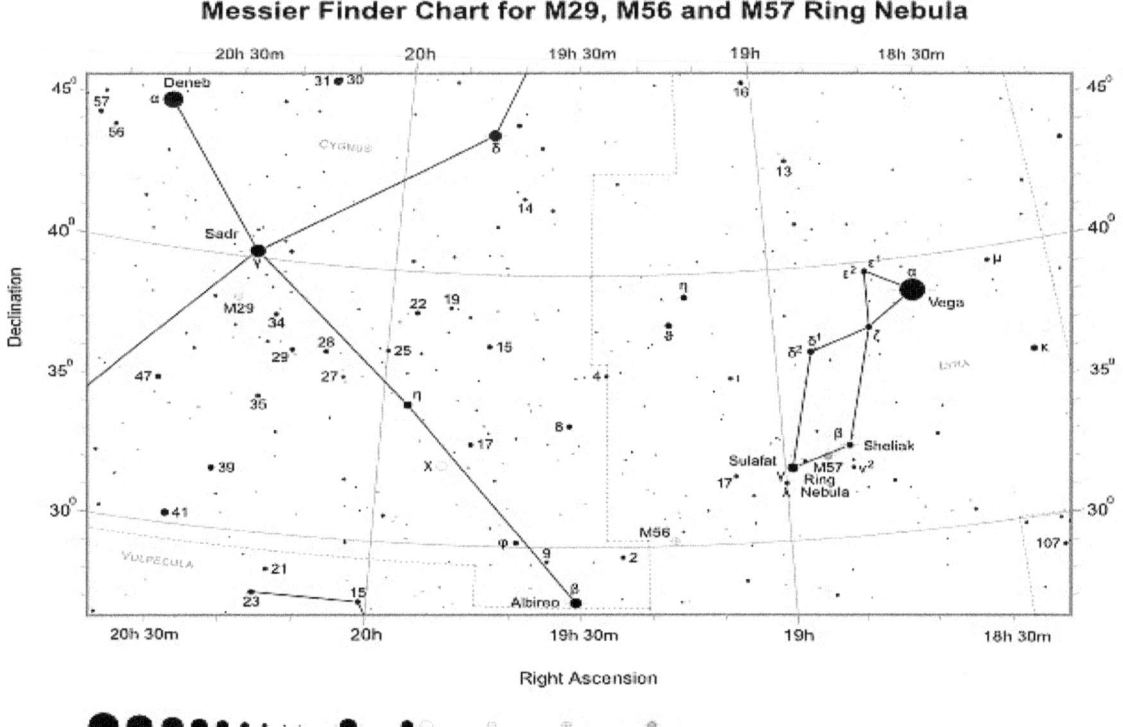

Messier Finder Chart for M29, M56 and M57 Ring Nebula

Star magnitudes: -1 0 1 2 3 4 5 6 7

Double star   Variable stars   Open cluster   Globular cluster   Planetary nebula

Messier 57 - M57 - The Ring Nebula (Planetary Nebula)

M57 The Ring Nebula is a showpiece planetary nebula located in constellation of Lyra. It is probably the most well-known, studied and photographed object of its kind and a perennial favorite with amateur astronomers. It is relatively bright at mag. +8.8 and easy to locate; about 40% along an imaginary line connecting Sheliak (β Lyr - mag. +3.5) to Sulafat (γ Lyr - mag. +3.2). For Northern Hemisphere observers the Ring Nebula is high in the sky during summer months. From southern latitudes it appears much lower down.

M57 is a difficult 10x50 binocular object appearing at best as a faint out of focus star. It is certainly much easier to locate when using larger 20x80 binoculars. Small telescopes fair better. A 100mm (4-inch) telescope reveals M57 as a small grey puffed out slightly elliptical shaped patch of light, but noticing the ring shape with it centre hole is challenging even when using averted vision. When viewed through a 200mm (8-inch) telescope, the shape is much clearer with finer detail also visible.

Very large amateur scopes show more intricate detail but the 15th magnitude central star, at the heart of the Ring Nebula, is difficult to spot. However, it's easy to image.

The Ring Nebula was discovered by Antoine Darquier de Pellepoix in January 1779.

M57 Data Table

Messier	57
NGC	6720
Name	Ring Nebula
Object Type	Planetary Nebula
Constellation	Lyra
Distance (kly)	2.3
Apparent Mag.	8.8
RA (J2000)	18h 53m 35s
DEC (J2000)	33d 01m 43s
Apparent Size (arcmins)	1.4 x 1.0
Radius (light years)	0.5

Messier # 057

Date:	Time:
Site:	
Temp:	Wind:	Hum:
Clouds:	Moon:
Scope:	
EP:	Mag:
NELM:	See/Trans:
Type:	# Stars:
Mag:	Age:
Const:	
Notes:	

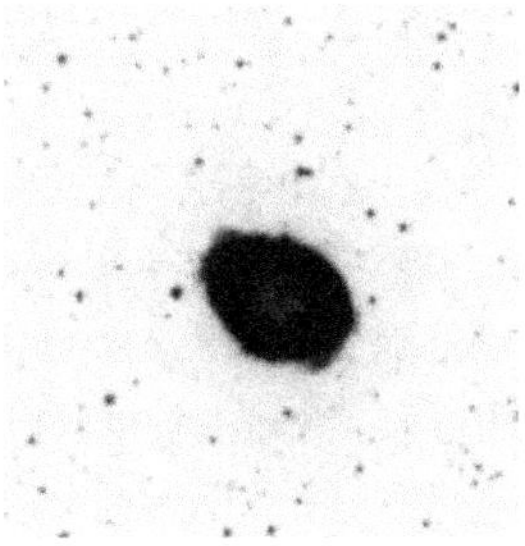

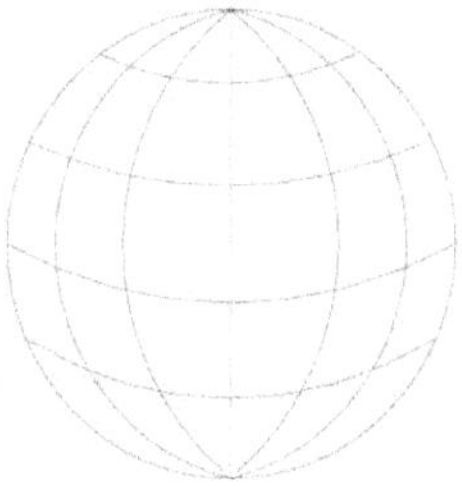

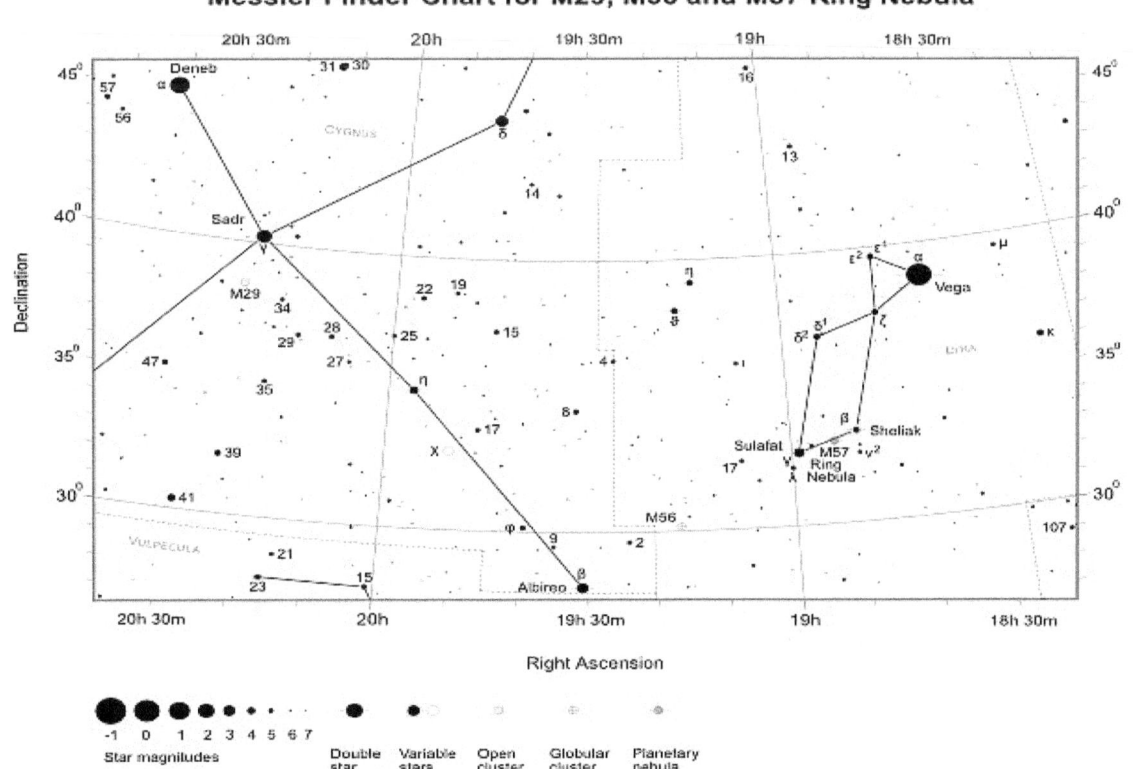

Messier Finder Chart for M29, M56 and M57 Ring Nebula

Declination

Right Ascension

Star magnitudes
-1 0 1 2 3 4 5 6 7

Double star
Variable stars
Open cluster
Globular cluster
Planetary nebula

Messier 58 - M58 - Barred Spiral Galaxy

M58 is a magnitude +9.8 barred spiral galaxy that's one of the brightest members of the famous Virgo cluster. At a distance of 68 million light-years and despite being one of the furthest objects in the Messier catalogue it's bright enough to be visible in large binoculars and small telescopes. Although unknown when discovered, M58 was the most distant known object of all at that time. It's best seen during the months of March, April and May.

The centre of the Virgo cluster is positioned approx. halfway along a line connecting Denebola (β Leo - mag. +2.1) to Vindemiatrix (ε Vir - mag. +2.8). M58 is located on the southern side of this line, just over halfway along. Positioned one degree east of M58 is M59 and M60 with M89 located one degree to the northwest of M58.

The galaxy was one of Charles Messier original discoveries. He found it on April 15, 1779; the same night he located elliptical galaxies M59 and M60, which were discovered a few days earlier by Johann Gottfried Koehler. Along with M91, M95 and M109, it's one of four barred spiral galaxies in the Messier catalogue.

M58 is a fine galaxy for backyard observers. It's visible in larger binoculars (e.g. 20x80s) as a small faint near circular haze of light. A 100mm (4-inch) scope will easily show the bright nucleus of M58. With a 200mm (8-inch) telescope under dark skies and good seeing conditions it's possible to see hints of the central bar structure. Larger scopes show more subtle details but not a great deal more. The galaxy covers 6.0 x 4.8 arc minutes of apparent sky. Despite not being terrifically detailed, M58 is large and bright enough to be impressive through most backyard scopes. Some 30 arc minutes southwest of M58 is a curious pair of interacting galaxies NGC 4567 and NGC 4568, popularly called the Siamese Twins.

M58 has an active galactic nucleus, which contains a super massive black hole and some starburst activity. Two supernovae have been observed in M58; a type II supernova at mag. +13.5 in 1998 and a mag. +12.2 type I supernova a year later.

The galaxy is one of the earliest recognized spiral galaxies and was listed by Lord Rosse as one of 14 "spiral nebulae" discovered to 1850.

M58 Data Table

Messier	58		**DEC (J2000)**	11d 49m 05s
NGC	4579		**Apparent Size (arcmins)**	6.0 x 4.8
Object Type	Barred spiral galaxy		**Radius (light years)**	58,000
Classification	SAB(rs)6		**Number of Stars**	400 Billion
Constellation	Virgo		**Notable Feature**	M58 is one of the brightest galaxies in the Virgo cluster
Distance (kly)	68,000			
Apparent Mag.	9.8			
RA (J2000)	12h 37m 44s			

Messier # 058

Date:	Time:
Site:	
Temp:	Wind:	Hum:
Clouds:	Moon:
Scope:	
EP:	Mag:
NELM:	See/Trans:
Type:	# Stars:
Mag:	Age:
Const:	
Notes:	

Messier Finder Chart for M49, M58, M59, M60, M84, M85, M86, M87, M88, M89, M90, M91, M98, M99 and M100

Also shown M53, M64 Black Eye Galaxy, M65 and M66

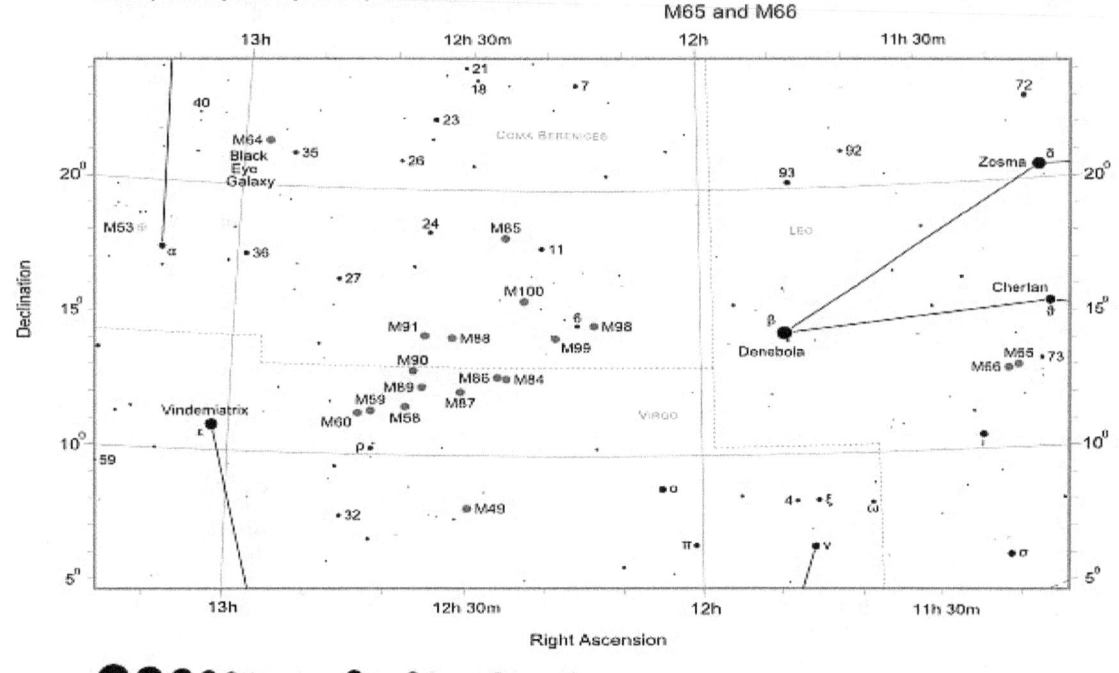

Star magnitudes -1 0 1 2 3 4 5 6 7

Double star   Variable stars   Galaxy   Globular cluster

Messier 59 – M59 – Elliptical Galaxy

M59 is an elliptical galaxy in Virgo and a member of the Virgo cluster that was discovered by Johann Gottfried Koehler on April 11, 1779 while observing a comet in that region of sky. On the same day, he also discovered neighboring galaxy M60, a slightly larger and brighter version of M59. Also comet chasing at that time was Charles Messier who independently found both galaxies four days after Koehler. During his search, Messier also discovered M58 another nearby Virgo cluster galaxy that was missed by Koehler. Of the three, Messier described M60 as the brightest galaxy with M59 and M58 being fainter and of similar magnitude.

At apparent magnitude +9.8, M59 is a challenging small telescope object. It can be spotted with small 80mm (3.1-inch) scopes or even large binoculars, but dark skies are a must. Even then it only appears as a hazy patch, that's better seen with larger amateur instruments. The galaxy is located 60 Million light years distant. It displays an apparent size of 5.4 x 3.7 arc minutes, which corresponds to a spatial diameter of 95,000 light-years. Despite been one of the larger elliptical galaxies in the Virgo cluster, M59 is considerably less massive and less luminous than the other great cluster ellipticals, M49, M60 and M87.

A good proportion of the Messier Virgo Cluster galaxies can be found along or near an imaginary line connecting Denebola (β Leo - mag. +2.1) to Vindemiatrix (ε Vir - mag. +2.8). M59 is no exception and it's positioned about 5 degrees from Vindemiatrix. Located 0.4 degrees east of M59 is M60 with M58 one degree west of M59.

The galaxies are best seen during the months of March, April and May.

When viewed through medium sized 150mm (6-inch) or 200mm (8-inch) telescopes, M59 appears as an elongated diffuse patch of light. The central core appears condensed and bright with a large faint nebulous halo surrounding it. Through the eyepiece, M59 looks slightly smaller and fainter than M60, but otherwise rather similar in appearance with both galaxies visible in the same low power eyepiece field of view.

In total, M59 contains somewhere between 1500 and 2400 globular clusters. This is considerably less than M49, M60 and M87 but still more than ten times the number in our own Milky Way Galaxy. To date, only one supernova has been recorded in M59 (1939B). It peaked in 1939 at magnitude of +11.9.

M59 Data Table

Messier	59		**DEC (J2000)**	11d 38m 48s
NGC	4621		**Apparent Size (arcmins)**	5.4 x 3.7
Object Type	Elliptical galaxy			
Classification	E5		**Radius (light years)**	47,500
Constellation	Virgo		**Number of Stars**	200 Billion
Distance (kly)	60000		**Notable Feature**	Member of the Virgo Cluster of galaxies
Apparent Mag.	9.8			
RA (J2000)	12h 42m 02s			

Messier # 059

Date:	Time:
Site:	
Temp:	Wind:	Hum:
Clouds:	Moon:
Scope:	
EP:	Mag:
NELM:	See/Trans:
Type:	# Stars:
Mag:	Age:
Const:	
Notes:	

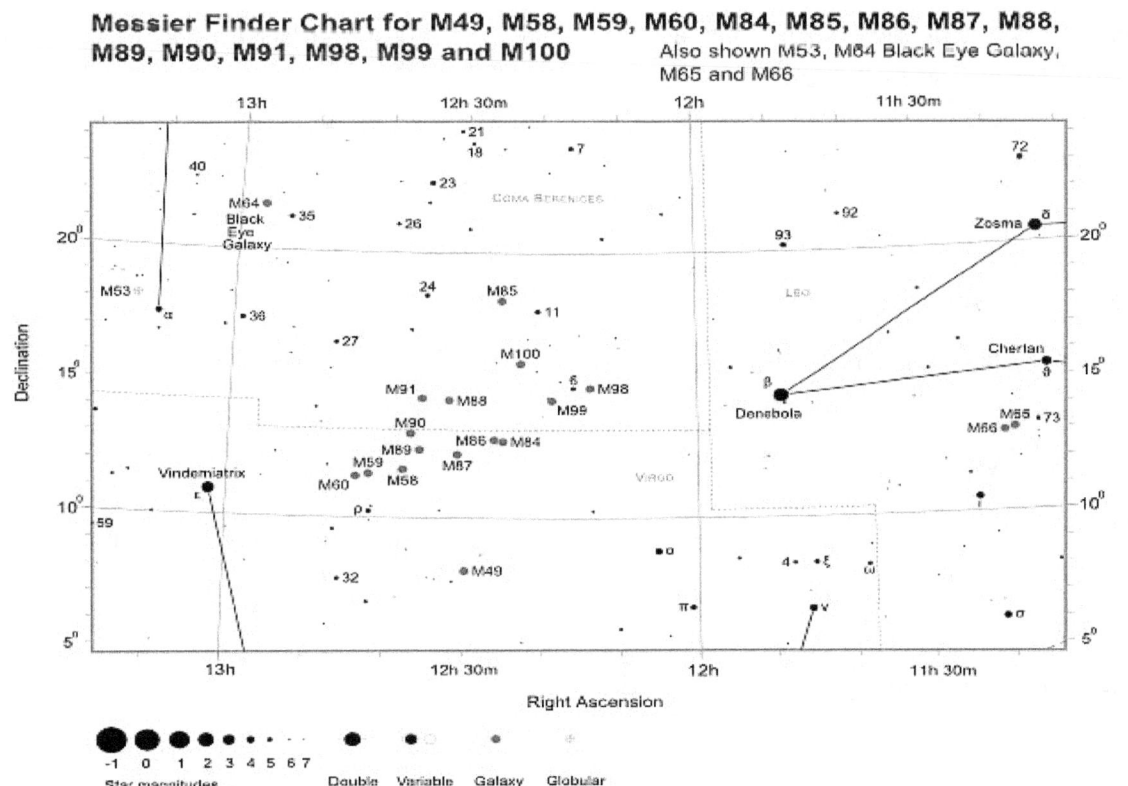

Messier Finder Chart for M49, M58, M59, M60, M84, M85, M86, M87, M88, M89, M90, M91, M98, M99 and M100 Also shown M53, M64 Black Eye Galaxy, M65 and M66

Messier 60 – M60 – Elliptical galaxy

M60 is an elliptical galaxy and a member of the Virgo cluster of galaxies. With an apparent magnitude of +9.2 it's the third brightest of the giant elliptical galaxies in the Virgo cluster. Only M49 (mag. +8.4) and M87 (mag. +8.7) appear more luminous from our perspective. M60 is visible with small scopes or large binoculars, but as with most galaxies it's better seen with greater aperture.

On April 11, 1779 while comet chasing, Johann Gottfried Koehler discovered M60 together with its slightly smaller and fainter neighbor M59. Also searching around the same time and the same part of the sky was none other than Charles Messier, who independently found both M59 and M60 four days after Koehler. During his search, Messier also discovered M58, another nearby Virgo cluster galaxy that was missed by Koehler. Of the three galaxies, Messier described M60 as the brightest with M59 and M58 being fainter and of similar magnitude.

Locating M60 is relatively easy. Start by imagining a line from Vindemiatrix (ε Vir - mag. +2.8) heading in the direction of Denebola (β Leo - mag. +2.1). About 4.5 degrees along this line is M60 with M59 positioned 0.4 degrees west of M60. Moving another degree in the same westerly direction arrives at M58.

M60 is estimated to lie 55 million light-years from Earth. It spans 7.6 x 6.2 arc minutes of apparent sky, which corresponds to a spatial diameter of 120,000 light-years. The galaxy contains about 400 billion stars and is best seen during the months of March, April and May.

Through a small 80mm (3.1-inch) telescope M60 appears as a fuzzy patch of light that's slightly brighter towards the core. It's marginally brighter and larger but otherwise rather similar in appearance to M59. When viewed through a 200mm (8-inch) scope, M60 is better seen, appearing diffuse with a bright nucleus but not a lot more detail apparent. On good nights, also visible is NGC 4647 (mag. +11.4) a small elliptical galaxy companion galaxy that lies 4 arc minutes northwest of M60 and in the same field of view.

The central black hole of M60 is 4.5 billion solar masses, one of the largest ever found. To date, only one supernova (SN 2004W) has been observed in the galaxy.

M60 Data Table

Messier	60
NGC	4649
Object Type	Elliptical galaxy
Classification	E2
Constellation	Virgo
Distance (kly)	55,000
Apparent Mag.	9.2
RA (J2000)	12h 43m 40s
DEC (J2000)	11d 33m 07s

Apparent Size (arcmins)	7.6 x 6.2
Radius (light years)	60,000
Number of Stars	400 Billion
Notable Feature	Central black hole of 4.5 billion solar masses is one of the largest ever found.

Messier # 060

Date:	Time:
Site:	
Temp:	Wind:	Hum:
Clouds:	Moon:
Scope:	
EP:	Mag:
NELM:	See/Trans:
Type:	# Stars:
Mag:	Age:
Const:	
Notes:	

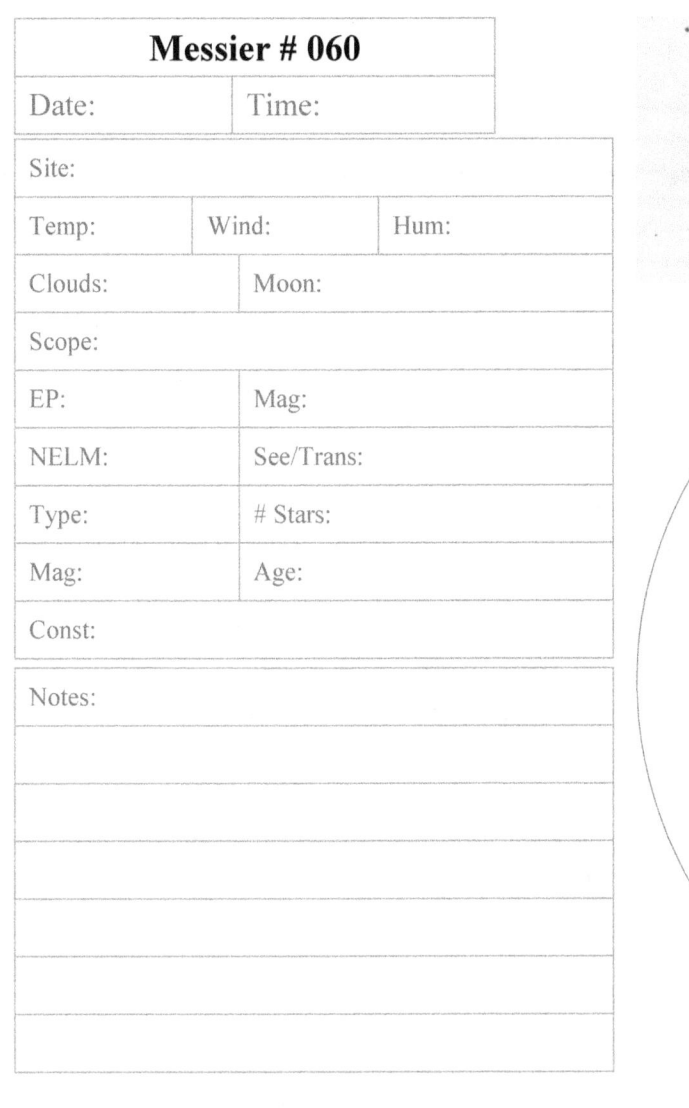

Messier Finder Chart for M49, M58, M59, M60, M84, M85, M86, M87, M88, M89, M90, M91, M98, M99 and M100

Also shown M53, M64 Black Eye Galaxy, M65 and M66

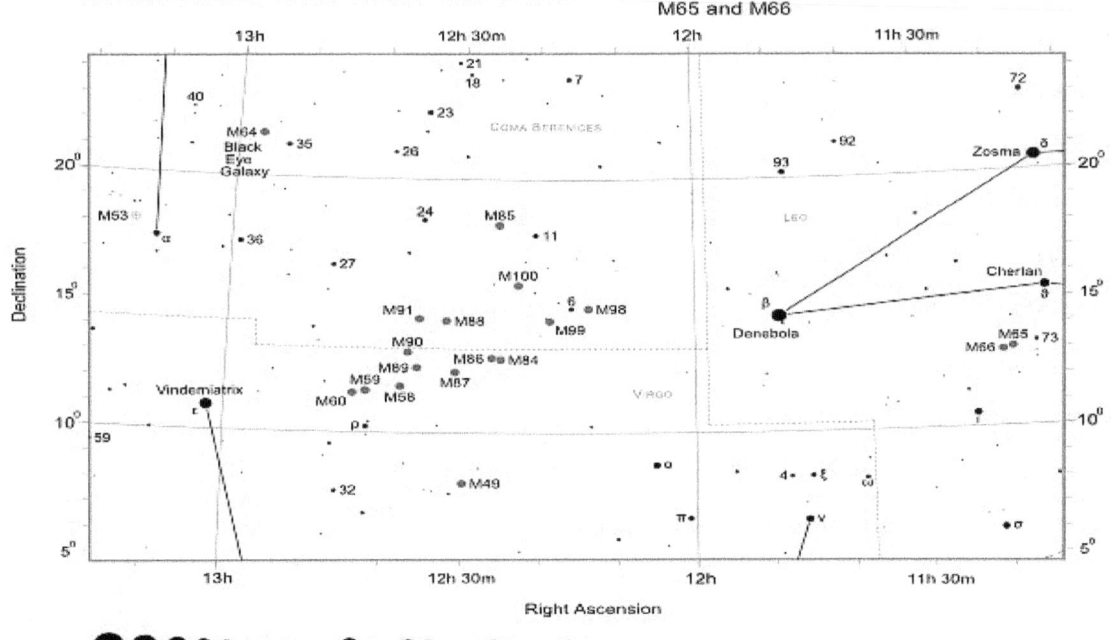

Messier 61 – M61 – Barred Spiral Galaxy

M61, mag. +9.9, is a face-on barred spiral galaxy that belongs to the Virgo cluster of galaxies. It was discovered by Barnabus Oriani while comet chasing on May 5, 1779. Ironically, Charles Messier observed M61 on the same day but mistakenly though he had seen a comet! A few days later he realized his mistake.

M61 is reasonably large galaxy with a diameter of 100,000 light-years, similar to own Milky Way. It has an apparent size of 6.5 x 5.9 arc minutes and is located 52.5 million light-years distant. The galaxy appears visually compact and is one of the finest "small" barred spiral galaxies in the sky for large backyard scopes.

It's located 8 degrees northwest of binary star Porrima (γ Vir - mag. +2.7), 1.25 degrees north-northeast of 16 Vir (mag. +5.0) and best seen during the months of March, April and May.

On dark nights, M61 can be spotted with small scopes as a faint patch of light. Through a 200mm (8-inch) instrument, it appears small with a stellar like core surrounded by a faint halo. Larger scopes of the order of 300mm (12-inch) display the structure in intricate detail with delicate spiral arms visible, surrounding a bright well-defined nucleus.

M61 is classification as a starburst galaxy due to high rates of star formation across its disk. This may be due to interactions with nearby satellite galaxies NGC 4292 and NGC 4303B. It's also a good galaxy for supernovae hunters. To date, seven have been observed with the brightest reaching magnitude +12 in 1964.

M61 Data Table

Messier	61
NGC	4303
Object Type	Spiral galaxy
Classification	SAB(rs) bc
Constellation	Virgo
Distance (kly)	52,500
Apparent Mag.	9.9
RA (J2000)	12h 21m 55s
DEC (J2000)	04d 28m 24s
Apparent Size (arcmins)	6.5 x 5.9
Radius (light years)	50,000
Number of Stars	>300 Billion
Notable Feature	Member of the Virgo cluster of galaxies

Messier # 061

Date:		Time:	
Site:			
Temp:	Wind:		Hum:
Clouds:		Moon:	
Scope:			
EP:		Mag:	
NELM:		See/Trans:	
Type:		# Stars:	
Mag:		Age:	
Const:			

Notes:

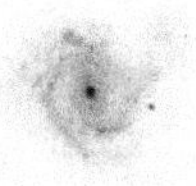

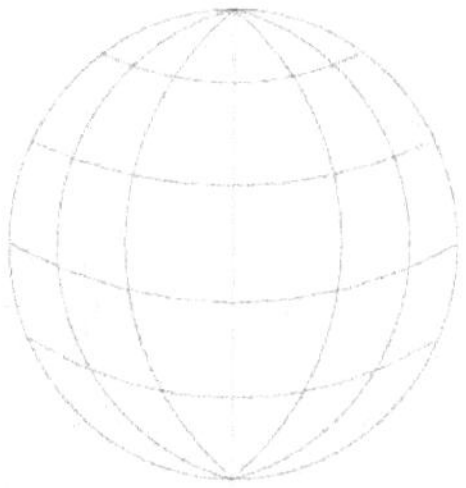

Messier Finder Chart for M61
Also shown M49 and M104 Sombrero Galaxy

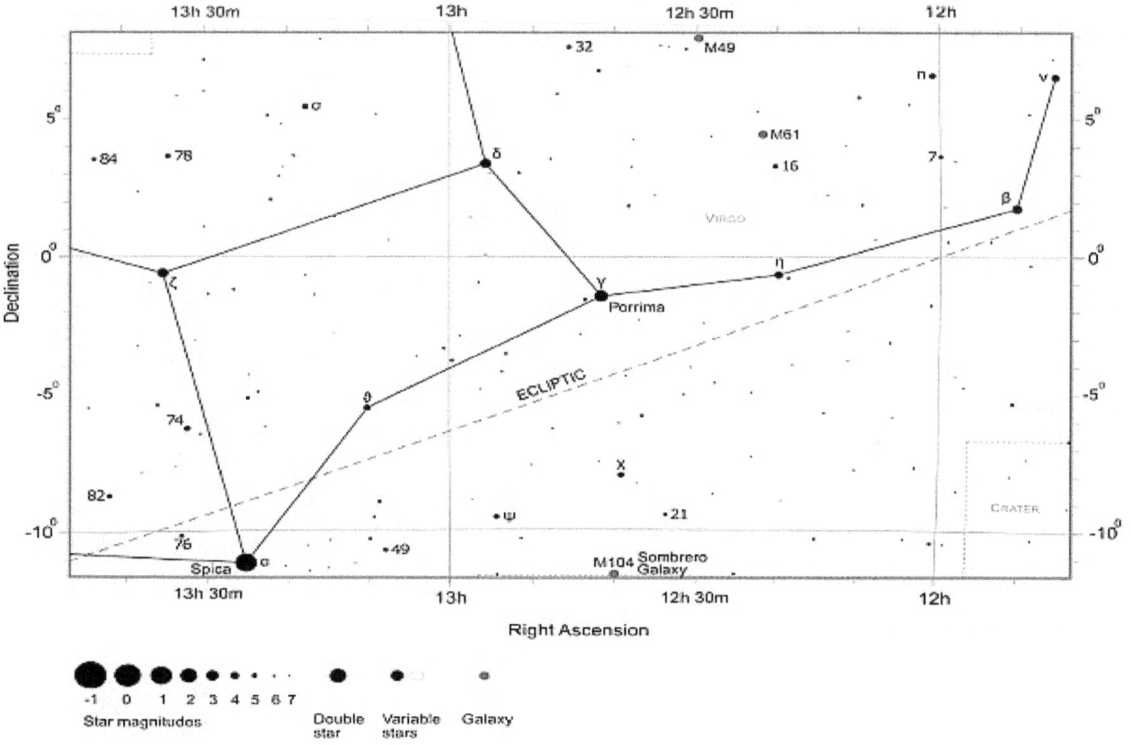

Star magnitudes   -1 0 1 2 3 4 5 6 7

Double star   Variable stars   Galaxy

Right Ascension

Messier 62 - M62 - Globular Cluster

M62 is a magnitude +6.5 globular cluster in the constellation of Ophiuchus at the border with Scorpius. It's located close to the centre of the Milky Way which may be the reason why it's one of the most irregular shaped globulars known. At a distance of only 6,100 light years from the galactic centre, M62 is subject to large deforming tidal forces. From Earth, the globular is much further away at 22,200 light-years.

M62 is visible with binoculars as a faint small fuzzy ball of light. However, since it's located amongst the rich star fields of the Milky Way it can easily be missed. It's best seen from tropical and southern hemisphere latitudes during the months of May, June and July. For mid latitude northern hemisphere based observers M62 is a tricky object. With a declination of 30 degrees south it's doesn't rise particularly high above the southern horizon, therefore never well situated for observation. The globular was discovered by Charles Messier on June 7, 1771. However, he didn't accurately measure its position until 1779 when it was added to his catalogue. William Herschel first resolved M62 into stars describing it as a miniature version of M3. Finding M62 can be challenging as there are no bright stars right next to it. It can be located by imagining a right angle triangle formed by connecting Antares (α Sco - mag. +1.0) with epsilon Sco (ε - mag. +2.3) and M62. Antares is 7.5 degrees northwest of M62 and ε Sco is 4.75 degrees southwest of M62.

Positioned 4.5 degrees north of M62 is slightly fainter but larger globular cluster M19 (mag. +7.2).

When viewed through a 80mm (3.1-inch) telescope, M62 appears as a diffuse ball of light with a bright core. On nights of good seeing and transparency a 200mm (8-inch) scope at medium to high magnifications will begin to resolve the outer edges. Averted vision helps significantly but larger size amateur scopes are required for better resolution to obtain a deeper view. Also noticeable is the oval shape of the cluster with the long axis orientated in the north-south direction. The center of the cluster appears condensed and it's a more compact globular than neighbor M19.

In total M62 covers 15 arc minutes of sky which corresponds to a spatial diameter 96 light-years. Visually it appears smaller, covering 8 to 10 arc minutes of apparent sky. The globular contains a high number of variable stars of which at least 89 are know, many of them RR Lyrae type. It's also likely that at some point in its history M62 has undergone a core collapse (like M15, M30 and M70).

M62 Data Table

Messier	62		Apparent Size (arcmins)	15 x 15
NGC	6266		Radius (light years)	48
Object Type	Globular cluster			
Constellation	Ophiuchus		Age (years)	11,780M
Distance (kly)	22.2		Number of Stars	150,000
Apparent Mag.	6.8		Notable Features	Very irregular in shape. Contains a high number of variable stars (at least 89)
RA (J2000)	17h 01m 12.5s			
DEC (J2000)	-30d 06m 44s			

Messier # 062

Date:	Time:
Site:	
Temp:	Wind:	Hum:
Clouds:	Moon:
Scope:	
EP:	Mag:
NELM:	See/Trans:
Type:	# Stars:
Mag:	Age:
Const:	
Notes:	

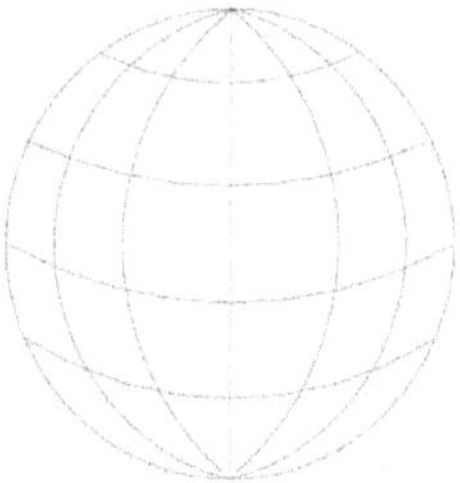

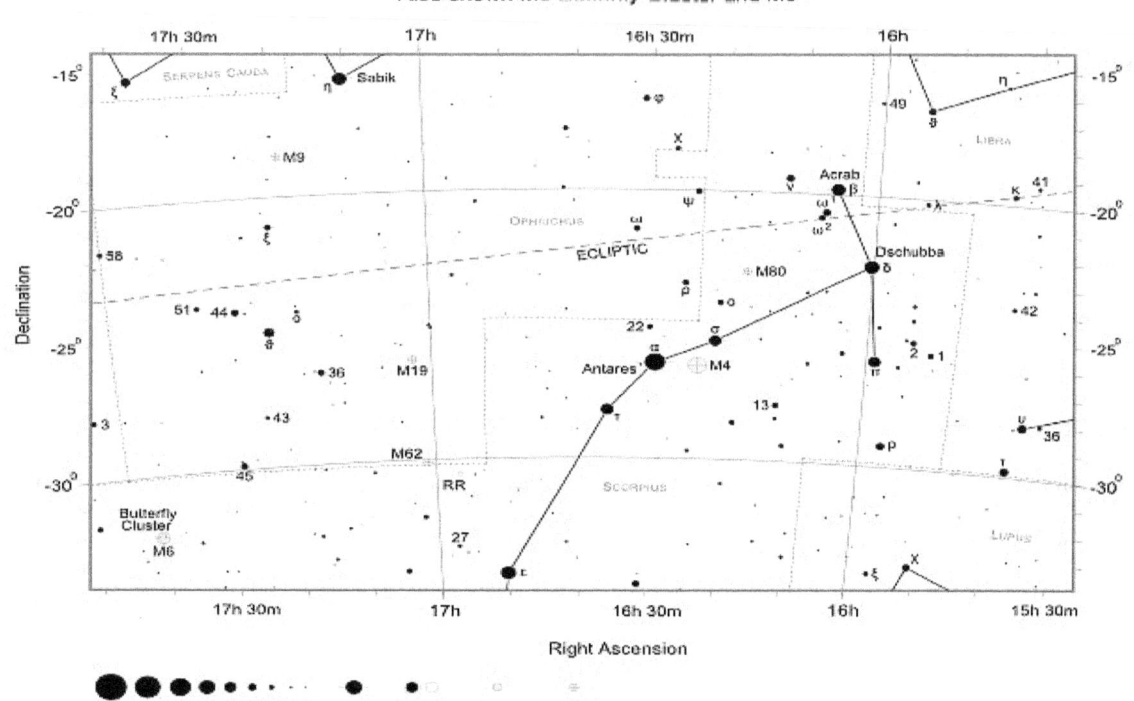

Messier Finder Chart for M4, M19, M62 and M80
Also shown M6 Butterfly Cluster and M9

Star magnitudes: -1 0 1 2 3 4 5 6 7

Double star Variable stars Open cluster Globular cluster

Messier 63 - M63 - The Sunflower Galaxy (Spiral Galaxy)

M63, the Sunflower Galaxy is a beautiful and spectacular spiral galaxy in the constellation of Canes Venatici. Along with M51 "The Whirlpool Galaxy", M94 and M106 it's one of four Messier galaxies located in the constellation. Of these, M51 is a much studied interacting grand design spiral that claims the title of finest galaxy in Canes Venatici. However, M63 is only marginally fainter and not far behind its illustrious neighbor. In addition, M63 and M51 are gravitationally bound and along with at least 6 other smaller galaxies they form the M51 Group of galaxies.

M63 is best seen from the Northern Hemisphere during the months of March, April or May.

The galaxy was the very first discovery made by Pierre Mechain, the great friend of Charles Messier, who found it on June 14, 1779. On the same day, Messier included it in his catalogue. Although Canes Venatici is a faint constellation locating M63 is easy since we can use the famous Plough or Big Dipper asterism of Ursa Major as a starting point. First locate Alkaid (η UMa - mag. +1.9) the eastern and end star of the handle of the Plough. Imagine a line from Alkaid moving in a southwesterly direction for about 14 degrees. At the end of this line is the brightest star in Canes Venatici, Cor Caroli (α CVn - mag. +2.9). M63 is positioned about 2/3rds of the way along this line.

At magnitude +8.9, the Sunflower Galaxy is a challenging binocular object, appearing at best as an out of focus star. Through a 80mm (3.1 inch) refractor it's recognizable as a galaxy but shows no detail unless viewed with medium or larger amateur scopes. A 150 mm (6 inch) instrument at high power reveals a bright core surrounded by a smooth oval shaped patch of nebulosity. A scope of 200mm (8-inch) aperture will start to show the spiral structure including the short arm segments. More subtle details such as dust lanes are visible in even larger scopes.

M63 has an apparent size of 12.6 x 7.2 arc minutes and is located 37 Million light years from Earth; equating to an actual diameter of 135,000 light years. In the early 1800s, Lord Rosse identified spiral structure within the galaxy, making this one of the first galaxies in which such structure was identified. It was listed as one of 14 "spiral nebulae" discovered to 1850.

In 1971, a magnitude +11.8 supernova appeared in one of the arms of M63.

M63 Data Table

Messier	63
NGC	5055
Name	Sunflower Galaxy
Object Type	Spiral galaxy
Classification	SA (rs)bc
Constellation	Canes Venatici
Distance (kly)	37,000
Apparent Mag.	8.9

RA (J2000)	13h 15m 49s
DEC (J2000)	42d 01m 50s
Apparent Size (arcmins)	12.6 x 7.2
Radius (light years)	67,500
Number of Stars	>400 Billion
Notable Feature	Member of the M51 Group of galaxies

Messier # 063

Date:	Time:
Site:	
Temp:	Wind:	Hum:
Clouds:	Moon:
Scope:	
EP:	Mag:
NELM:	See/Trans:
Type:	# Stars:
Mag:	Age:
Const:	
Notes:	

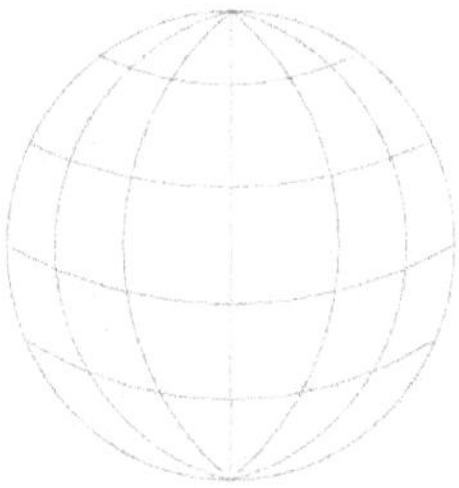

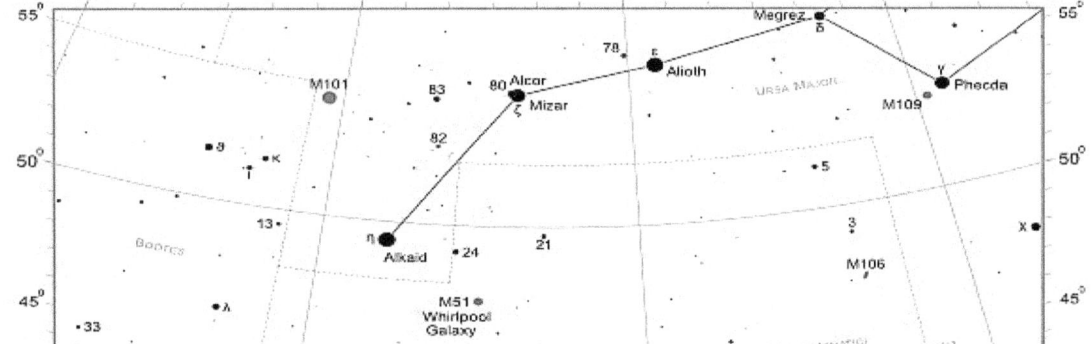

Messier Finder Chart for M51 Whirlpool Galaxy, M63 Sunflower Galaxy, M94, M101 and M106

Also shown M109

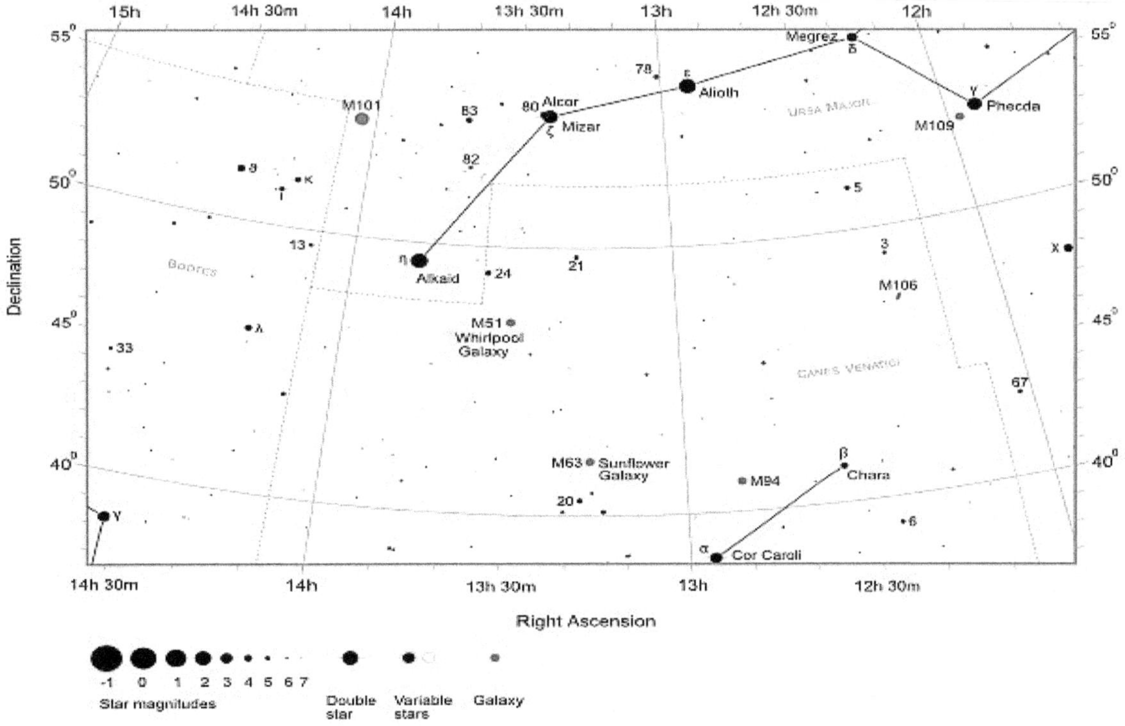

Messier 64 - M64 - Black Eye Galaxy (Spiral Galaxy)

M64 is a beautiful spiral galaxy known as the Black Eye Galaxy due to a spectacular dark band of absorbing dust in front of the nucleus, resulting in a smudged appearance. With an apparent magnitude of +8.8, it can be glimpsed with good binoculars on dark nights, appearing as a faint slightly irregular patch of light.

The Black Eye Galaxy is located in the constellation of Coma Berenices and was discovered by English astronomer Edward Pigott on March 23, 1779. Twelve days later Johann Elert Bode independently found it and Charles Messier adding it to his catalogue on March 1, 1780. The dark dust feature was discovered by William Herschel in 1785, comparing it to a black eye.

It's located 5 degrees northwest of Diadem (α Com - mag. +4.3) on an imaginary line connecting stars, 35 Com (mag. +4.9) and 40 Com (mag. +5.5), with M64 positioned one degree northeast of 35 Com.Arcturus, the brightest star in the northern section of the sky and fourth brightest overall is located 19 degrees east and a little south of M64.

The galaxy is 24 Million light-years distant and has an apparent size of 10.0 x 5.4 arc minutes, which corresponds to an actual linear diameter of 70,000 light-years. It's estimated to contain 100 billion stars and is best seen from northern latitudes during the months of March, April and May.

The Black Eye Galaxy is an extremely rewarding telescope object that's one of the brightest and easily observed galaxies anywhere in the sky. When viewed through a 100mm (4-inch) telescope, it appears irregular in shape with general uneven brightness and a large bright core. On nights of excellent seeing, the characteristic standout dark dust lane can be glimpsed with a telescope of this size, but it's easier with larger scopes. When viewed through 150mm (6-inch) instruments, the oval-shaped is accentuated and the dark dust lane easy to spot. A 200mm (8-inch) scope reveals the dark patch, a sharp condensed bright core surrounded by a large outer envelope of wispy nebulosity - a wonderful sight!

To date, no supernova has ever been recorded in M64.

M64 Data Table

Messier	64		**DEC (J2000)**	21d 40m 58s
NGC	4826		**Apparent Size (arcmins)**	10.0 x 5.4
Name	Black Eye Galaxy		**Radius (light years)**	35,000
Object Type	Spiral galaxy			
Classification	SA(rs)ab		**Number of Stars**	100 Billion
Constellation	Coma Berenices		**Notable Feature**	Also known as Sleeping Beauty Galaxy or sometimes Evil Eye Galaxy
Distance (kly)	24000			
Apparent Mag.	8.8			
RA (J2000)	12h 56m 44s			

Messier # 064

Date:		Time:	
Site:			
Temp:	Wind:		Hum:
Clouds:		Moon:	
Scope:			
EP:		Mag:	
NELM:		See/Trans:	
Type:		# Stars:	
Mag:		Age:	
Const:			
Notes:			

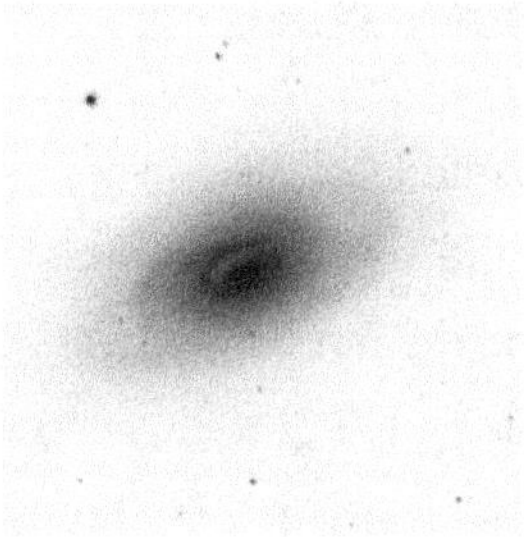

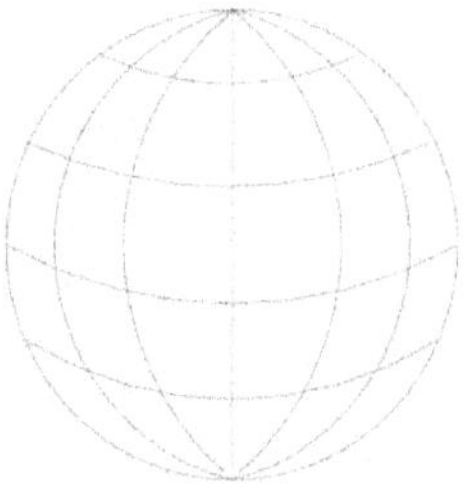

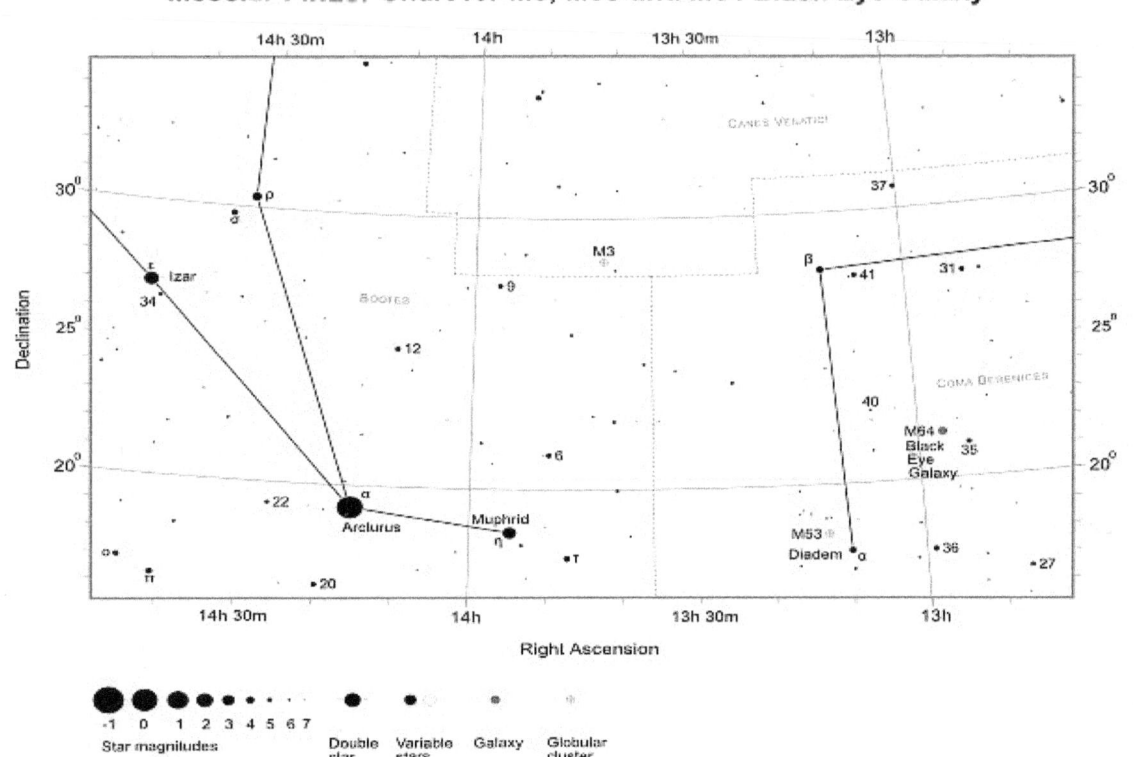

Messier Finder Chart for M3, M53 and M64 Black Eye Galaxy

Star magnitudes   -1 0 1 2 3 4 5 6 7

Double star   Variable stars   Galaxy   Globular cluster

Messier 65 – M65 – Spiral Galaxy

M65 is a magnitude +9.6 spiral galaxy located 35 million light years from Earth in the constellation of Leo. It was discovered by Charles Messier on March 1, 1780, the night he also discovered M66. Messier described the galaxy as a "very faint nebula without stars." Apart from M81 and M82 in Ursa Major, M65 and M66 are probably the most sought after galaxy pair for amateur astronomers. With a third galaxy, NGC 3628, lying nearby the trio forms the heart of a small group of celebrated galaxies known as "The Leo Triplet" or "M66 group". All three can be observed or photographed in the same field of view and as a result are a sought after grouping for amateur observers and astrophotographers alike.M65 is located in the eastern section of the relatively large and bright zodiac constellation of Leo "the Lion", with the Lion's heart in western skylore marked by the constellations brightest and only first magnitude star, Regulus (α Leo – mag +1.4). Positioned approximately 16 degrees to the northeast of Regulus is the star Chertan (θ Leo - mag. +3.3). Along with the Zosma (δ Leo - mag. +2.6) to the north and Denebola (β Leo - mag. +2.1) to the east, Chertan forms a prominent right-angled triangle. Now move 2 degrees south of Chertan to arrive at magnitude +5.3 star 73 Leo. M65 lies 0.75 degrees east of 73 Leo with M66 a further 0.33 degrees southeast of M65. Located 0.5 degrees to the north of the Messier pair is the 3rd member of the famous Leo triplet, NGC 3628.

The galaxies are best seen during the months of March, April and May.

With a magnitude of +9.6 and a high surface brightness, M65 is visible in binoculars from a dark site. Although dim, 10x50 binoculars display a brighter centre surrounded by fainter edges. Also visible in the same binocular field of view is M66. In total, M65 spans 8.7 x 2.4 arc minutes of apparent sky, appearing almost side on when viewed from our perspective. A small telescope such as an 80mm (3.1-inch) refractor nicely shows the elongation as a thin slivery disk of light with a brighter core. Subtle details can be made out of the galaxy hazy structure through medium sized telescopes of the order of 200mm (8-inch) aperture with even larger amateur telescopes revealing beautiful dark lanes within the galaxy's structure.M65 is marginally the faintest of the Leo Triplet and similar in appearance to M66. Although NGC 3628 is similar in magnitude to M66 it's also the hardest to pick out due to its larger apparent size and lower surface brightness. Once you have identified all three, it's a worthwhile taking time to compare these wonderful galaxies through the eyepiece.

M65 Data Table

Messier	65	**DEC (J2000)**	13d 05m 37s
NGC	3623	**Apparent Size (arcmins)**	8.7 x 2.4
Object Type	Spiral galaxy	**Radius (light years)**	45,000
Classification	SAB (rs) a	**Number of Stars**	200 Billion
Constellation	Leo	**Notable Feature**	Member of the Leo Triplet along with M66 and NGC 3628
Distance (kly)	35,000		
Apparent Mag.	9.6		
RA (J2000)	11h 18m 56s		

Messier # 065

Date:	Time:	
Site:		
Temp:	Wind:	Hum:
Clouds:	Moon:	
Scope:		
EP:	Mag:	
NELM:	See/Trans:	
Type:	# Stars:	
Mag:	Age:	
Const:		

Notes:

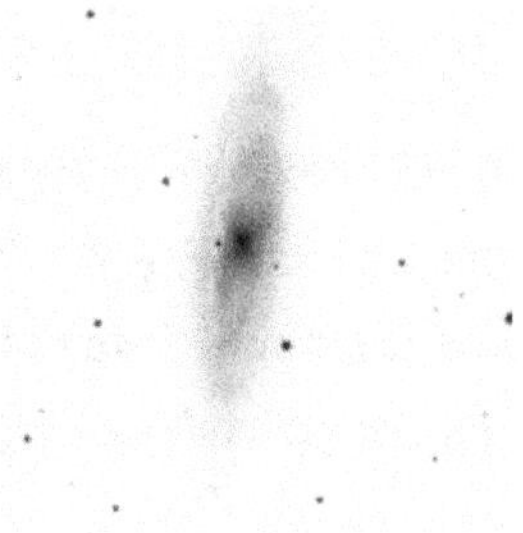

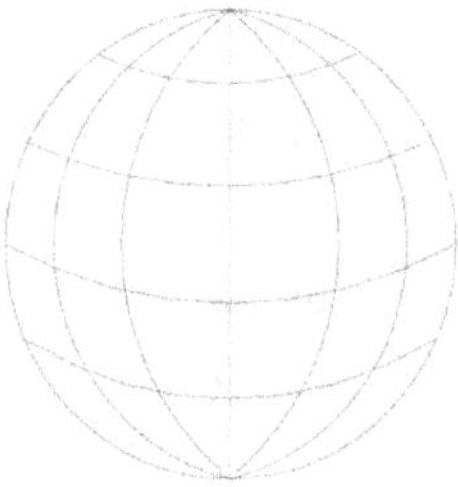

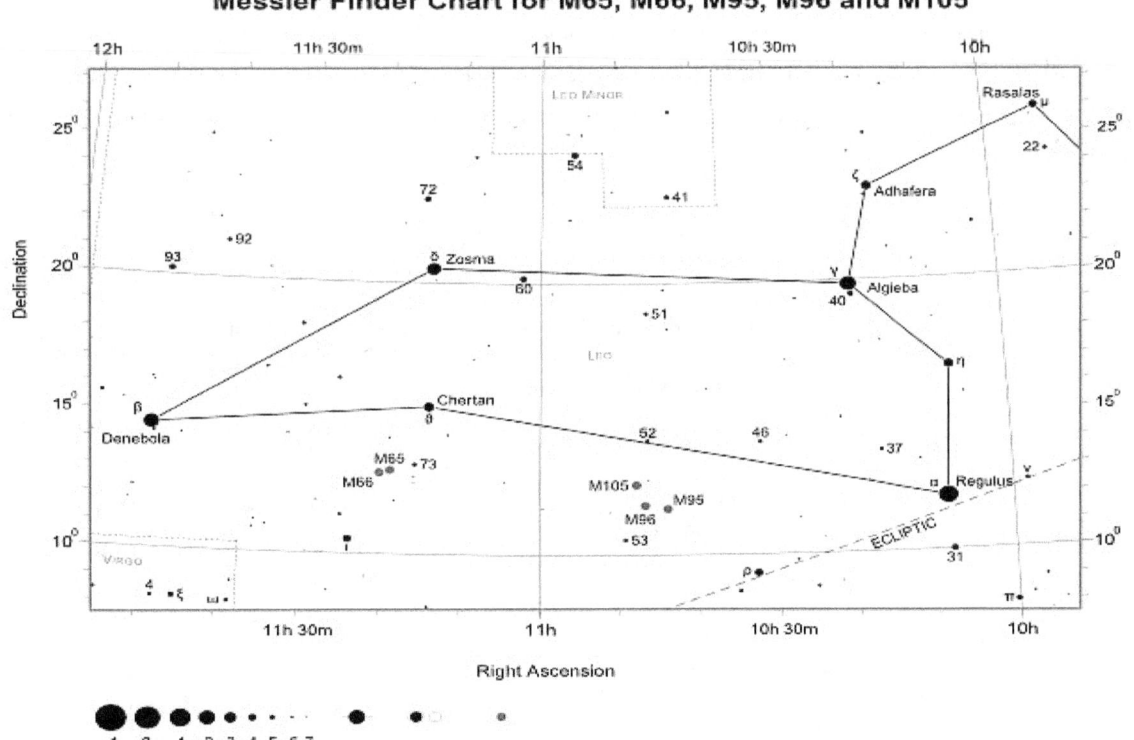

Messier Finder Chart for M65, M66, M95, M96 and M105

Messier 66 - M66 - Intermediate Spiral Galaxy

M66 is a superb bright intermediate spiral galaxy located in the constellation of Leo. It's the brightest of a trio of galaxies that form - along with M65 and NGC 3628 - the well known and popular "Leo Triplet" or "M66 group". All three objects can be observed with small telescopes in the same low power field of view. With the exception of the M81/M82 pair in Ursa Major, the Leo Triplet is arguably the most sought after galaxy grouping for amateur astronomers.

Charles Messier discovered both M66 (mag. +8.9) and M65 (mag. +9.6) on March 1, 1780. The third member of the triplet, NGC 3628, has a debatable apparent magnitude. Some texts record it as the brightest member of the three while others the faintest! For our purposes we estimate NGC 3628 to be brighter than M65 and almost as bright as M66. However, what is clear is that NGC 3628 suffers from low surface brightness and therefore is the most difficult member of the trio to spot. Messier missed it completely and it was not until April 8, 1784 when it was finally discovered by William Herschel. To find the triplet, look to the eastern part of Leo. This zodiac constellation is relatively large and bright and somewhat looks like the Lion it represents. The brightest star in Leo and its only first magnitude star is Regulus (α Leo – mag +1.4). Positioned approximately 16 degrees northeast of Regulus is Chertan (θ Leo - mag. +3.3). Along with the Zosma (δ Leo - mag. +2.6) to the north and Denebola (β Leo - mag. +2.1) to the east, Chertan forms a prominent right-angled triangle. Located 2 degrees south of Chertan is 73 Leo (mag. +5.3). M65 lies 0.75 degrees east of this star with M66 a further 0.33 degrees southeast of M65. Located 0.5 degrees north of the Messier pair is NGC 3628.

The galaxies are best seen during the months of March, April and May. From a dark site, M66 (and M65) are faintly visible as fuzzy patches of light with 10x50 binoculars. Through larger 20x80mm binoculars or a small 80mm (3.1-inch) telescope, M66 appears as a thin oval smudge of light with a brighter centre. Since M66 and M65 are separated by only 20 arc minutes of apparent sky, both galaxies fit neatly in the same low magnification field of view.

A medium size telescope of 150mm (6-inch) aperture shows M66 as a slightly edge on silvery disk of light with a bright stellar like nucleus. A 250mm (10-inch) scope hints at a spiral structure with the brighter nebulosity wrapped in a fainter halo. On the other hand, M65 appears fainter and more edge on when compared to M66. In total, M66 measures 9.1 x 4.1 arc minutes in apparent size. At a distance of 36 million light years from Earth, this corresponds to an actual diameter of 95,000 light years. The galaxy is estimated to contain about 200 billion stars. So far, four supernovae have been recorded in M66 (1973R, 1989B, 1997bs and 2009hd). The brightest of the four was 1989B which reached magnitude +12.2 on February 1, 1989.

M66 Data Table

Messier	66	**Apparent Mag.**	8.9	**Number of Stars**	200 Billion
NGC	3627	**RA (J2000)**	11h 20m 15s		
Object Type	Intermediate spiral galaxy	**DEC (J2000)**	12d 59m 28s	**Notable Feature**	Member of the Leo Triplet along with M65 and NGC 3628
Classification	SAB (s)	**Apparent Size (arcmins)**	9.1 x 4.1		
Constellation	Leo	**Radius (light years)**	47,500		
Distance (kly)	36,000				

Messier # 066

Date:	Time:
Site:	
Temp:	Wind:	Hum:
Clouds:	Moon:
Scope:	
EP:	Mag:
NELM:	See/Trans:
Type:	# Stars:
Mag:	Age:
Const:	

Notes:

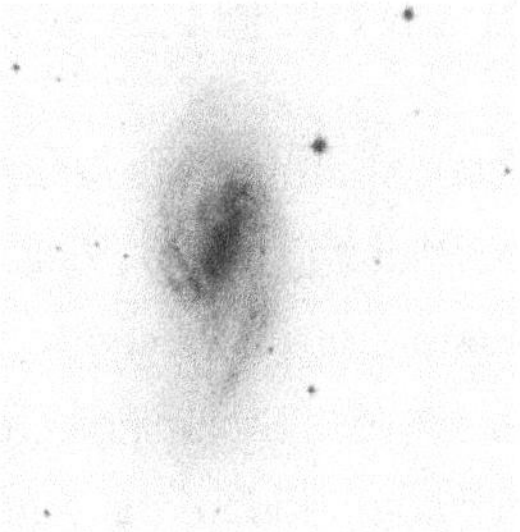

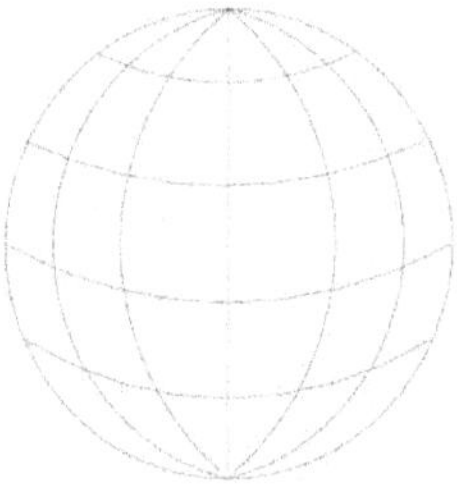

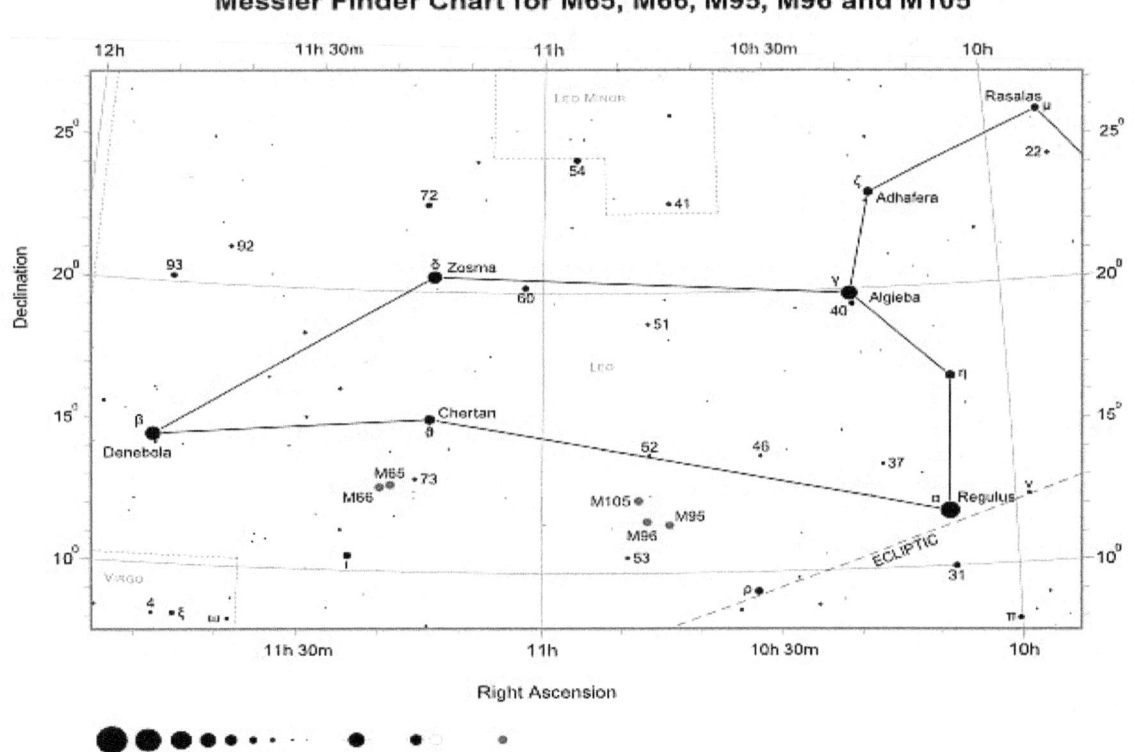

Messier Finder Chart for M65, M66, M95, M96 and M105

Right Ascension

Declination

LEO MINOR

Rasalas

Adhafera

Zosma

Algieba

Chertan

Denebola

Regulus

M65
M66
M105
M96

VIRGO

LEO

ECLIPTIC

Star magnitudes -1 0 1 2 3 4 5 6 7

Double star   Variable stars   Galaxy

Messier 67 - M67 - Open Cluster

Messier 67 is a very old open cluster located in the constellation of Cancer. Estimations of the cluster age vary between 3.2 and 5 billion years with recent valuations indicating it to be nearer to 4.0 billion years, making M67 one of the oldest known open clusters. Only a handful of open clusters are believed to be older but none of them are as close to us as M67. As a result of its age, it contains a variety of star types including many Sun like stars, red giants and white dwarfs and is easily the most ancient Messier open cluster. For comparison, the other Messier object in Cancer the great open cluster M44 "The Praesepe" is 600 million years old and brilliant open cluster M45 "The Pleiades" is a very youthful 100 million years of age. The constellation of Cancer "the Crab" is a faint zodiac constellation that is bordered by much brighter Leo to the east and Gemini to the west. To the north is faint Lynx, with Canis Minor and Hydra located on the southern side. First locate the heart of Cancer, a grouping of four faint stars. They are Asellus Australis (δ Cnc - mag. +3.9), Asellus Borealis (γ Cnc - mag. +4.7), η Cnc (mag. +5.3) and θ Cnc (mag. +5.3) with M44 contained within this grouping. Then identify another faint star in Cancer, α Cnc (mag. +4.3), which is positioned a further 9 degrees to the southeast of M44. M67 is located 1.75 degrees west of this star. Alternatively, M67 can be found about half way along an imagining line connecting first magnitude stars Regulus (α Leo - mag. +1.4) and Pollux (β Gem - mag. +1.1).M67 was discovered sometime before 1779 by German astronomer Johann Gottfried Koehler. He described the cluster as being rather conspicuous and nebula like in appearance although with his basic telescope, he was unable to resolve it into stars. Charles Messier then independently rediscovered M67, resolved it into stars and cataloged it on April 6, 1780 as a "Cluster of small stars with nebulosity, below the southern claw of Cancer. The position determined from the star Alpha [Cancri].".

M67 is best seen from the Northern Hemisphere during the months of February, March and April. With an apparent magnitude of +6.1, M67 is at the boundary of naked eye visibility. It's easily seen in 10x50 binoculars, appearing as an elongated "blur" of light with several component stars visible under good seeing. Through telescopes M67 is a beautiful object. A 80mm (3.1-inch) telescope reveals a large concentrated misty patch of light with a sprinkling of bright stars visible. Medium sized scopes fair even better. When viewed through 150mm (6-inch) or 200mm (8-inch) scopes, M67 is resolved into dozens of stars mainly clustered towards the centre, superimposed on a hazy background. Higher magnifications helps to resolve fainter stars. A larger 300mm (12-inch) scope at about 100x magnification displays about 100 stars, at the same time removing the background mist seen in small scopes.M67 is a rich cluster of at least 200 faint stars in the constellation of Cancer that has a diameter of 30 arc minutes, equivalent to that of the full Moon. It lies at a distance of 2700 light-years from Earth and with an estimated age of 4 Billion years is one of the oldest known and most studied open clusters. Although often overshadowed by M44, its famous neighbor to the north, M67 is a wonderful cluster in its own right that provides a marvelous contrast to the sprawling Beehive cluster.

M67 Data Table

Messier	67	DEC (J2000)	11d 48m 43s
NGC	2682	Apparent Size (arcmins)	30 x 30
Object Type	Open cluster	Radius (light years)	10
Constellation	Cancer	Age (years)	4000M
Distance (kly)	2.7	Number of Stars	>200
Apparent Mag.	6.1	Other Name	Collinder 204
RA (J2000)	08h 51m 20s		

Messier # 067

Date:	Time:

Site:

Temp:	Wind:	Hum:

Clouds:	Moon:

Scope:

EP:	Mag:

NELM:	See/Trans:

Type:	# Stars:

Mag:	Age:

Const:

Notes:

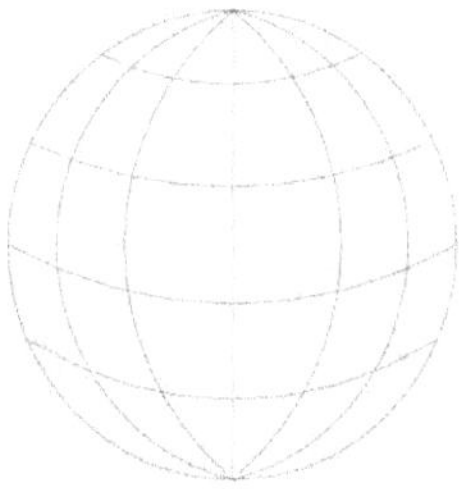

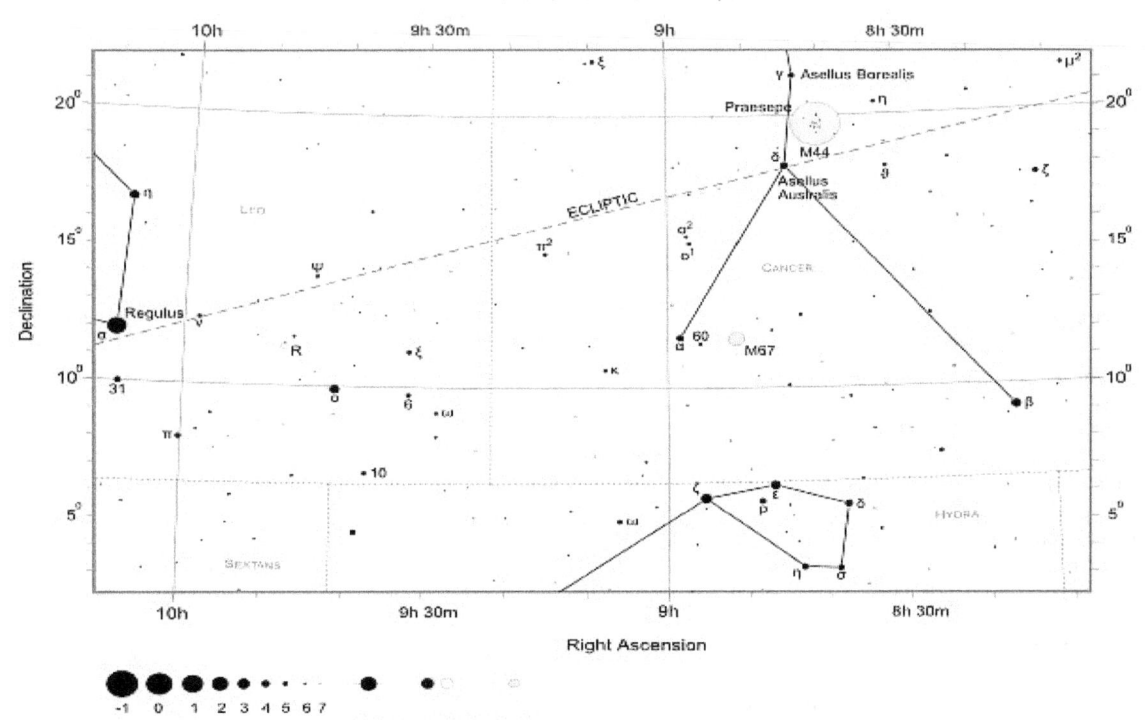

Messier Finder Chart for M67
Also shown M44 Praesepe

Star magnitudes
-1 0 1 2 3 4 5 6 7

Double star   Variable stars   Open cluster

Messier 68 - M68 - Globular Cluster

M68 is a mag. +7.8 medium sized globular cluster located in eastern Hydra that was discovered by Charles Messier on April 9, 1780. Although not as spectacular as great globulars such as Omega Centauri, 47 Tucanae or M13, it's easily visible with 7x50 or 10x50 binoculars and appears obviously non-stellar. The globular is well seen through medium and large sized amateur scopes.

Hydra is the night sky's largest constellation. However, despite it's immense apparent size it contains only one reasonably bright star, Alphard (α Hya) at mag. +2.0. Despite this, locating M68 is quite easy as it's positioned just south of the relatively bright quadrangle of Corvus (Crv) and 3.5 degrees southeast of star β Crv (mag. +2.6).

With a declination of -26.7 degrees, M68 is best seen from the Southern Hemisphere during the months of March, April and May. From northern temperate locations it appears low down and doesn't climb very high above the southern horizon at best.

Through a small 80mm (3.1-inch) telescope, M68 reveals a brighter centre surrounded by a fuzzy halo that gradually fades to the edges. Under nights of good transparency and seeing conditions the outer parts of M68 are resolved in a 200mm (8-inch) scope. The core is not compact and large amateur scopes of the order of 300mm (12-inch) will resolve stars across the full face of the cluster. In total, M68 spans 11 arc minutes of apparent sky, although it appears somewhat smaller through the eyepiece.

M68 is located approximately 33,000 light years distance. It has a spatial diameter of about 105 light-years and is estimated to contain more than 100,000 stars.

M68 Data Table

Messier	68
NGC	4590
Object Type	Globular cluster
Constellation	Hydra
Distance (kly)	33.3
Apparent Mag.	7.8
RA (J2000)	12h 39m 28s
DEC (J2000)	-26d 44m 34s
Apparent Size (arcmins)	11 x 11
Radius (light years)	53
Age (years)	11,200M
Number of Stars	>100,000
Notable Feature	Relatively metal poor cluster

Messier # 068

Date:	Time:	
Site:		
Temp:	Wind:	Hum:
Clouds:	Moon:	
Scope:		
EP:	Mag:	
NELM:	See/Trans:	
Type:	# Stars:	
Mag:	Age:	
Const:		

Notes:

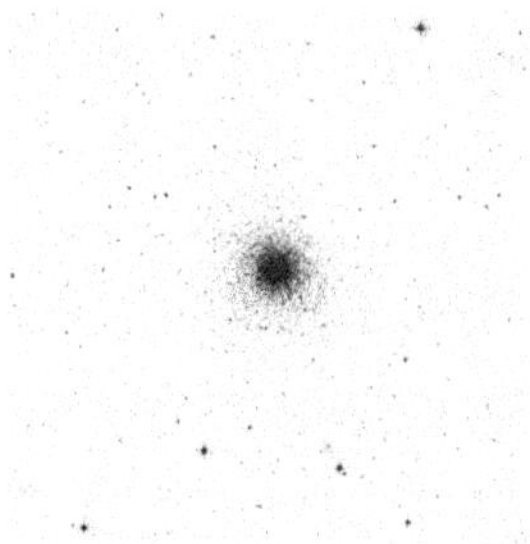

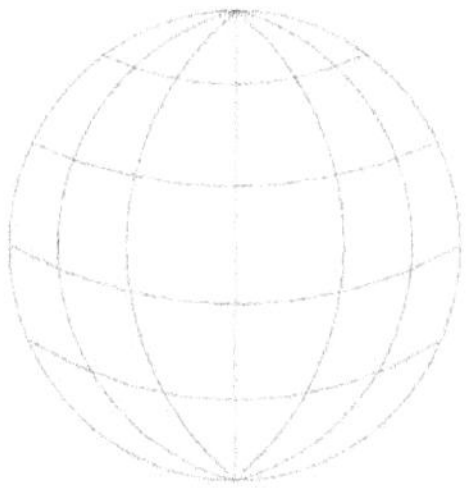

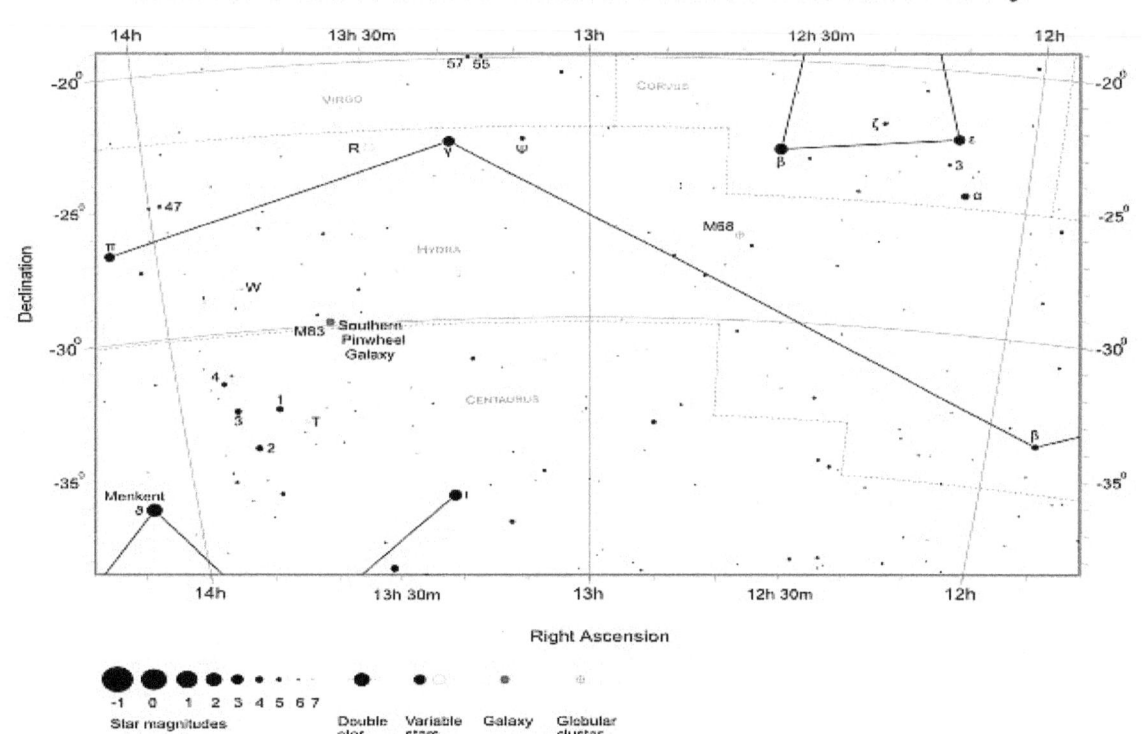

Messier Finder Chart for M68 and M83 Southern Pinwheel Galaxy

Declination

Right Ascension

Star magnitudes -1 0 1 2 3 4 5 6 7

Double star   Variable stars   Galaxy   Globular cluster

Messier 69 - M69 - Globular Cluster

M69 is a globular cluster located inside the bright "Teapot" asterism of Sagittarius. It shines at magnitude +7.6 and therefore within the range of good quality 7x50 or 10x50 binoculars, although faint and only star like in appearance. The cluster is best seen from the Southern Hemisphere during the months of June, July and August. However, from northern temperate latitudes it's a difficult object as it never climbs high above the southern horizon.

M69 is located 29,700 light-years from Earth and was discovered by Charles Messier on August 31, 1780. On this night he also discovered M70, a physically close neighbor of M69; spatially they are separated by just 1,800 light years. Any potential observers located on planets orbiting stars inside M69 would have a spectacular view of M70 and vice-versa. Of course, this is assuming that the many thousands of bright stars visible in their own globular won't block the view of the other.

Finding M69 is easy once one is familiar within the teapot asterism of Sagittarius. Start by focusing on the base of the teapot and image a line connecting Kaus Australis (ε Sgr - mag. +1.8) with Ascella (ζ Sgr - mag. +2.6). Positioned 1.5 degrees along this line and 2 degrees north is M69, with two 5th magnitude stars located just south of the cluster.

When viewed through a small 80mm (3.1-inch) scope, M69 appears small with a faint halo surrounding a bright centre. It looks somewhat like a comet but even at high powers is not resolvable. Of the three Messier globulars (M54, M69 and M70) inside the "Teapot", M69 is the brightest. All three are fine clusters in their own right but they are small in apparent size and therefore difficult to resolve with amateur scopes.

A 200mm (8-inch) scope at high powers hints at resolving M69. Averted vision also helps but even larger scopes are recommended. The central cluster concentration is also apparent with a gradually tailing of brightness to the outer edges. In total, M69 has an apparent size of almost 10 arc minutes but through amateur scopes it appears much smaller. It's estimated to contain 125,000 stars and is extremely poor in variable stars. Outstanding twentieth century American astronomer Harlow Shapley could not find any and to date only 8 variables have been discovered.

M69 Data Table

Messier	69
NGC	6637
Object Type	Globular cluster
Constellation	Sagittarius
Distance (kly)	29.7
Apparent Mag.	7.6
RA (J2000)	18h 31m 23s
DEC (J2000)	-32d 20m 53s

Apparent Size (arcmins)	9.8 x 9.8
Radius (light years)	42
Age (years)	13,060M
Number of Stars	125,000
Notable Feature	One of the most metal rich globulars known.

Messier # 069

Date:	Time:
Site:	
Temp:	Wind:	Hum:
Clouds:	Moon:
Scope:	
EP:	Mag:
NELM:	See/Trans:
Type:	# Stars:
Mag:	Age:
Const:	
Notes:	

**Messier Finder Chart for M8 Lagoon Nebula,
M20 Trifid Nebula, M21, M22, M28, M54, M69, M70**

Also shown M6 Butterfly Cluster,
M7, M18, M23, M24 Star Cloud
and M55

Star magnitudes
-1 0 1 2 3 4 5 6 7

Double star   Variable stars   Open cluster   Globular cluster   Bright nebulae

Messier 70 - M70 - Globular Cluster

M70 is an eighth magnitude globular cluster located in Sagittarius that's faintly visible with binoculars, appearing "star" like. It's much easier to spot with small telescopes where despite being small with little detail visible, it appear obviously non-stellar. To resolve M70 into stars large amateur scopes are required.

Charles Messier discovered M70 on August 31, 1780, describing it as a "nebula without star". On the same night he also discovered M69, another close by globular (both apparently and spatially). M70 has an extremely dense core and is believed at some time previously to have suffered a core collapse, similar to Messier globulars M15, M30 and possibly M62. It was William Herschel who first resolved M70 into stars, describing it as a miniature version of M3.

M70 is located 29,300 light years from Earth. Spatially, it's separated by only 1,800 light-years from M69 with both objects located close to the galactic centre. They are best seen from the Southern Hemisphere during the months of June, July and August. However, from northern temperate latitudes they are never well positioned, at best climbing just a few degrees above the southern horizon.

Finding M70 is easy once familiar within the teapot asterism of Sagittarius. Locate the two stars that make up the base of the teapot, Kaus Australis (ε Sgr - mag. +1.8) and Ascella (ζ Sgr - mag. +2.6). Imagine a line connecting these two stars. M70 is positioned almost exactly halfway along this line.

At magnitude +8.0, M70 is much easier to spot with larger 11x70 or 20x80 binoculars than with standard size models. When viewed through 80mm (3.1-inch) scopes, the globular appears as a faint diffuse ball of light with a slightly brighter central region. Telescope apertures of 250mm (10-inch) or larger are required to start resolving some of the outer stars. In total, M70 covers 8 arc minutes of apparent sky but through amateur scopes, visually it appears much smaller than this. The cluster is estimated to be 12.8 billion years old and contains 75,000 stars.

M70 made headlines in 1995 when Alan Hale and Thomas Bopp observed it and discovered the great comet Hale-Bopp nearby.

M70 Data Table

Messier	70	**Apparent Size (arcmins)**	8 x 8
NGC	6681	**Radius (light years)**	34
Object Type	Globular cluster	**Age (years)**	12,800M
Constellation	Sagittarius	**Number of Stars**	75,000
Distance (kly)	29.3	**Notable Feature**	Believed to have previously suffered a core collapse
Apparent Mag.	8.0		
RA (J2000)	18h 43m 13s		
DEC (J2000)	-32d 17m 31s		

Messier # 070

Date:		Time:	
Site:			
Temp:	Wind:		Hum:
Clouds:		Moon:	
Scope:			
EP:		Mag:	
NELM:		See/Trans:	
Type:		# Stars:	
Mag:		Age:	
Const:			
Notes:			

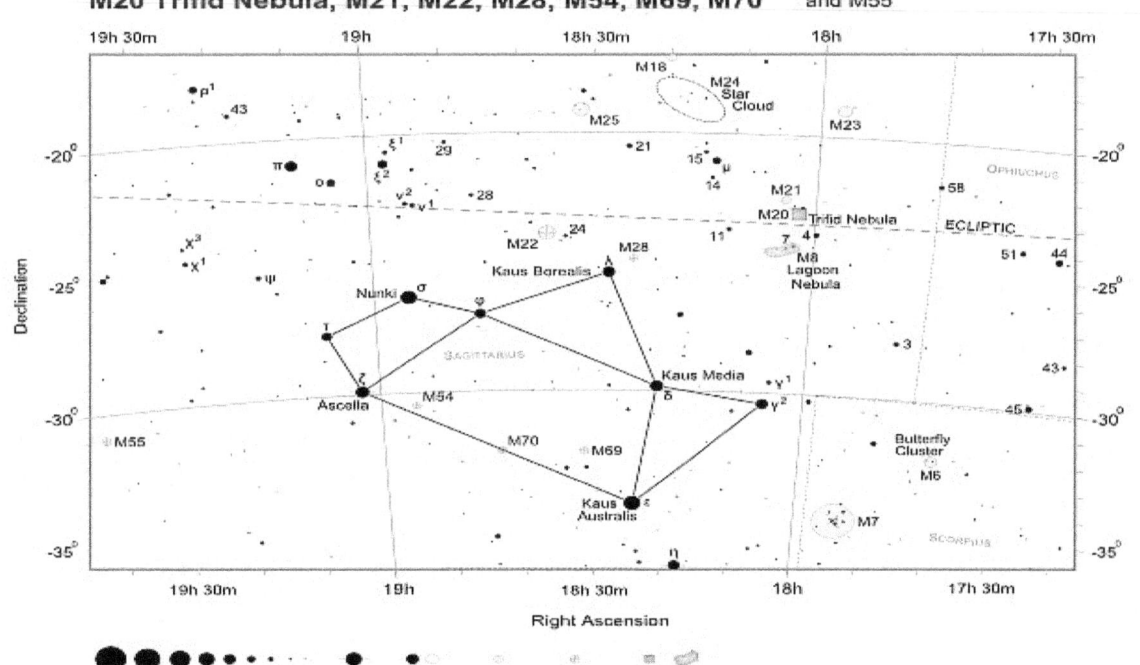

Messier Finder Chart for M8 Lagoon Nebula, M20 Trifid Nebula, M21, M22, M28, M54, M69, M70

Also shown M6 Butterfly Cluster, M7, M18, M23, M24 Star Cloud and M55

Messier 71 - M71 - Globular Cluster

M71 is a very loose but attractive globular cluster located in the small constellation of Sagittarius, "the Arrow". At magnitude +7.1, it's not visible to the naked eye but can be seen with binoculars. The cluster is best placed in the sky during the months of June, July and August.

M71 was discovered by wealthy Swiss landowner Philippe Loys de Cheseaux in 1746 and then re-discovered by Dresden based astronomer Johann Koehler in 1775 and Pierre Méchain in June 1780. Méchain informed his friend Messier who searched for the cluster, found it and finally catalogued it on October 4, 1780. He described it as "very faint...containing no stars...the least light extinguishes it".

Of course, the view through modern backyard telescopes is considerably better, but lets first find M71. The finder chart below depicts its position in Sagittarius. The globular is very easy to locate as it lies midway between gamma Sge (γ Sge - mag. +3.5) and delta Sge (δ Sge - mag. +3.7). These stars form part of the distinctive, although not particularly bright, "Arrow" asterism. Two degrees southwest of γ Sge is M71.

Through binoculars, M71 appears as a reasonably large fuzzy patch of light. It's a faint object for small telescopes and is best seen with averted vision. For example, through a 80mm (3.1-inch) scope little or no detail is discernible but increase the aperture just a small amount and what a difference! Through a 150mm (6-inch) or 200mm (8-inch) scope the globular appears large and loose with many 11th and 12th magnitude stars resolvable across the cluster surface. Comparisons are often made with M11, the larger and brighter Wild Duck open cluster and it's easy to see why M71 was earlier questioned (and sometimes still is) as an open cluster.

Larger telescopes still reveal many more stars. A 300mm (12-inch) telescope resolves over 100 stars across a seven arc minute diameter disk. It's also apparent that M71 lacks the typically strong central condensation of normal globular clusters.M71 is of the smallest and youngest globular clusters know and also one of the nearest at only 13,000 light years distant. It has a spatial diameter of only 27 light-years and is estimated to be 9.5 billion years old. In total, it contains at least 20,000 stars and is a nice globular cluster with a remarkable history, well worth a look.

M71 Data Table

NGC	6838	Radius (light years)	13.5
Object Type	Globular Cluster	Age (years)	9,500M
Constellation	Sagittarius	Number of Stars	>20,000
Distance (kly)	13	Other Name	Collinder 409
Apparent Mag.	7.1	Notable Feature	The irregular variable star Z Sge (Z Sagittae) is a member of this cluster
RA (J2000)	19h 53m 46s		
DEC (J2000)	18d 46m 42s		
Apparent Size (arcmins)	7.2 x 7.2		

Messier # 071

Date:	Time:
Site:	
Temp:	Wind:	Hum:
Clouds:	Moon:
Scope:	
EP:	Mag:
NELM:	See/Trans:
Type:	# Stars:
Mag:	Age:
Const:	
Notes:	

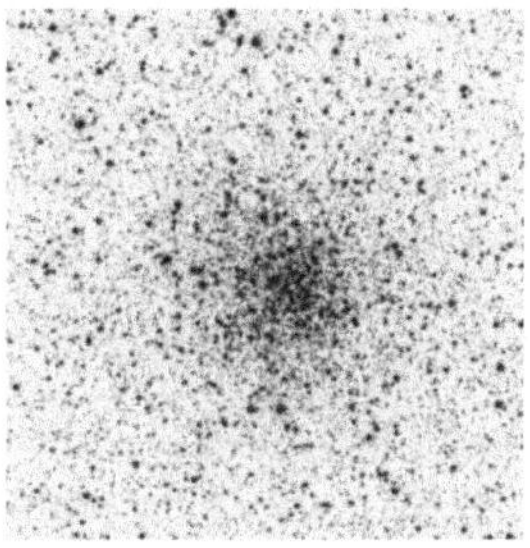

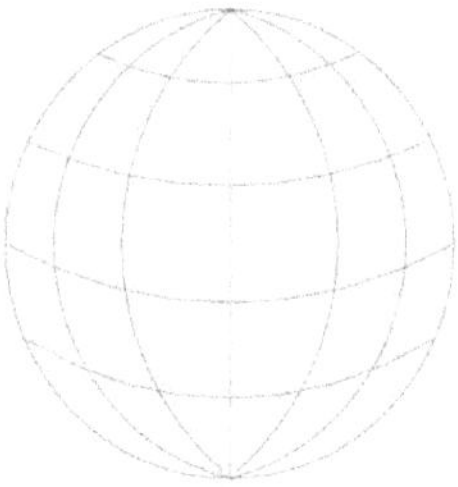

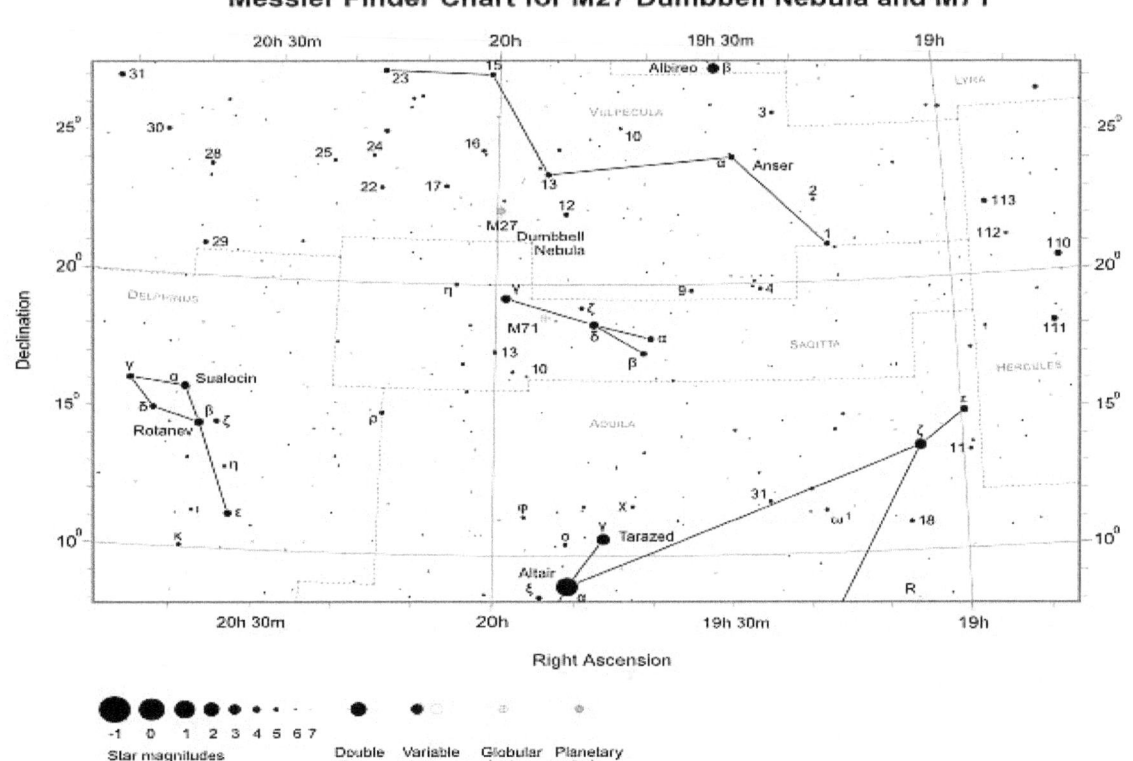

Messier Finder Chart for M27 Dumbbell Nebula and M71

Star magnitudes: -1 0 1 2 3 4 5 6 7

Double star   Variable stars   Globular cluster   Planetary nebula

Messier 72 - M72 - Globular Cluster

M72 is a distant, magnitude +9.3, globular cluster located in the constellation of Aquarius. At 55,000 light years from Earth, it's way beyond the galactic center and consequently one of the faintest and smallest (apparent size) of Messier's globulars. Hence, M72 appears as a very faint diffuse ball of light in small telescopes that's easy to miss.M72 was discovered by Pierre Méchain on August 29, 1780. His friend Charles Messier found it a few days later on October 4th and included it in his catalogue. With their instruments - good quality at the time although low quality by today's commercially standards - both Méchain and Messier had difficulty in deciding exactly what M72 was. In the end they believed it to be a faint nebula rather than a star cluster. A few years later, British astronomer Sir William Herschel using a larger instrument was able to determine the true nature of M72, resolving it into stars. Of note, on the same night as he catalogued M72, Messier discovered M73. This curious item, located just over a degree east of M72, was described by Messier as a cluster of four stars with some nebulosity. However, it's now believed to be just an asterism of four stars with no associated nebulosity.

Locating M72:- There are several ways to locate this object but none are particularly easy; amateur astronomers often regard M72 as one of the more difficult Messier objects to find. Since it's positioned in the southern part of Aquarius towards the constellation boundary, one method is to start with neighboring Capricornus. Once familiar with the stars and general shape of Capricornus, focus on theta (θ Cap - mag. +4.1). Then image a line from θ Cap traveling in a northwesterly direction for 9 degrees. Once there you will reach a slightly brighter star, magnitude +3.8 Albali (ε Aqr). Just over halfway along this line is M72.The globular is best seen during the months of July, August and September and ideally from tropical, Southern Hemisphere regions where it appears high in the sky.

Due to its vast distance and faintness, even large 80mm binoculars struggle to pick out M72, appearing as just a star like point of light. Small 100mm (4-inch) telescopes fair little better, with only the brighter core region readily visible. Overall this is a tough cluster to resolve with amateur instruments and even a 250mm (10-inch) telescope only hints at resolving the extreme edges of M72. However, what is noticeable is the even brightness across the face of M72. Very large amateur telescopes of 400mm (16-inch) aperture or more show M72 looking rather like M13 as seen with a small scope.M72 has a spatial diameter of 104 light-years and is believed to contain at least 100,000 stars. Of these, 42, a considerable number are know variables (mostly RR Lyrae stars). It's estimated to be 9.5 Billion years old.

In summary, distant globular M72 is a challenging object for amateur astronomers. It's visible in small scopes but to get the most from this far off cluster, dark skies together with a medium / large size telescope are essential. This not very compact cluster is difficult to resolve but intrinsically luminous, hence why we can seen it at all from such a far distance with small scopes.

M72 Data Table

Messier	72		**DEC (J2000)**	-12d 32m 13s
NGC	6981		**Apparent Size (arcmins)**	6.6 x 6.6
Object Type	Globular cluster		**Radius (light years)**	52
Constellation	Aquarius		**Age (years)**	9,500M
Distance (kly)	54		**Number of Stars**	>100,000
Apparent Mag.	9.3			
RA (J2000)	20h 53m 28s			

Messier # 072

Date:	Time:
Site:	
Temp:	Wind:	Hum:
Clouds:	Moon:
Scope:	
EP:	Mag:
NELM:	See/Trans:
Type:	# Stars:
Mag:	Age:
Const:	

Notes:

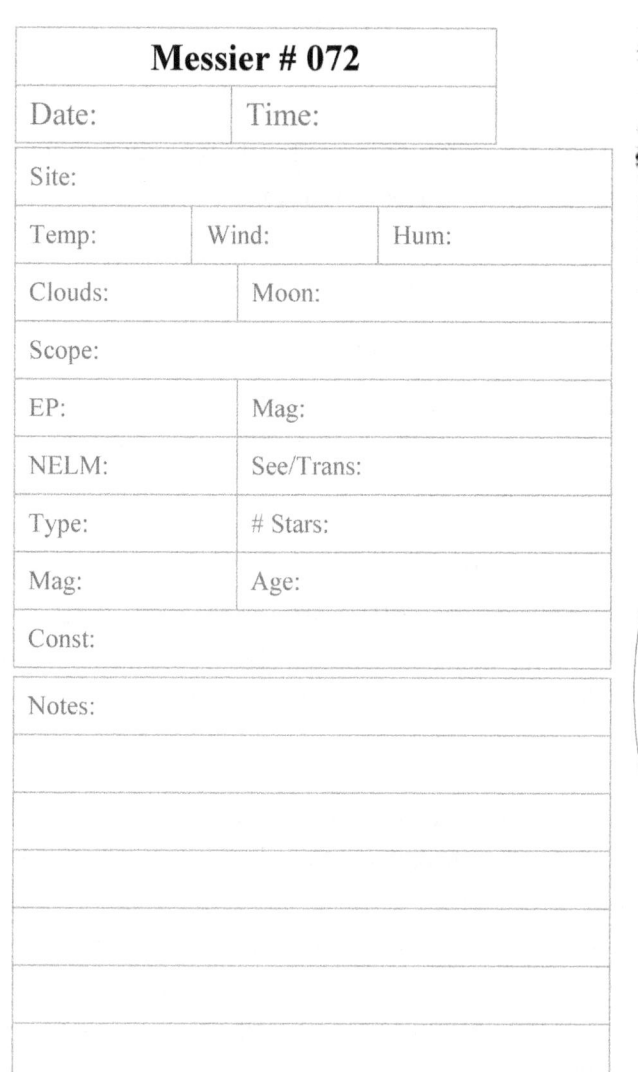

Messier Finder Chart for M30, M72 and M73
Also shown M75

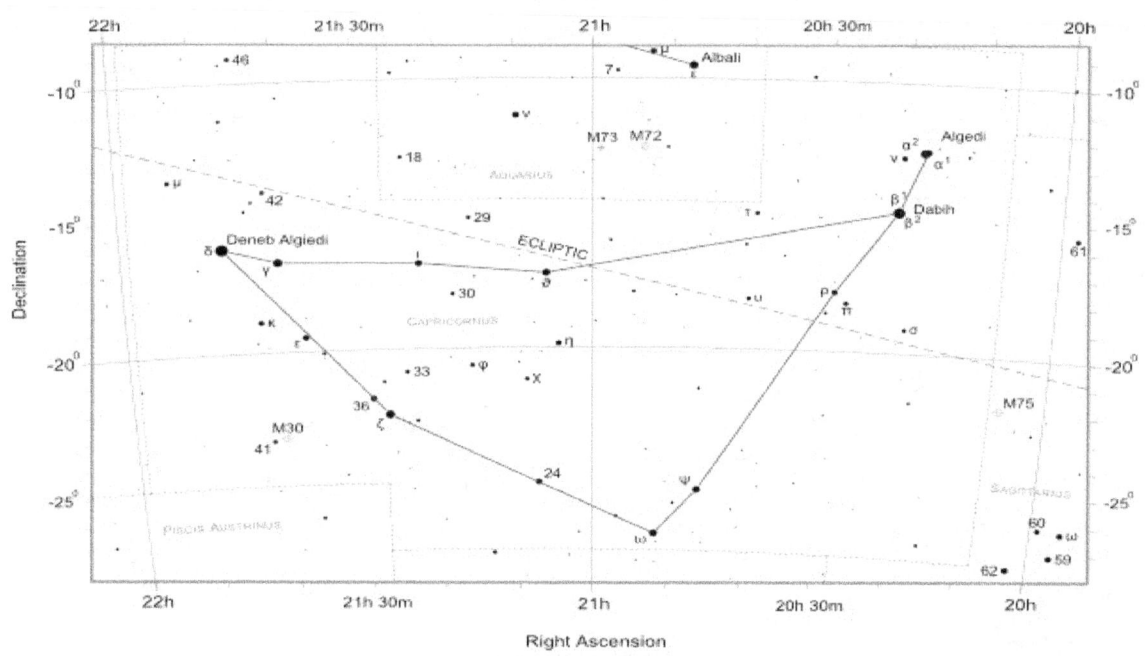

Right Ascension

Messier 73 - M73 - Asterism

M73 is a curious entry in the Messier catalogue. It's a small group of four faint stars shaped like a "Y" in the constellation of Aquarius. The grouping was discovered by Messier himself on October 4, 1780, the night he also catalogued M72. He described the object as a cluster of three or four stars with some nebulosity. Subsequent observations by William and John Herschel revealed no nebulosity. Nevertheless, John Herschel included M73 in his General Catalogue of clusters, nebulae, and galaxies (GC) and John Dreyer added M73 to the New General Catalogue (NGC).

It's now known that no such nebulosity exists and recent measurements suggest that M73 is not a true open cluster at all but just an asterism.M73 is located in southern Aquarius close to the border with Capricornus. The asterism is positioned just over 1 degree east of 9th magnitude distant globular cluster M72. Finding M73 - also M72 - can be quite challenging; Aquarius and Capricornus are both relatively faint constellations with few bright stars.

One possible route to locate M73 is with the stars of Capricornus. Locate theta (θ Cap - mag. +4.1) and then image a line from this star in a northwesterly direction for 9 degrees until you reach marginally brighter star, Albali (ε Aqr - mag. +3.8). Just over halfway along this line is M72 and located 1 1/3 degrees to the east of M72 is M73.The asterism is best seen during the months of July, August and September and ideally from tropical, Southern Hemisphere regions where it appears high in the sky.

M73 has an apparent magnitude of +9.0 and hence is a difficult 7x50 or 10x50 binocular object. Larger 20x80 binoculars fair better, although since the four stars are between 10th and 12th magnitude in brightness, the asterism appears as a poor faint point of light. Through 100mm (4-inch) telescopes the "Y" shape of the group is visible with three of the stars blue-white in color, the other orange. What's noticeable is that one star is fainter than the other three (component magnitudes are +10.4, +11.3, +11.7 and +12.3). The apparent diameter of M73 is only 2.8 arc minutes and hence it's possible to push up the magnification as high as the seeing conditions allow. A 150mm (6-inch) telescope or larger telescope will easily show the asterism in full detail. It's often been claimed that the telescopic view of M73 appears fuzzy suggesting of the same appearance Messier himself observed. Scientifically there has not been a great deal of interest in M73. The cluster was previously treated as a sparsely populated open cluster, where the stars are loosely bound to each other by mutual gravitational attraction. However, evidence has always been thin on the ground and there has been a small debate on whether the stars are an asterism or an open cluster. The current belief is that the cluster is an asterism with the four stars located at different distances and moving in different directions. A rough estimate of the average distance of M73 is 2500 light years. Despite being only a chance alignment of stars, analysis of asterisms is useful to astronomers studying how open clusters are pulled apart by Milky Way gravitational forces.

M73 Data Table

Messier	73		**RA (J2000)**	20h 58m 56s
NGC	6994		**DEC (J2000)**	-12d 38m 08s
Object Type	Asterism		**Apparent Size (arcmins)**	2.8 x 2.8
Constellation	Aquarius		**Number of Stars**	4
Distance (kly)	~2.5		**Other Name**	Collinder 426
Apparent Mag.	9.0			

Messier # 073

Date:	Time:
Site:	

Temp:	Wind:	Hum:

Clouds:	Moon:
Scope:	
EP:	Mag:
NELM:	See/Trans:
Type:	# Stars:
Mag:	Age:
Const:	

Notes:

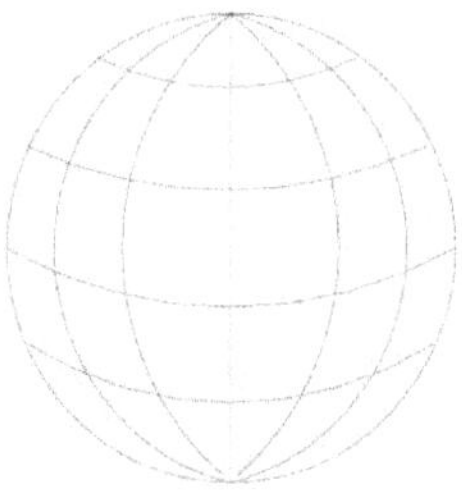

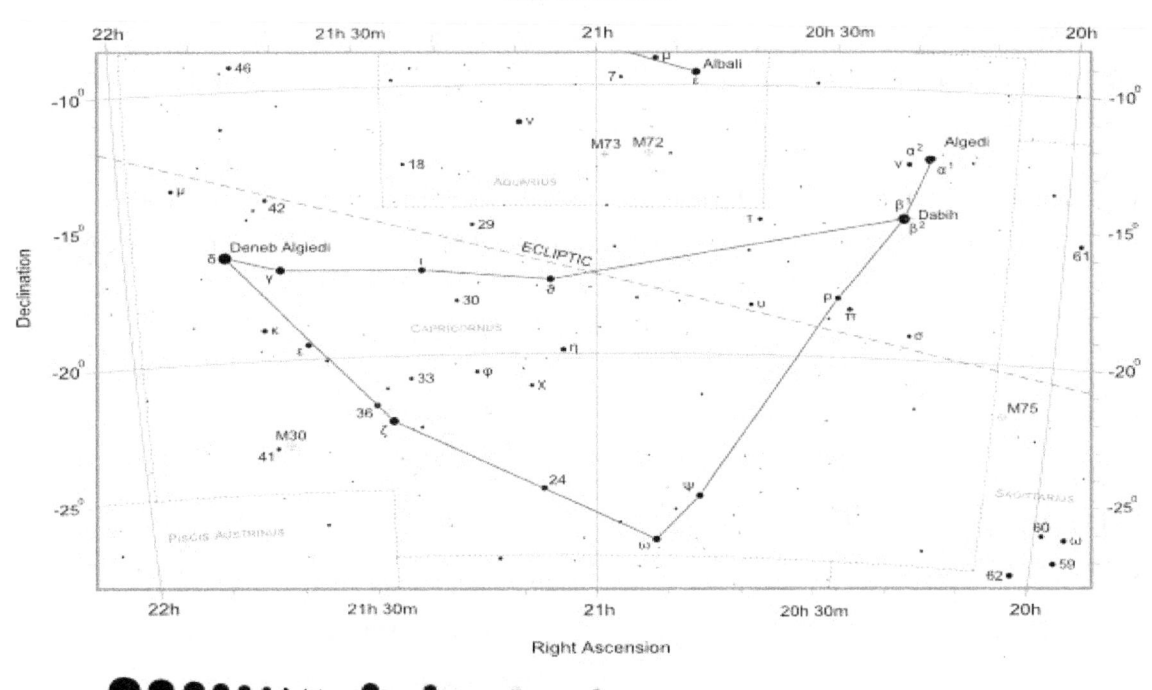

Messier Finder Chart for M30, M72 and M73
Also shown M75

Right Ascension

Declination

22h · 21h 30m · 21h · 20h 30m · 20h

46 · Albali · μ · 7 · ν · M73 · M72 · AQUARIUS · 18 · μ · 42 · 29 · ECLIPTIC · Deneb Algiedi · δ · γ · ι · 30 · δ · τ · υ · ρ · π · σ · α² · α¹ · Algedi · ν · β γ · Dabih · β² · 61 · CAPRICORNUS · κ · ε · η · 33 · φ · χ · 36 · ζ · M30 · 41 · 24 · ψ · M75 · ω · 60 · ω · SAGITTARIUS · 62 · 59 · PISCIS AUSTRINUS

-10⁰ · -15⁰ · -20⁰ · -25⁰

Star magnitudes   -1 0 1 2 3 4 5 6 7

Double star   Variable stars   Globular cluster   Asterism

Messier 74 - M74 - Spiral Galaxy

M74 is a beautiful face-on spiral galaxy located in the constellation of Pisces. It is an archetypal example of a Grand Design spiral galaxy that was discovered by Pierre Méchain sometime during September 1780. Méchain then reported his discovery to Charles Messier, who subsequently determined its position and catalogued it on October 18, 1780. With a magnitude of +9.4, M74 is not a faint galaxy but it does suffer from low surface brightness and can be difficult to locate even with just a hint of light pollution. As a result, it is widely regarded as one of the more difficult Messier objects.

M74 is located in the barren star fields of Pisces. Despite Pisces being the 14th largest constellation in the sky (889 sq. degrees), it contains no bright stars. The brightest star Eta Piscium (η Psc) is only of magnitude +3.6. Luckily this star is a perfect marker; M74 lies only 1.5 degrees to the east-northeast. To pinpoint Eta Piscium, start in the adjacent constellation of Aries. First locate the constellations two brightest stars, Hamal (mag. +2.0) and Sheratan (mag. +2.6). Then imagine a line connecting Hamal to Sheratan and extend this line in a southwest direction for just over 7.5 degrees. This leads directly to Eta Piscium. The galaxy is best seen during the months of October, November and December.

Observationally, M74 is a very difficult binocular object. In 10x50 binoculars it's extremely challenging with dark, light pollution free skies a must and even then it may still be invisible. At best it appears only as a very faint smudge of light. Larger 20x80 binoculars make the task easier with M74 appearing as a dim small round patch of nebulosity without detail. The view through a small 100mm (4-inch) telescope at low magnification is slightly improved; the core region of the galaxy is visible surrounded by a very faint fuzzy halo. With a 200mm (8-inch) telescope the halo is more pronounced and appears larger. The halo of course is the beautiful spiral arms, evidence of which can be glimpsed in larger telescopes of the order of 300mm (12-inch) aperture or greater. In appearance, M74 looks like a smaller version of M33 The Triangulum Galaxy but with a more defined nucleus.

The galaxy is located 32.5 million light-years from Earth and has a diameter of about 95,000 light-years, similar to that of the Milky Way. It has a large apparent size of 10.5 x 9.5 arc minutes. In 2002 and again in 2003, supernovae were discovered in M74. This first supernova brightened up to mag. +12.3, the second to mag. +13.2.

Interestedly, M74 was the subject of a mistake made by Sir John Herschel in his General Catalogue (GC), published in 1864. In it Herschel mysteriously catalogued M74 as a globular cluster, an error that was also carried over into the subsequent New General Catalogue (NGC).

M74 is a beautiful spectacular face-on spiral galaxy with a low surface brightness. It therefore can be a difficult object to locate visually, but is rewarding once found. The full glory and majesty of M74 is revealed in images and photographs.

M74 Data Table

Messier	74
NGC	628
Object Type	Spiral galaxy
Classification	SA(s)c
Constellation	Pisces
Distance (kly)	32500

Apparent Mag.	9.4
RA (J2000)	01h 36m 42s
DEC (J2000)	15d 47m 03s
Apparent Size (arcmins)	10.5 x 9.5
Radius (light years)	47,500

Messier # 074

Date:	Time:
Site:	
Temp:	Wind:	Hum:
Clouds:	Moon:
Scope:	
EP:	Mag:
NELM:	See/Trans:
Type:	# Stars:
Mag:	Age:
Const:	

Notes:

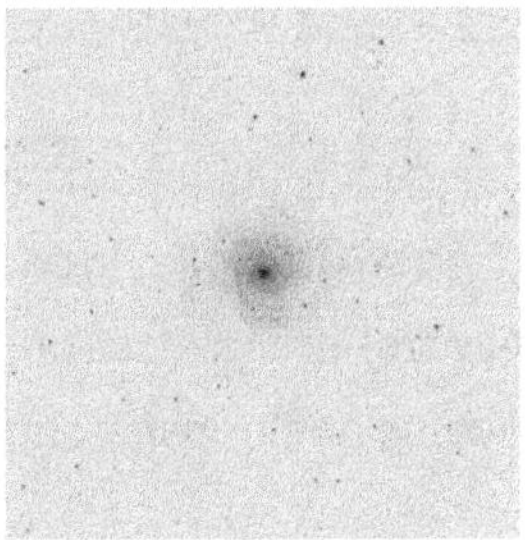

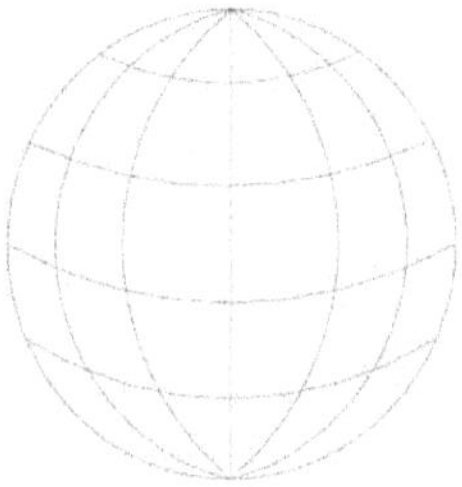

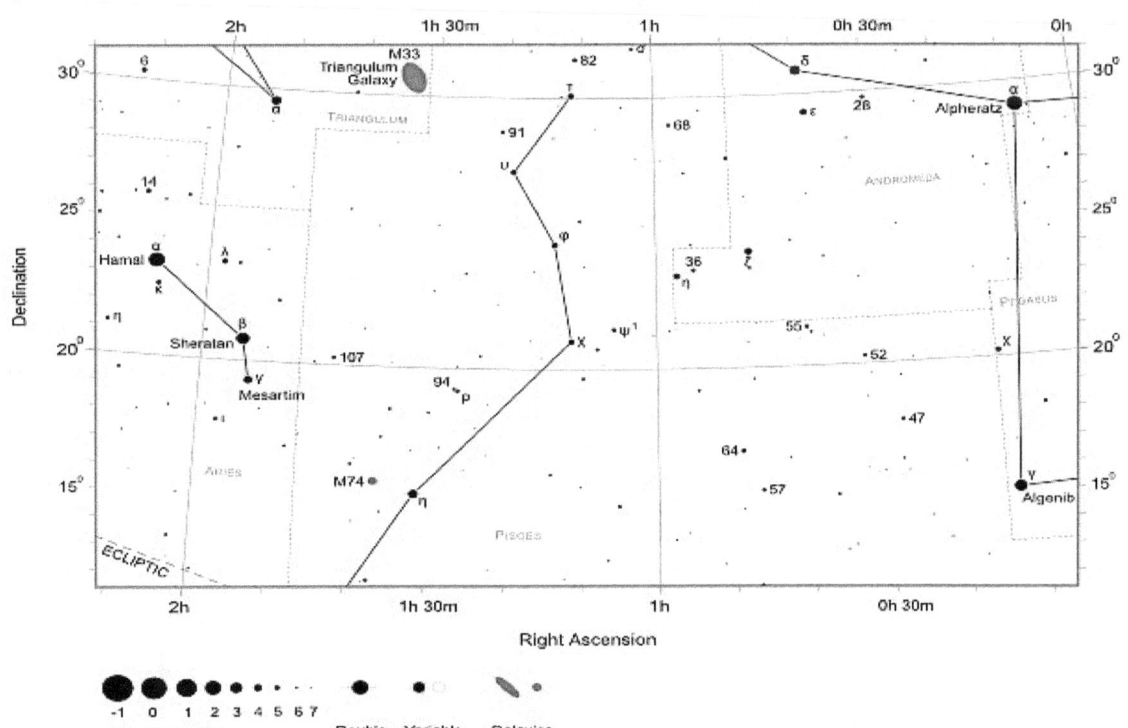

Messier Finder Chart for M74
Also shown M33 Triangulum Galaxy

Declination

Right Ascension

Star magnitudes -1 0 1 2 3 4 5 6 7

Double star   Variable stars   Galaxies

Messier 75 - M75 - Globular Cluster

M75 (mag. +8.7) is a very distant and compact globular cluster located in eastern Sagittarius. At 67,500 light years from Earth it's one of the more remote Messier globulars and hence appears faint and small from our perspective. It's located far beyond the galactic centre (46,700 light-years) and almost on the opposite side of the galaxy to us. Despite this, M75 is intrinsically bright and on nights of good seeing and transparency can be glimpsed with a pair of 7x50 or 10x50 binoculars.

M75 was discovered by Pierre Méchain on the night of August 27, 1780. Charles Messier observed it soon after and added it to his catalogue a few weeks later. It was William Herschel who first resolved M75 into stars describing it (like M62 and M70) as a "miniature version of M3."

M75 is located right at the Sagittarius-Capricornus border. It's positioned about 23 degrees northeast of the centre of the Sagittarius "Teapot" asterism and 5.5 degrees north and a little east of a small group of four faint naked-eye stars (59 Sgr - mag. +4.5, 60 Sgr - mag. +4.8, 62 Sgr - mag. +4.4 and ω Sgr - mag. +4.7). The globular is best seen during the months of June, July and August from the Southern Hemisphere and the tropics.

For binocular observers M75 appears stellar in nature. A 100mm (4-inch) telescope at high magnifications will start to pick out a certain degree of fuzziness. The cluster however appears small, visually spanning only 3 arc minutes in diameter. It's roughly comparable in size and brightness to the Messier globulars located within the teapot (M54, M69 and M70). To begin resolution, larger telescopes of the order of 250mm (10-inch) or more are required.

In total M75 covers 6.8 arc minutes of apparent sky, which corresponds to a spatial diameter of 134 light-years. It's very old at 13 Billion years plus and is estimated to contain 400,000 stars. Classified as class I, M75 is one of the more densely concentrated globular clusters known. However, despite been faint and small it's a remarkable globular given its incredible, extragalactic-like distance.

M75 Data Table

Messier	75
NGC	6864
Object Type	Globular cluster
Constellation	Sagittarius
Distance (kly)	67.5
Apparent Mag.	8.7
RA (J2000)	20h 06m 05s
DEC (J2000)	-21d 55m 17s

Apparent Size (arcmins)	6.8 x 6.8
Radius (light years)	67
Age (years)	>13,000M
Number of Stars	400,000
Notable Feature	Distant globular and one of the most densely concentrated clusters known

Messier # 075

Date:		Time:	
Site:			
Temp:	Wind:		Hum:
Clouds:		Moon:	
Scope:			
EP:		Mag:	
NELM:		See/Trans:	
Type:		# Stars:	
Mag:		Age:	
Const:			
Notes:			

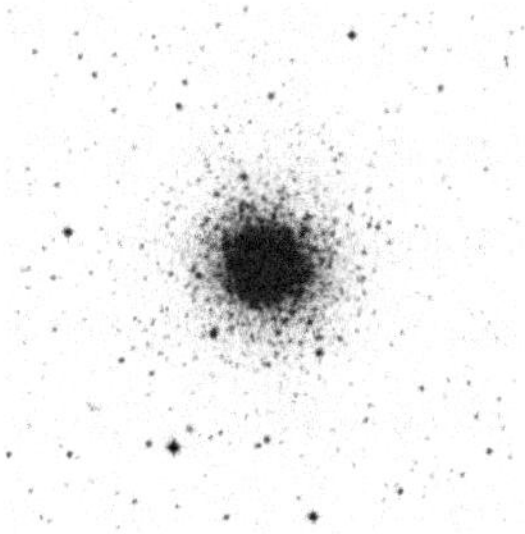

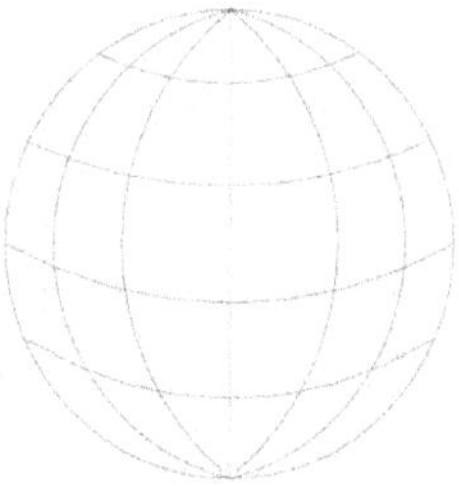

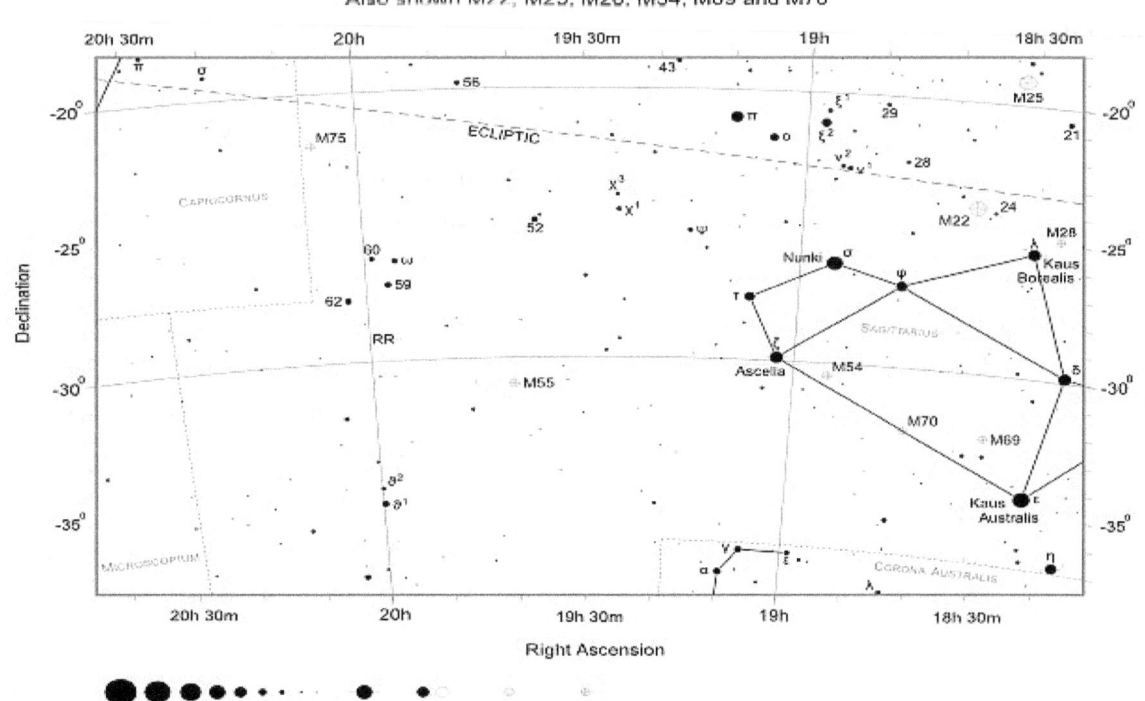

Messier Finder Chart for M55 and M75
Also shown M22, M25, M28, M54, M69 and M70

Right Ascension

Declination

Star magnitudes -1 0 1 2 3 4 5 6 7

Double star   Variable stars   Open cluster   Globular cluster

Messier 76 - M76 - The Little Dumbbell Nebula (Planetary Nebula)

M76 or "The Little Dumbbell Nebula" is a planetary nebula located in Perseus. At magnitude +10.1 and spanning 2.7 x 1.8 arc minutes, its one of the faintest and smallest objects in Messier's catalogue. The nebula was discovered by Pierre Méchain on September 5, 1780 and first recognized as a planetary nebula by American astronomer Heber Doust Curtis in 1918. However, Isaac Roberts suggested it was similar to M57 (Ring Nebula) in 1891.

M76 itself looks like a miniature version of the famous Dumbbell Nebula (M27) in Vulpecula, from which it derives its name. Interestingly, it was assigned two NGC numbers - NGC 650 and 651 - since wrongly suspected of consisting of two separate emission nebulae. The structure is now classed as a bipolar planetary nebula.

The Little Dumbbell Nebula maybe faint but not difficult to locate; it's positioned just south of the prominent "W" asterism of Cassiopeia and only a degree north-northwest of Phi Persei (φ Per - mag. +4.0). It's best seen from the Northern Hemisphere during the months of October, November and December. From latitudes 40N or more it's circumpolar and hence never sets. However from southern temperate latitudes, M76 is a difficult object that never climbs high above the northern horizon.

When viewed through a small 80mm (3.1-inch) telescope or large 20x80 binoculars, M76 appears faint, small and diffuse. It can be glimpsed with direct vision but better seen when using averted vision and/or a nebula filter. Since faint it's a difficult object at best appearing as an odd shape glow.

The nebula is a fine sight through small and medium size scopes. A 200mm (8-inch) telescope will show the two almost equal sized lobes separated by dark lane. If conditions permit, push up the magnitude to between 20x and 30x per inch of aperture to bring out more subtle details. Of the two lobes, the southern one appears slightly brighter.

The Little Dumbbell Nebula is located approx. 2,500 light-years from Earth, which corresponds to a spatial diameter of about 2 light-years.

M76 Data Table

Messier	76
NGC	650, 651
Name	Little Dumbbell Nebula
Object Type	Planetary nebula
Constellation	Perseus
Distance (kly)	2.5
Apparent Mag.	10.1
RA (J2000)	01h 42m 18s

DEC (J2000)	51d 34m 16s
Apparent Size (arcmins)	2.7 x 1.8
Radius (light years)	1.0
Notable Feature	Considered to be one of the most difficult objects to see in Messier's list

Messier # 076

Date:	Time:
Site:	
Temp:	Wind:	Hum:
Clouds:	Moon:
Scope:	
EP:	Mag:
NELM:	See/Trans:
Type:	# Stars:
Mag:	Age:
Const:	
Notes:	

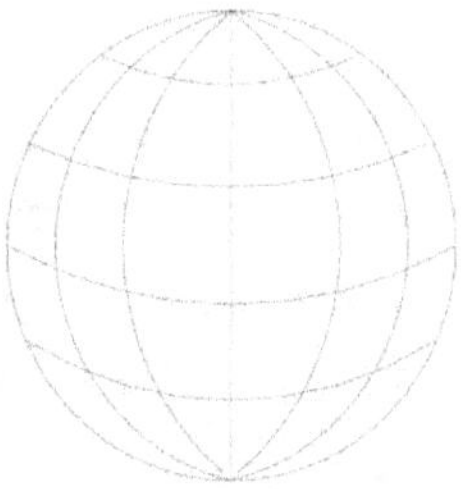

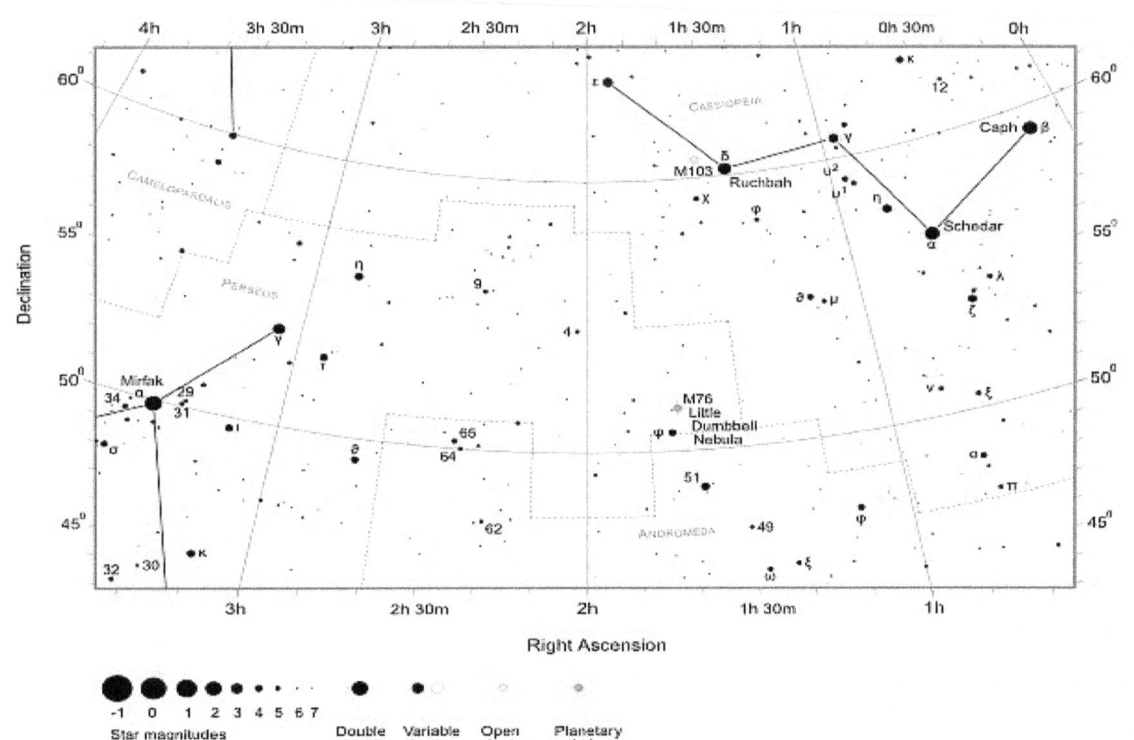

Messier Finder Chart for M76 Little Dumbbell Nebula
Also shown M103

Messier 77 - M77 - Barred Spiral Galaxy

M77 is a notable 9th magnitude face-on barred spiral galaxy in the constellation of Cetus. The nucleus of the galaxy contains at its core an Active Galaxy Nucleus (AGN), which although obscured from view by astronomical dust at visible wavelengths is detectable due to strong emissions in the infrared, ultraviolet, and X-ray regions of the electromagnetic spectrum. This type of galaxy is known as a Seyfert galaxy, named after 20th century American astronomer Carl Seyfert who first identified the class in 1943, of which M77 is the brightest example. They are believed to be home to a super massive black hole of between 10 and 100 million solar masses.

M77 was discovered by Pierre Méchain on October 29, 1780. He described the object as a nebula and subsequently reported it to Charles Messier who then added it to his catalogue. Messier and later William Herschel both described M77 - incorrectly - as a star cluster. The galaxy is located about 1 degree southeast of magnitude 4.1 star delta Cet (δ Cet) and a few degrees to the northeast of famous long period variable star Mira. The constellation Cetus in Greek mythology represented a "Sea Monster" although today it's often referred to as "the Whale" or "the Shark". It's the fourth largest constellation in the sky covering 1231 square degrees and is bordered by Aries, Pisces, Aquarius, Sculptor, Fornax, Eridanus and Taurus.

Although it shines at magnitude +9.1, M77 is a compact galaxy with a bright center and therefore a relatively easy target for large binoculars (15x70s or 20x80s) and small telescopes. It can even be spotted with 10x50 binoculars from dark sites with excellent seeing conditions. A 80mm (3.1-inch) scope shows the galaxy as a condensed ball of fuzzy light with a slightly brighter central core. A 100mm (4-inch) telescope enhances the view, with the galaxy displaying an oval shaped halo surrounding a bright central region. When viewed through a 200mm (8-inch) scope the details are pronounced with the central core of the galaxy evident and almost stellar like. The large surrounding diffuse halo is prominent. A bright star lies nearby to the east of M77.M77 is a magnificent galaxy and one of the largest in Messier's catalogue. It's estimated to be at least 47 Million light years distant, with an apparent size of 7.1 x 6.0 arc minutes this corresponds to an actual diameter of 100,000 light years. The galaxy is probably even larger still, as evidence suggests of faint outer spiral extensions. The stars in the inner region of M77 are young while those away from the centre tend to be older. It is one of the first recognized spiral galaxies, and listed by Lord Rosse as one of 14 "spiral nebulae" discovered to 1850. When Vesto Slipher was working on galaxy spectra, M77 was one of two galaxies in which he detected large red shifts (the other being M104, the "Sombrero").This interesting barred spiral galaxy is best seen during the months of September, October and November.

M77 Data Table

Messier	77	**Apparent Size (arcmins)**	7.1 x 6.0
NGC	1068	**Radius (light years)**	50,000
Object Type	Barred spiral galaxy	**Number of Stars**	>300 Billion
Classification	(R)SA(rs)b	**Other Name**	Arp 37
Constellation	Cetus	**Notable Features**	Brightest Seyfert galaxy and one of the largest Messier galaxies.
Distance (kly)	47,000		
Apparent Mag.	9.1		
RA (J2000)	02h 42m 41s		
DEC (J2000)	-00d 00m 47s		

Messier # 077

Date:	Time:
Site:	
Temp:	Wind:	Hum:
Clouds:	Moon:
Scope:	
EP:	Mag:
NELM:	See/Trans:
Type:	# Stars:
Mag:	Age:
Const:	
Notes:	

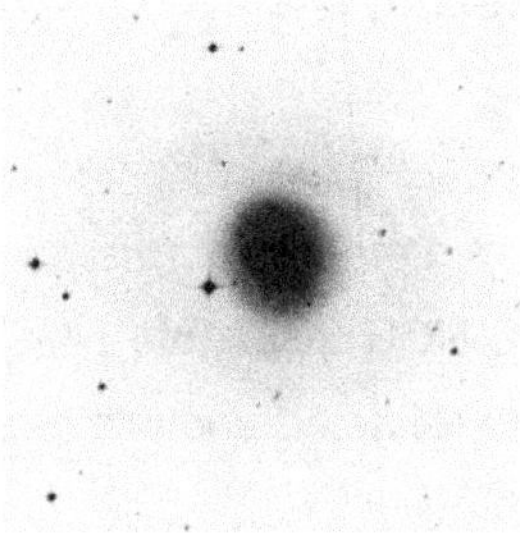

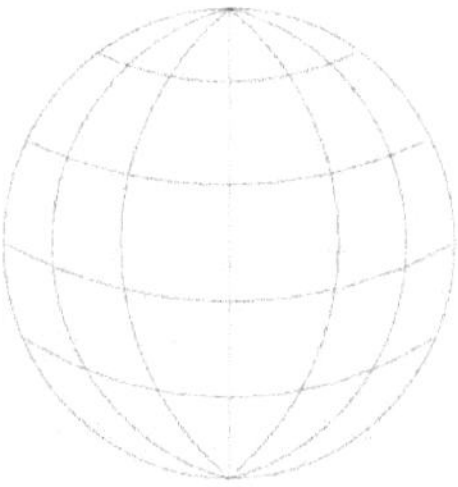

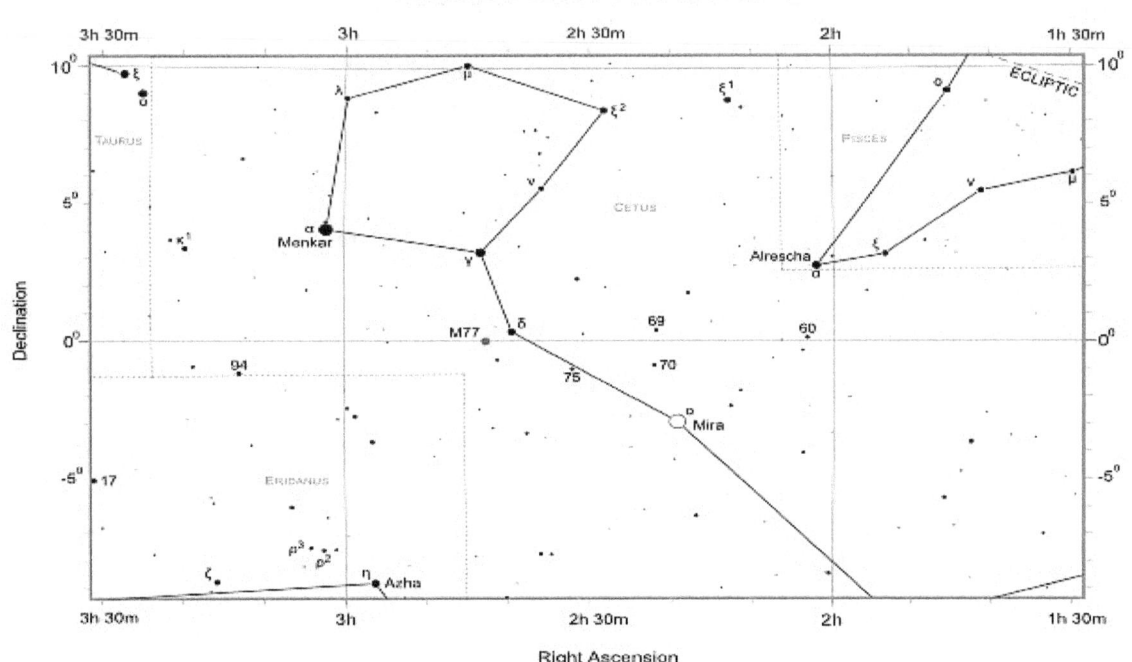

Messier Finder Chart for M77

Star magnitudes: -1 0 1 2 3 4 5 6 7

Double star Variable stars Galaxy

Messier 78 - M78 - Reflection Nebula

M78 is an often forgotten nebula in the constellation of Orion. It's a reflection nebula that's part of the Orion Molecular Cloud Complex, a large cloud of gas and dust centered on the famous Orion Nebula (M42) and De Mairan's Nebula (M43). Also included in this grouping are NGC 2064, NGC 2067 and NGC 2071 and other nebulae. Reflection nebulae like M78 are clouds of interstellar dust that shine due to reflected and scattered light from nearby stars.

M78 was discovered by Pierre Méchain in early 1780 with Charles Messier adding it to his catalogue on December 17, 1780. Although only of 8th magnitude, finding M78 is easy since it's positioned just a few degrees northeast of Orion's famous belt. The three bright stars that make up the belt are Alnitak (ζ Ori - mag. +1.72), Alnilam (ε Ori - mag. +1.69) and Mintaka (δ Ori - mag. +2.25). Positioned 2.5 degrees northeast of Alnitak is M78.

The nebula is visible in 10x50 binoculars, appearing as a small, faint, hazy patch of light. A small telescope of aperture 80mm (3.1-inch) or large 20x80 binoculars reveal a little more detail, especially the brighter northern part of the nebula, which appears as a comet like fan shape. Also visible, surrounded by the nebula, are a pair of 10th magnitude stars that are responsible for making the cloud of dust visible. Of all known reflection nebulae, M78 is the brightest in the sky. With a 200mm (8-inch) or large scope it's possible to notice brighter areas and twists in the nebula especially when using averted vision.

About 45 variable stars of the T Tauri type, young stars still in the process of formation are known to exist in M78. Faint nebulosity, NGC 2071, lies very close to the north edge of M78.

M78 has an apparent size of 8.0 x 6.0 arc minutes. At a distance of 1,600 light years this corresponds to a maximum spatial diameter of about 4 light years. The nebula is best seen during the months of November, December and January

M78 Data Table

Messier	78
NGC	2068
Object Type	Reflection Nebula
Constellation	Orion
Distance (kly)	1.6
Apparent Mag.	8.2
RA (J2000)	05h 46m 46s

DEC (J2000)	00d 04m 45s
Apparent Size (arcmins)	8.0 x 6.0
Radius (light years)	2.0
Notable Feature	Part of Orion Molecular Cloud Complex

Messier # 078

Date:	Time:
Site:	
Temp:	Wind:	Hum:
Clouds:	Moon:
Scope:	
EP:	Mag:
NELM:	See/Trans:
Type:	# Stars:
Mag:	Age:
Const:	
Notes:	

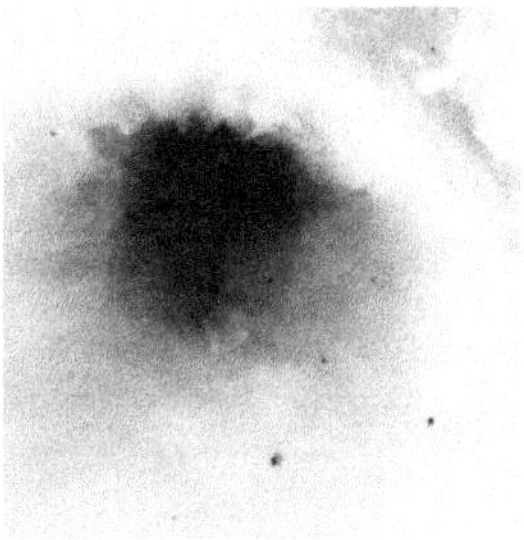

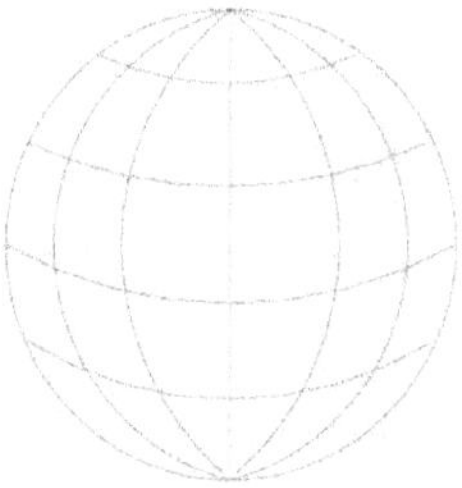

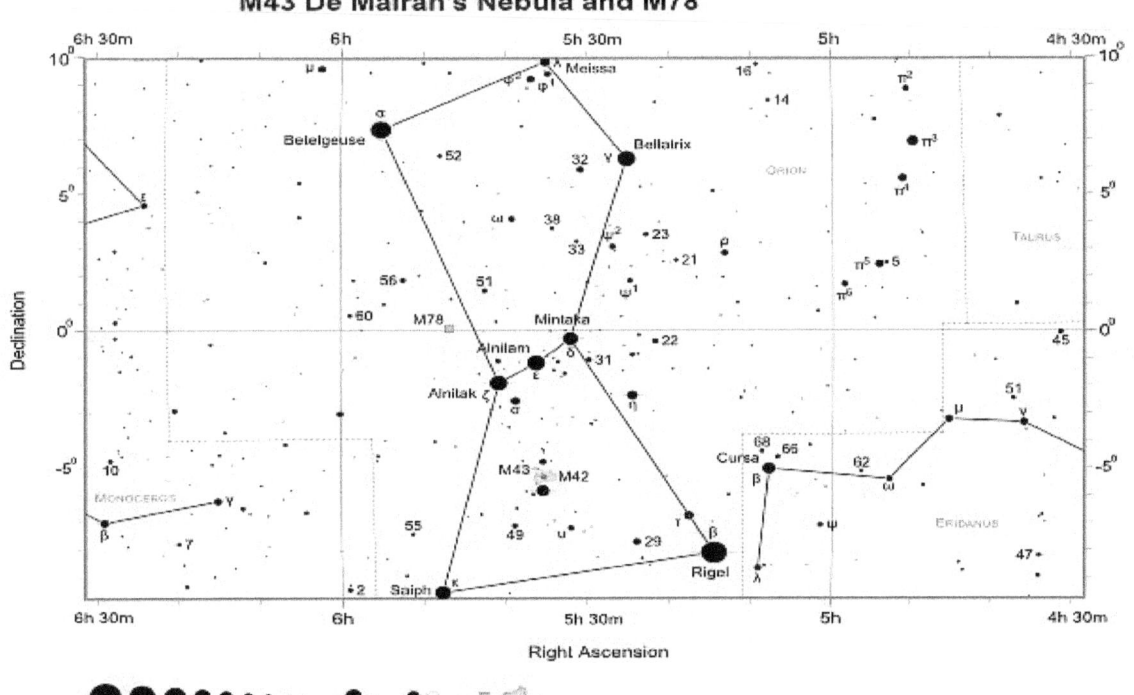

Messier Finder Chart for M42 Great Orion Nebula, M43 De Mairan's Nebula and M78

Messier 79 - M79 - Globular Cluster

M79 is an intriguing eighth magnitude globular cluster located in the constellation of Lepus. At a distance of 41,000 light-years from Earth and 60,000 light-years from the Milky Way centre, it's believed to be an extragalactic globular and a native of the nearby Canis Major Dwarf galaxy. The only other extragalactic globular cluster in Messiers catalogue is M54, which belongs to the Sagittarius Dwarf Elliptical galaxy.

Unusual for globulars, M79 is located opposite the galactic center and therefore best seen during the Southern Hemisphere summer and Northern Hemisphere winter months. With a declination of -24.5 degrees south it never rises particular high above the southern horizon from northern temperate latitudes. However, it's one of the finest globulars that can be seen during this time of year.

The constellation of Lepus is located south of Orion and west of Canis Major. It contains few deep sky objects within the range of amateur scopes and M79 is the only Messier object found within its boundaries. Locating M79 is easy; its positioned 20 degrees southwest of Sirius (α CMa - mag. -1.46), the brightest star in the night sky. An imaginary line connecting Arneb (α Lep - mag. +2.6) with Nihal (β Lep - mag. +2.8) and extending southwards for about the same distance again leads to M79. About 0.5 degrees southwest of M79 lies the magnitude +5.1 double star HD 35162 (HIP 25045) with its 7th magnitude companion, separated by 3 arc minutes.

At magnitude +8.1, M79 is visible as a fuzzy spot through 7x50 or 10x50 binoculars. A small 80mm (3.1-inch) telescope reveals the cluster as a somewhat yellow looking unresolved hazy comet-like ball of light. It has a bright core and therefore stands up well to light polluted skies and a certain amount of moonlight. Larger scopes of minimum 250mm (10-inch) aperture are required to begin resolution of the outer edges.

In total M79 spans 8.7 arc minutes in diameter, which corresponds to a spatial diameter of 104 light-years. The globular contains 150,000 stars and is estimated to be 11.7 billion years old.

M79 Data Table

Messier	79
NGC	1904
Object Type	Globular cluster
Constellation	Lepus
Distance (kly)	41
Apparent Mag.	8.1
RA (J2000)	05h 24m 11s
DEC (J2000)	-24d 31m 27s

Apparent Size (arcmins)	8.7 x 8.7
Radius (light years)	52
Age (years)	11,700M
Number of Stars	150,000
Notable Feature	Likely belongs to the Canis Major Dwarf Galaxy

Messier # 079	
Date:	Time:

Site:		
Temp:	Wind:	Hum:
Clouds:	Moon:	
Scope:		
EP:	Mag:	
NELM:	See/Trans:	
Type:	# Stars:	
Mag:	Age:	
Const:		

Notes:

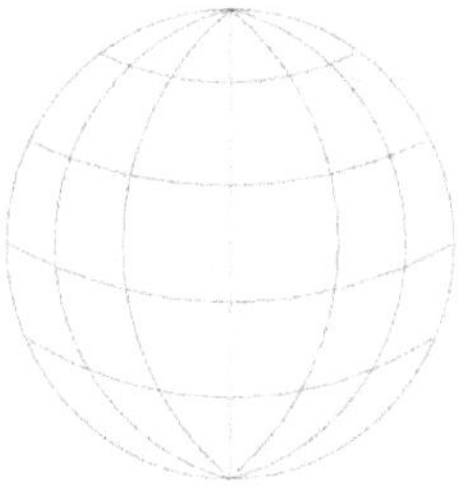

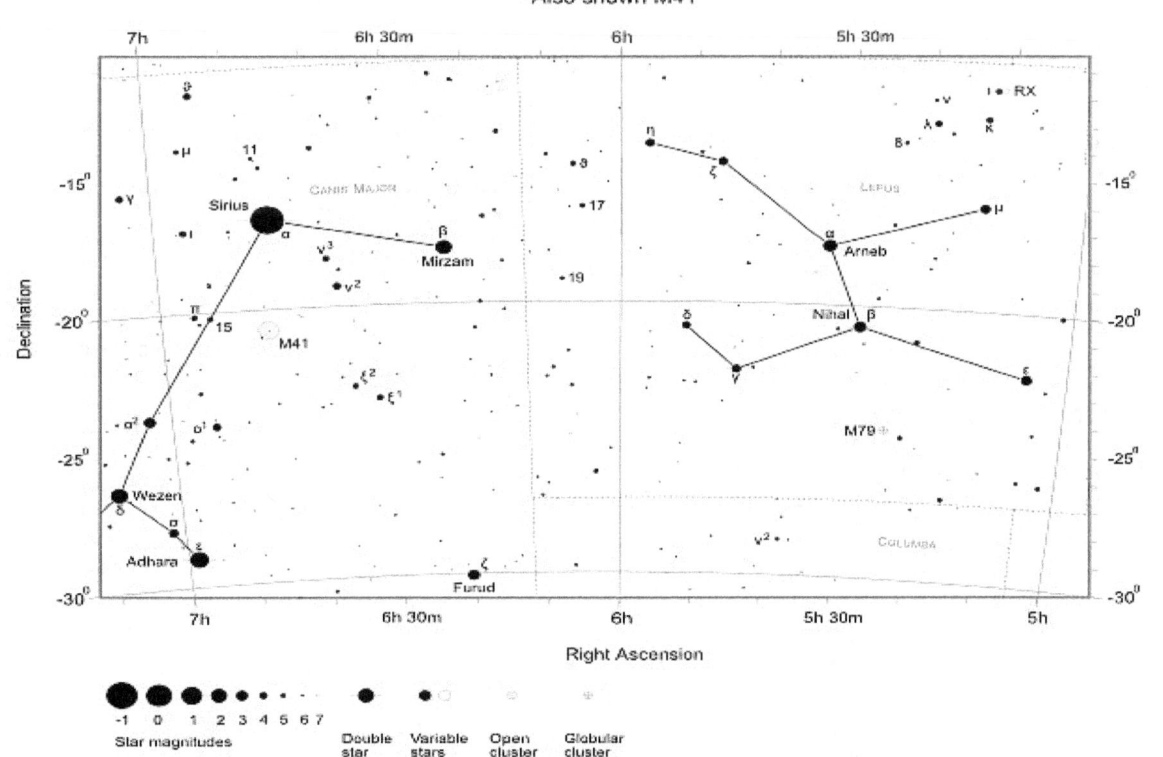

Messier Finder Chart for M79

Also shown M41

7h
6h 30m
6h
5h 30m
5h
-15°
-20°
-25°
-30°
Declination
Right Ascension
Canis Major
Lepus
Columba
Sirius
Mirzam
M41
Wezen
Adhara
Furud
Arneb
Nihal
M79
RX
11
15
17
19
Star magnitudes
-1 0 1 2 3 4 5 6 7
Double star
Variable stars
Open cluster
Globular cluster

Messier 80 - M80 - Globular Cluster

M80 is a small but compact globular cluster located in Scorpius. It shines at magnitude +7.5 and therefore within the range of popular 7x50 or 10x50 binoculars. At its core, M80 contains a large number of "blue stragglers", stars that appear much younger than the age of the globular cluster itself! The likely reason is they have probably lost part of their cooler outer layers due to close encounters with other stars. Since M80 contains more blue stragglers than average it implies exceptionally high core stellar interaction rates.

M80 was discovered by Charles Messier on January 4, 1781. Though not conspicuous, M80 is easy to locate as its positioned just 4 degrees northwest of brilliant red supergiant star Antares (α Sco - mag. +1.0). The globular is situated halfway along an imaginary line connecting Antares with Acrab (β Sco - mag. +2.6). Located just west of Antares is magnificent globular cluster M4.

The finder chart below shows the position of M80. The globular is best seen from tropical and Southern Hemisphere latitudes during the months of May, June and July.

When viewed with binoculars or small telescopes, M80 appears as a mottled ball of light. A small 80mm (3.1-inch) scope reveals a soft round structure that's not resolvable. Larger 200mm (8-inch) telescopes display a bright compact core and an outer halo that extends up to 5 arc minutes in diameter. On nights of good seeing and transparency the outer regions hint at resolution. Much better resolution is achieved with apertures of 300mm (12-inch) or greater, with the brightest member stars being of about 14th magnitude. On May 21, 1860 a bright nova (T Sco) reached magnitude +7.0 in M80 and for a short time it outshone the entire cluster.

M80 is located at a distance of about 32,600 light-years and contains at least 200,000 stars. In total it covers 10 arc minutes of apparent sky, which corresponds to a spatial diameter of 96 light-years. The cluster is estimated to be 12.54 billion years old.

M80 Data Table

Messier	80
NGC	6093
Object Type	Globular cluster
Constellation	Scorpius
Distance (kly)	32.6
Apparent Mag.	7.5
RA (J2000)	16h 17m 03s
DEC (J2000)	-22d 58m 30s

Apparent Size (arcmins)	10 x 10
Radius (light years)	48
Age (years)	12,540M
Number of Stars	>200,000
Notable Feature	Contains a relatively large number of blue stragglers

Messier # 080

Date:	Time:
Site:	
Temp:	Wind:	Hum:
Clouds:	Moon:
Scope:	
EP:	Mag:
NELM:	See/Trans:
Type:	# Stars:
Mag:	Age:
Const:	
Notes:	

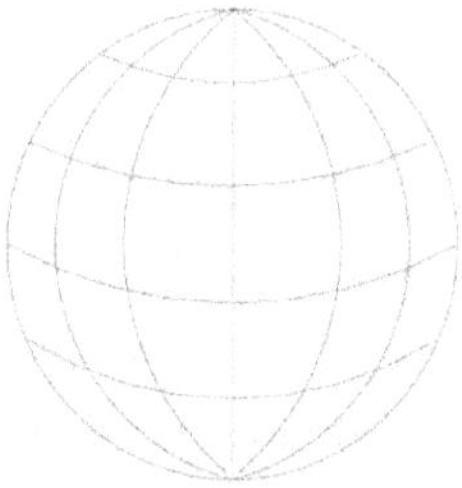

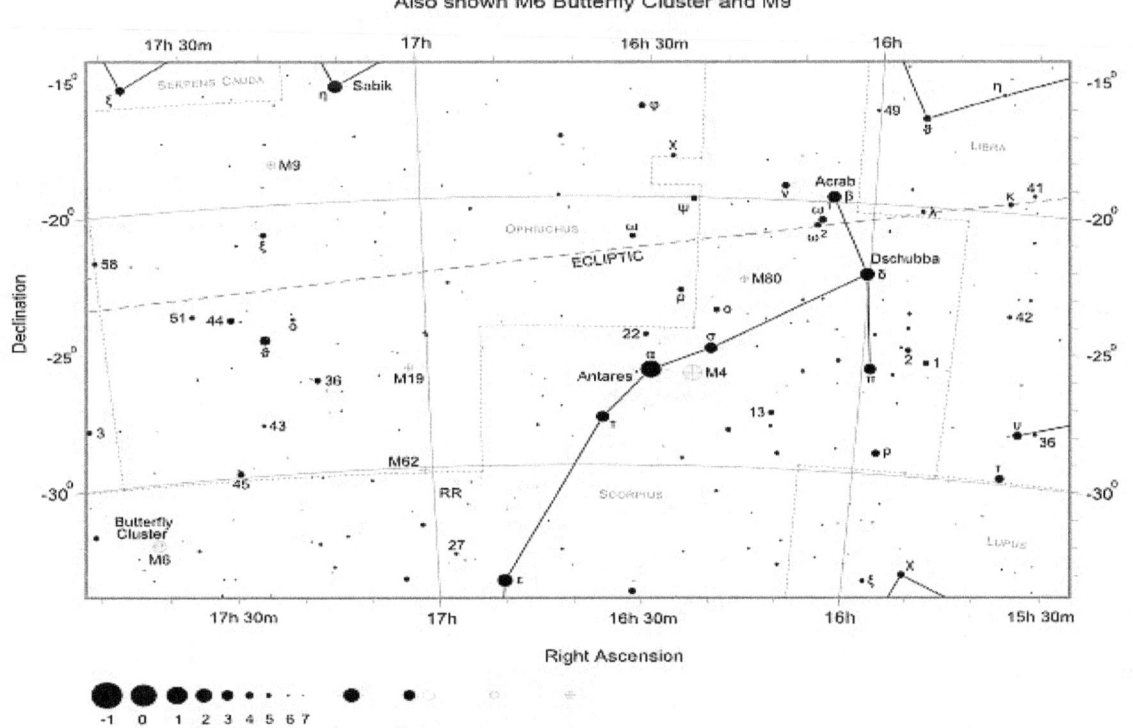

Messier Finder Chart for M4, M19, M62 and M80
Also shown M6 Butterfly Cluster and M9

Messier 81 - M81 - Bode's Galaxy (Spiral Galaxy)

M81 or Bode's galaxy is a large bright spiral galaxy located 11.8 million light-years from Earth in the constellation of Ursa Major. With an apparent magnitude of +6.9 it's easily visible with binoculars, a fine target for small telescope owners and a wonderful sight in larger scopes. The galaxy is a striking example of a grand design spiral; a galaxy that exhibits prominent near perfect and well defined spiral arms. In the same binocular and low magnification telescope field of view as M81 is M82, another prominent galaxy. At mag. +8.4, M82 is fainter (and smaller) than M81 and a very different type of galaxy. It's a starburst galaxy in which stars are forming at exceptionally high rates. Also known as the "Cigar" galaxy, M82 is the prototype object of its type and provides a striking compliment to the near perfect spiral shape of M81. Together the pair forms a popular visual and imaging target for amateur astronomers.

Both M81 and M82 were discovered by Johann Elert Bode on December 31, 1774. Pierre Mechain then independently rediscovered both galaxies in August 1779. He reported his observations to Charles Messier who added them to his catalogue on February 9, 1781.Finding M81 is not particularly difficult as the famous "Plough" asterism of Ursa Major can be used as the starting point. First focus on Dubhe (α UMa - mag. +1.8) the northwest corner star of the bowl of the Plough. The M81 / M82 pair is located 10 degrees northwest of this star with M82 positioned 38 arc minutes directly north of M81.

M81 appears a faint patch of light in 7x50 or 10x50 binoculars. There are rare reports that some eagle-eyed stargazers have managed, under exceptional conditions, to see M81 with the naked eye but this is sensational viewing by all accounts. It would mean M81 being by far the most distant permanent object that can be viewed without a telescope. However, M33 is commonly regarded as the holder of this record. A small 80mm (3-inch) scope at low power reveals M81 as a bright oval haze without detail. Also visible and in the same field of view is M82, which appears as a slim grey needle of uniform light. A 150mm (6-inch) scope at high power reveals M81 as a huge low surface brightness halo of nebulosity surrounding a bright core. In larger scopes more subtle details can be seen. The small unbarred spiral galaxy NGC 2976 lies 1.5 degrees southwest of M81 and can be seen without detail through a 200mm telescope.M81 has an apparent size of 27 x 14 arc minutes, which corresponds to a spatial diameter of 90,000 light-years. It's estimated to contain more than 250 billion stars and is the largest member of the Ursa Major or M81 group of galaxies that contains at least 34 members. M82 is also a member of this group, which is one of the closest groups of galaxies beyond our own Local Group. The galaxies are best seen from the Northern Hemisphere during the months of March, April and May. They are circumpolar and hence never set from locations north of 21 degrees north.

Only one supernova, SN 1993J, has been detected in M81. It was discovered by F. Garcia in Spain on March 28, 1993 and peaked at apparent magnitude +10.8, making it the brightest supernova in the northern part of the sky since 1954.

M81 Data Table

Messier	81		Apparent Mag.	6.9		Number of Stars	>250 Billion
NGC	3031		RA (J2000)	09h 55m 33s			
Name	Bode's Galaxy		DEC (J2000)	69d 03m 55s		Notable Feature	Largest member of the M81 Group of galaxies
Object Type	Spiral galaxy		Apparent Size (arcmins)	26.9 x 14.1			
Classification	SA(s)ab						
Constellation	Ursa Major		Radius (light years)	45,000			
Distance (kly)	11,800						

Messier # 081

Date:	Time:
Site:	
Temp:	Wind:	Hum:
Clouds:	Moon:
Scope:	
EP:	Mag:
NELM:	See/Trans:
Type:	# Stars:
Mag:	Age:
Const:	

Notes:

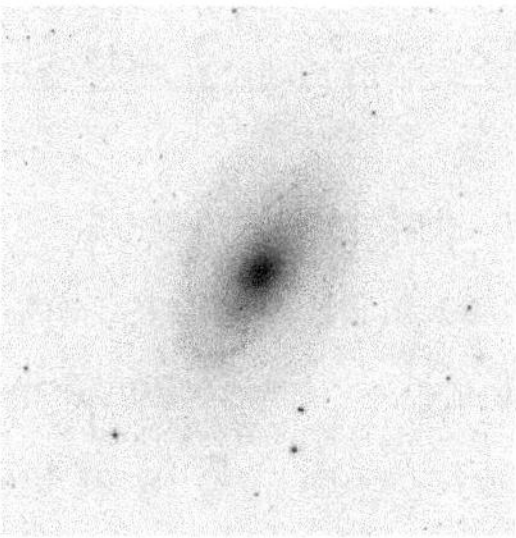

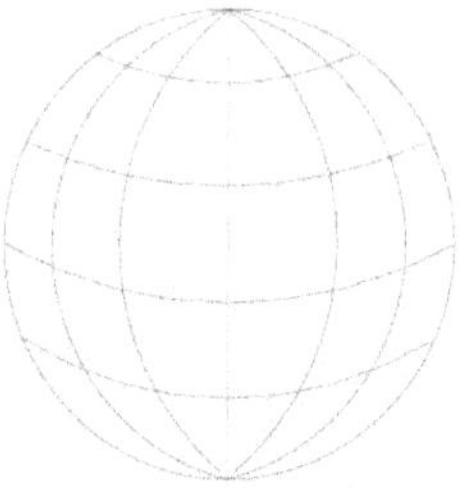

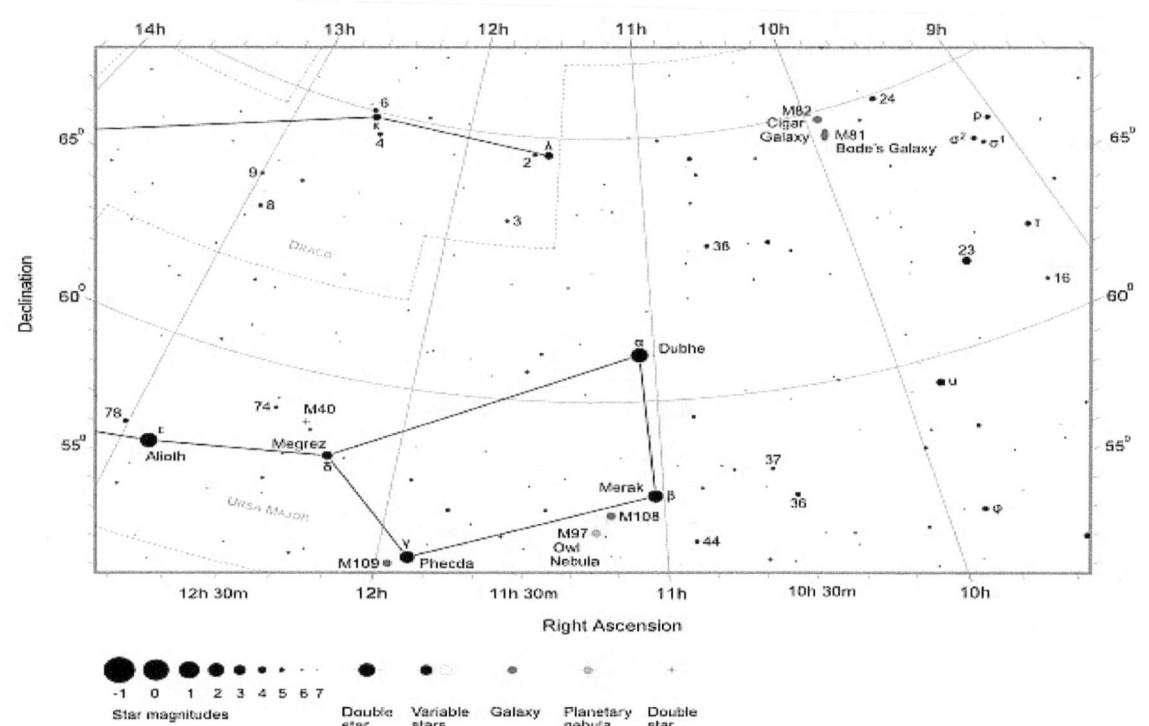

Messier Finder Chart for M81 Bode's Galaxy and M82 Cigar Galaxy
Also shown M40, M97 Owl Nebula, M108 and M109

Messier 82 - M82 - Cigar Galaxy (Starburst Galaxy)

M82 is a superb irregular galaxy of high surface brightness that's visible with binoculars and located in the constellation of Ursa Major. It has an apparent magnitude of +8.4 and is separated by only 38 arc minutes from M81 an even brighter and equally - if not more - stunning galaxy. However, these two fine galaxies are very different objects indeed. M81 (mag. +6.9) when viewed from Earth appears almost face-on and is of the grand spiral deign; a galaxy that exhibits prominent near perfect and well defined spiral arms. On the other hand, M82 (mag. +8.4) is edge on, appearing long and thin and as a result is often referred to as the "Cigar" galaxy. Unlike the perfectly formed spiral shape of M81, M82 is irregular and classified as the prototype starburst galaxy in which stars are forming at exceptionally high rates.

These two objects are the largest members of the Ursa Major or M81 group of galaxies, which at a distance of 11.7 million light years is one of the closest groups of galaxies beyond our own Local Group. In space M81 and M82 are physically close, separated by only about 150,000 light years and when seen through binoculars and telescopes at low magnifications they form a striking pair that appear in the same optical field of view. Both M81 and M82 were discovered by Johann Elert Bode on December 31, 1774. Pierre Mechain then independently rediscovered both galaxies in August 1779. He reported his observations to Charles Messier who added them to his catalogue on February 9, 1781.

Finding M82 is not particularly difficult as the famous "Plough" asterism of Ursa Major can be used as the starting point. First focus on Dubhe (α UMa - mag. +1.8) the northwest corner star of the bowl of the Plough. The M81 / M82 pair is located 10 degrees northwest of this star with M82 positioned directly north of M81.M82 appears as a faint thin rod of light in binoculars with M81 visible as a large diffuse hazy patch. It's noticeable the different sizes and shapes of these two great galaxies. A small 80mm (3.1-inch) telescope at low power shows M82 as a slim gray needle of uniform light whereas at high magnifications, a 150mm (6-inch) or 200mm (8-inch) instrument reveals dusty patches that cross the sharp surface of M82. The centre region appears brighter that the edges.

In total M82 covers 11.2 x 4.3 arc minutes of apparent sky. At a distance of 11.5 million light-years, this corresponds to an actual diameter of 38,000 light-years; less than half the 90,000 light-years of M81. Together the pair forms a popular visual and imaging target for amateur astronomers that are best seen from the Northern Hemisphere during the months of March, April and May. They are circumpolar and hence never set from locations north of 21 degrees north.

M82 is believed to contain at least 30 billion stars. On January 21, 2014, a bright Type 1a supernova (SN 2014J) was discovered in M82 by Stephen J. Fossey and his students at the University College London (UCL) observatory in London.

M82 Data Table

Messier	82	**Apparent Mag.**	8.4	**Stars**	
NGC	3034	**RA (J2000)**	09h 55m 51s	**Notable Features**	Prototype starburst galaxy. Member of the M81 Group of galaxies
Name	Cigar Galaxy	**DEC (J2000)**	69d 40m 43s		
Object Type	Starburst galaxy	**Apparent Size (arcmins)**	11.2 x 4.3		
Classification	Irregular				
Constellation	Ursa Major	**Radius (light years)**	19,000		
Distance (kly)	11,500				
		Number of	>30 Billion		

Messier # 082

Date:		Time:	
Site:			
Temp:	Wind:		Hum:
Clouds:		Moon:	
Scope:			
EP:		Mag:	
NELM:		See/Trans:	
Type:		# Stars:	
Mag:		Age:	
Const:			

Notes:

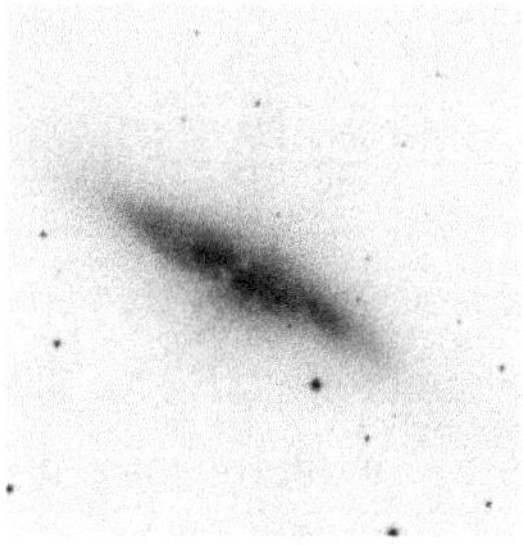

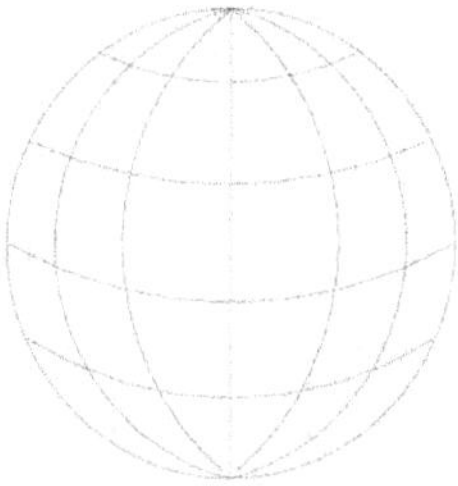

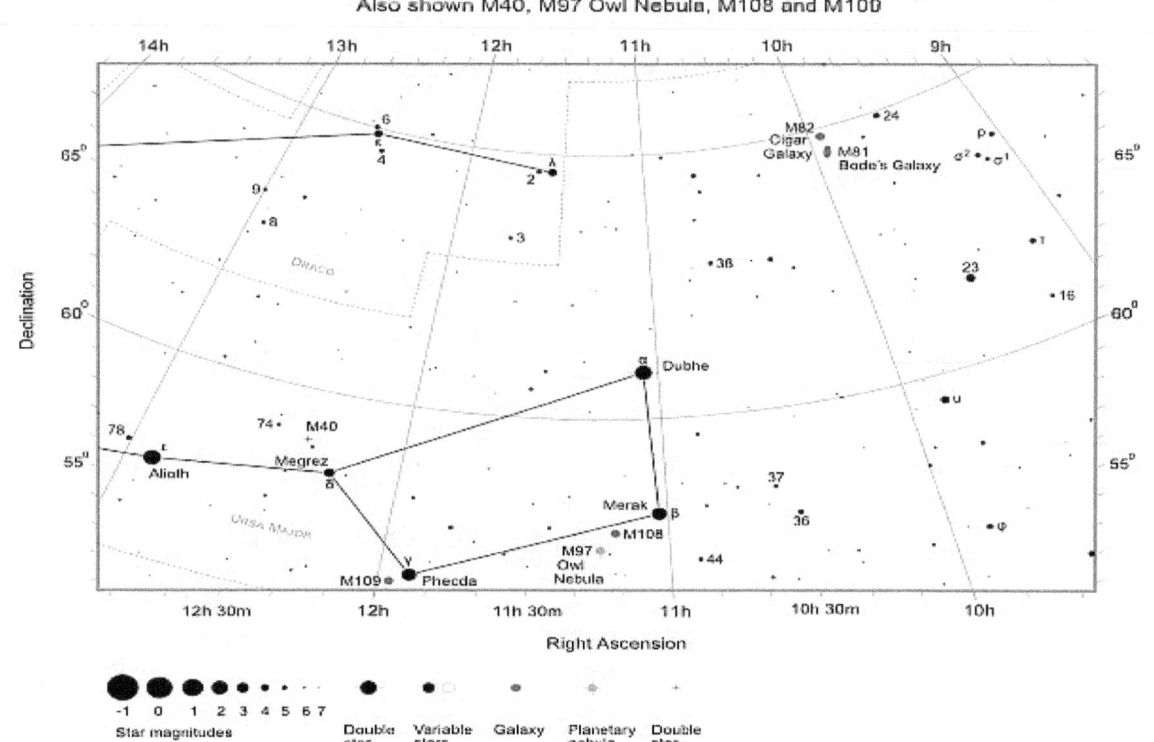

Messier Finder Chart for M81 Bode's Galaxy and M82 Cigar Galaxy
Also shown M40, M97 Owl Nebula, M108 and M109

Declination

Right Ascension

Star magnitudes   Double star   Variable stars   Galaxy   Planetary nebula   Double star

Messier 83 - M83 - Southern Pinwheel Galaxy (Barred Spiral Galaxy)

The Southern Pinwheel Galaxy also known as M83 is a barred spiral galaxy approximately 14.7 million light-years distant in the eastern section of the largest constellation of all, Hydra. It's one of the closest barred spirals, a showpiece galaxy and the finest barred spiral in the sky. With an apparent magnitude of +7.5, M83 is visible with 7x50 or 10x50 binoculars, appearing under dark skies as a patch of light with a brighter centre. It was discovered by Nicholas Louis de Lacaille at the Cape of Good Hope in South Africa on February 23, 1752 and added by Charles Messier to his catalogue on February 17, 1781.

With a declination of 30 degrees south, M83 is best seen from Southern Hemisphere or equatorial regions during the months of April, May and June. For mid-latitude northern hemisphere observers, the galaxy can be a difficult object; it's the southernmost galaxy in Messier's list and therefore never climbs particularly high above the southern horizon.

Despite being a relatively bright galaxy M83 can be tricky to locate as it's positioned in a part of the sky devoid of bright stars. It can be found by locating stars γ Hya (mag. +3.0) and π Hya (mag. +3.3). Imagine a line connecting these two stars and then move along the line until just short of the halfway mark. Located about 6 degrees south of this point is M83.

M83 is a superb telescope galaxy. A small 80mm (3.1-inch) scope reveals a bright nebula covering about 1/3 the diameter of the Moon with a brighter core. Don't be afraid to push up the magnification, this galaxy will take it. For example, through a 150mm (6-inch) telescope at high powers M83 is a splendid sight exhibiting a condensed bright nucleus, hints of the bar structure that's positioned within a large outer envelope of fainter nebulosity. Also visible on nights of good seeing are dark dust patches surrounding the nucleus. Through a larger 250mm (10-inch) scope, M83 is a stunning sight with well-formed spiral arms, numerous dust lanes and the distinct central bar nucleus visible.

M83 forms a small physical group, the M83 group, with peculiar radio galaxy Centaurus A (NGC 5128) and unusual galaxy NGC 5253 in Centaurus. In total five supernovae have been observed within M83 in the last 100 years (1923A, 1945B, 1950B, 1957D, 1968L and 1983N). Such an unusual high rate makes M83 a popular target for amateur supernovae hunters; next time you have the chance to observe M83, compare the stars around the galaxy with those on an archive photograph or image and if you observe a "new" star, you may have made a supernova discovery.

M83 Data Table

Messier	83
NGC	5236
Name	Southern Pinwheel Galaxy
Object Type	Barred spiral galaxy
Classification	SAB (s) c
Constellation	Hydra
Distance (kly)	14700
Apparent Mag.	7.5

RA (J2000)	13h 37m 00s
DEC (J2000)	-29d 52m 04s
Apparent Size (arcmins)	12.9 x 11.5
Radius (light years)	27,500
Number of Stars	40 Billion
Notable Feature	Six supernovae have been observed in M83

Messier # 083

Date:	Time:

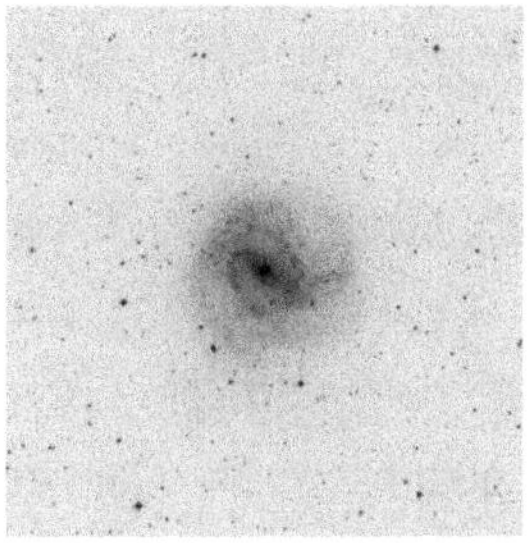

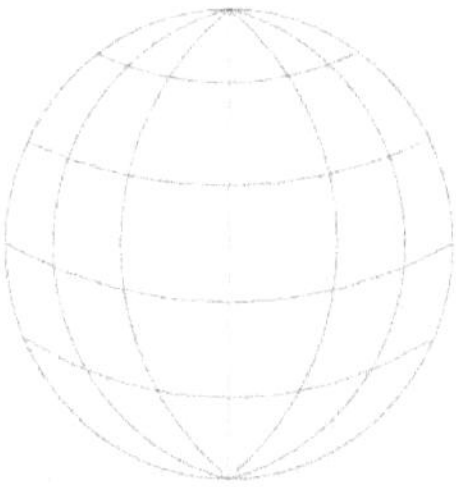

Site:

Temp:	Wind:	Hum:

Clouds:	Moon:

Scope:

EP:	Mag:
NELM:	See/Trans:
Type:	# Stars:
Mag:	Age:
Const:	

Notes:

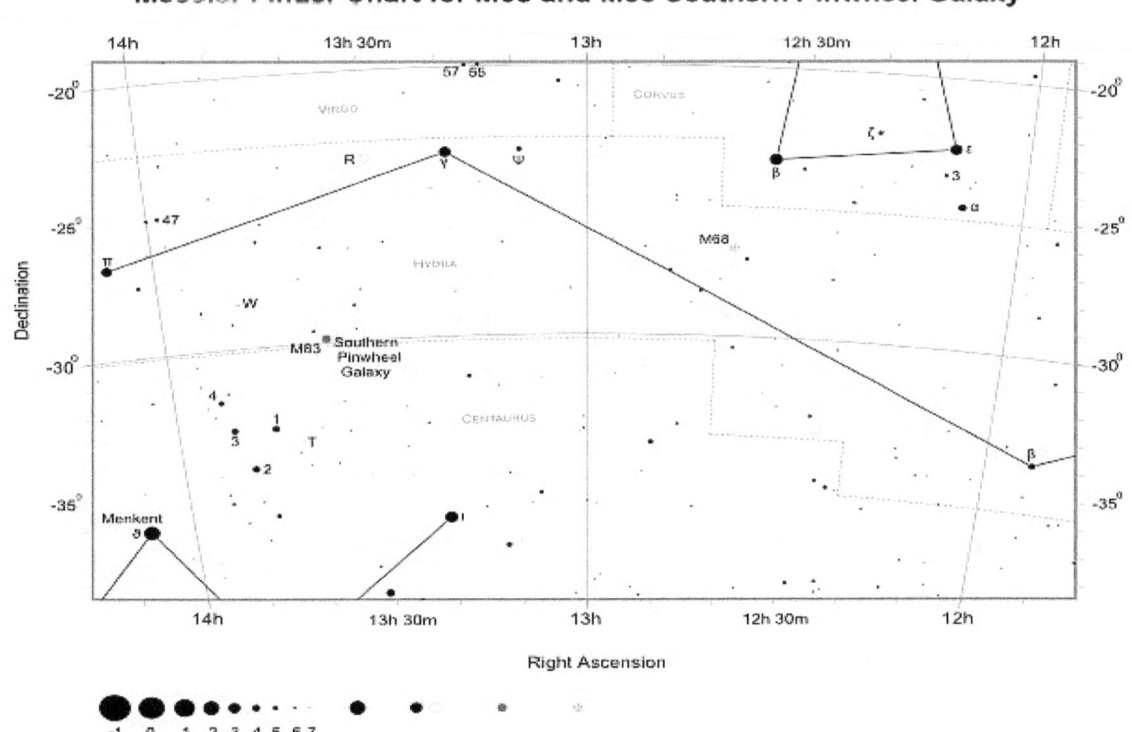

Messier Finder Chart for M68 and M83 Southern Pinwheel Galaxy

Messier 84 - M84 - Lenticular Galaxy (or Elliptical Galaxy)

M84 is a magnitude +9.4 lenticular or elliptical galaxy that belongs to the Virgo cluster of galaxies. Although one of the brighter members of the famous cluster, it's challenging to spot with popular 7x50 or 10x50 binoculars. Larger binoculars such as 20x80s or small telescopes make the task easier but as with most galaxies, dark skies are important. It's currently not clear what type of galaxy M84 is, it could either be a lenticular galaxy of type S0 seen face-on or an elliptical galaxy of type E1.

Charles Messier discovered M84 during one of his regular night sky patrols on March 18, 1781. He also discovered and catalogued another eight objects on the same day including M86, another giant lenticular or elliptical galaxy that's positioned just east of M84. The apparent size of M84 is 6.5 x 5.6 arc minutes and it's about 60 Million light years distant. This corresponds to a spatial diameter of 110,000 light-years.

M84 lies at the heart of the Virgo Cluster, close to the Virgo-Coma Berenices constellation border. It can be found by imagining a line connecting Denebola (β Leo - mag. +2.1) to Vindemiatrix (ε Vir - mag. +2.8). At the centre point of this line is M84 with M86 positioned 17 arc minutes east of M84.

The Virgo cluster galaxies are best seen during the months of March, April and May. Both M84 and M86 are visible together in the same low-power field of view. Small telescopes of the order of 80mm (3.1-inch) reveal both galaxies as small faint oval shapes of light with brighter centers. The other halos have low surface brightness, hence are better seen with medium or large sized scopes. Through a 200mm (8-inch) telescope under dark sky conditions, it's possible to also spot several more galaxies in the same field of view, including NGC 4435, NGC 4388, NGC 4402 and NGC 4438. Located about 1.5 degrees southeast of the M84/M86 pair is the giant elliptical galaxy M87.

Recent radio and Hubble Space Telescope observations have revealed two jets of matter shooting out from the centre of M84. The galaxy also has few young stars, indicating star formation is taking place at a slow rate. To date, three supernovae have been observed in M84 (SN 1957B, SN 1980I and SN 1991bg). The first one reached magnitude +13, the others magnitude +14. In total M84 contains about 400 billion stars.

M84 Data Table

Messier	84	**DEC (J2000)**	12d 53m 13s
NGC	4373	**Apparent Size (arcmins)**	6.5 x 5.6
Object Type	Lenticular galaxy (or Elliptical galaxy)	**Radius (light years)**	55,000
Classification	S0 (or E1)	**Number of Stars**	400 Billion
Constellation	Virgo	**Notable Feature**	Could be either a lenticular galaxy seen face-on or an elliptical galaxy
Distance (kly)	60000		
Apparent Mag.	9.4		
RA (J2000)	12h 25m 05s		

Messier # 084

Date:	Time:

Site:

Temp:	Wind:	Hum:

Clouds:	Moon:

Scope:

EP:	Mag:

NELM:	See/Trans:

Type:	# Stars:

Mag:	Age:

Const:

Notes:

Messier Finder Chart for M49, M58, M59, M60, M84, M85, M86, M87, M88, M89, M90, M91, M98, M99 and M100

Also shown M53, M64 Black Eye Galaxy, M65 and M66

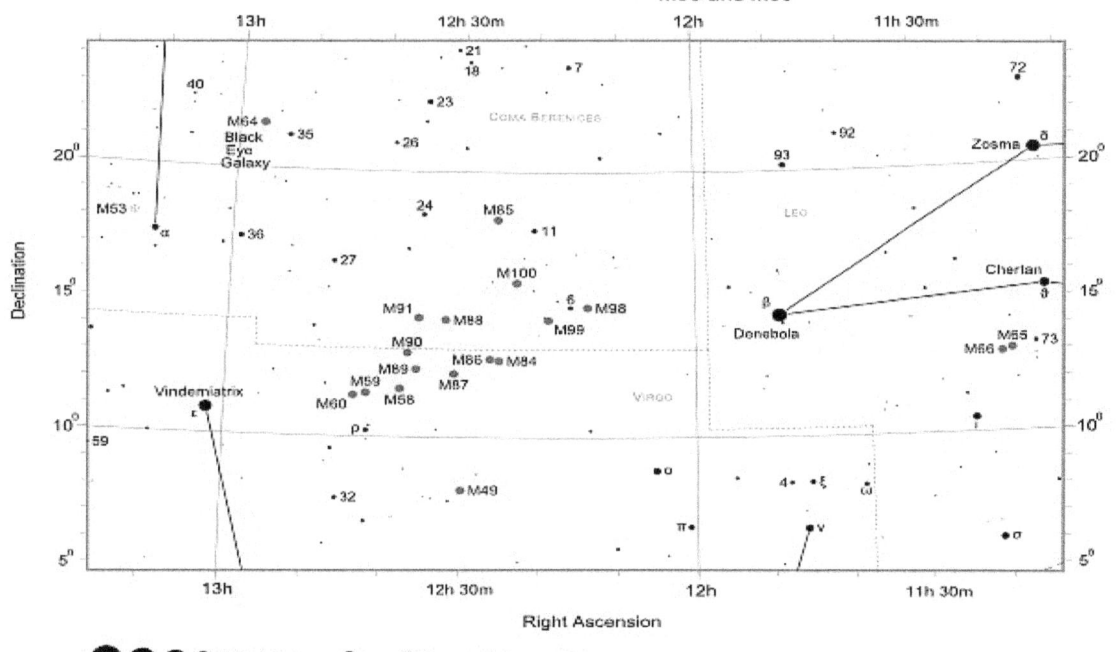

Messier 85 - M85 - Lenticular Galaxy (or Elliptical Galaxy)

M85 is a lenticular galaxy or an elliptical galaxy located in Coma Berenices that's a member of the Virgo cluster of galaxies. At magnitude +9.5 and covering 7.1 x 5.5 arc minutes it's similar in brightness and size to another Virgo cluster galaxy, M84. Spotting M85 with 7x50 or 10x50 binoculars is challenging due to its faintness, requiring good transparency and dark skies. A small 80mm (3.1-inch) scope shows a featureless ball of fuzz with a slightly brighter core. The view through amateur scopes in no way reflects the true nature of this distant enormous galaxy. It's located 60 million light years away making it one of the more remote objects in the Messier catalogue. The actual diameter of M85 is 125,000 light-years and is estimated to contain 400 billion stars. Long classified as a lenticular galaxy of type S0, recent observations of M85 have suggested that it could be an elliptical galaxy of type E1.

Pierre Méchain discovered M85 on March 4, 1781. He reported it to his friend Charles Messier who subsequently catalogued it on March 18, 1781. On the same night Messier discovered another seven galaxies, all of them Virgo Cluster members and also re-discovered bright globular cluster M92.The main crux of the Virgo cluster lies about halfway along an imaginary line connecting Denebola (β Leo - mag. +2.1) and Vindemiatrix (ε Vir - mag. +2.8), where most of the galaxies can be found. However, M85 is located at the very northern edge of the Virgo cluster, some 6 degrees northwest of the group centre and one degree northeast of star 11 Com (mag. +4.7).

It's best seen during the months of March, April and May.

Through a medium sized 150mm (6-inch) or 200mm (8-inch) scope, M85 appears as a bright, round, diffuse ball of light with a much brighter central core. However, even larger scopes don't show much more detail. Also visible in the same field of view is neighboring barred spiral galaxy NGC 4392. This galaxy is positioned 8 arc minutes to the east and is currently interacting with M85.

Supernova (SN 1960R) was observed in M85, reaching magnitude +11.7 in December 1960. But supernova hunters be warned, there's a foreground star south-southeast of the galaxy nucleus that often tricks many observers!

M85 Data Table

Messier	85	**RA (J2000)**	12h 25m 24s
NGC	4382	**DEC (J2000)**	18d 11m 27s
Object Type	Lenticular galaxy (or Elliptical galaxy)	**Apparent Size (arcmins)**	7.1 x 5.5
Classification	S0 (or E1)	**Radius (light years)**	62,500
Constellation	Coma Berenices	**Number of Stars**	400 Billion
Distance (kly)	60000		
Apparent Mag.	9.5		

Messier # 085

Date:	Time:

Site:

Temp:	Wind:	Hum:

Clouds:	Moon:

Scope:

EP:	Mag:
NELM:	See/Trans:
Type:	# Stars:
Mag:	Age:

Const:

Notes:

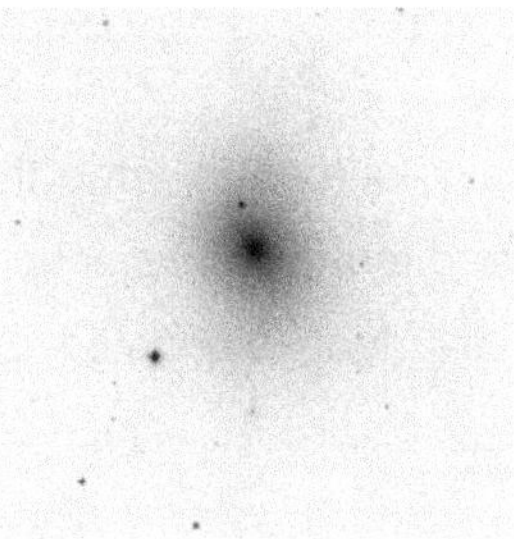

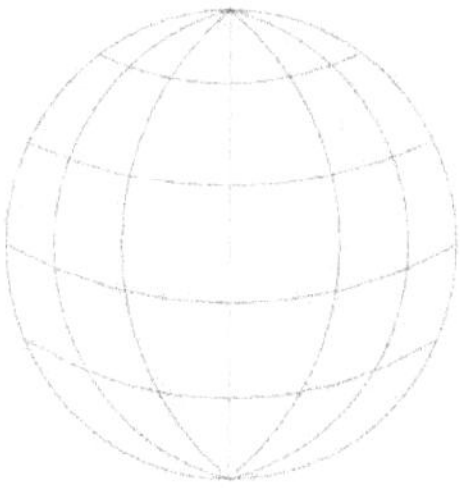

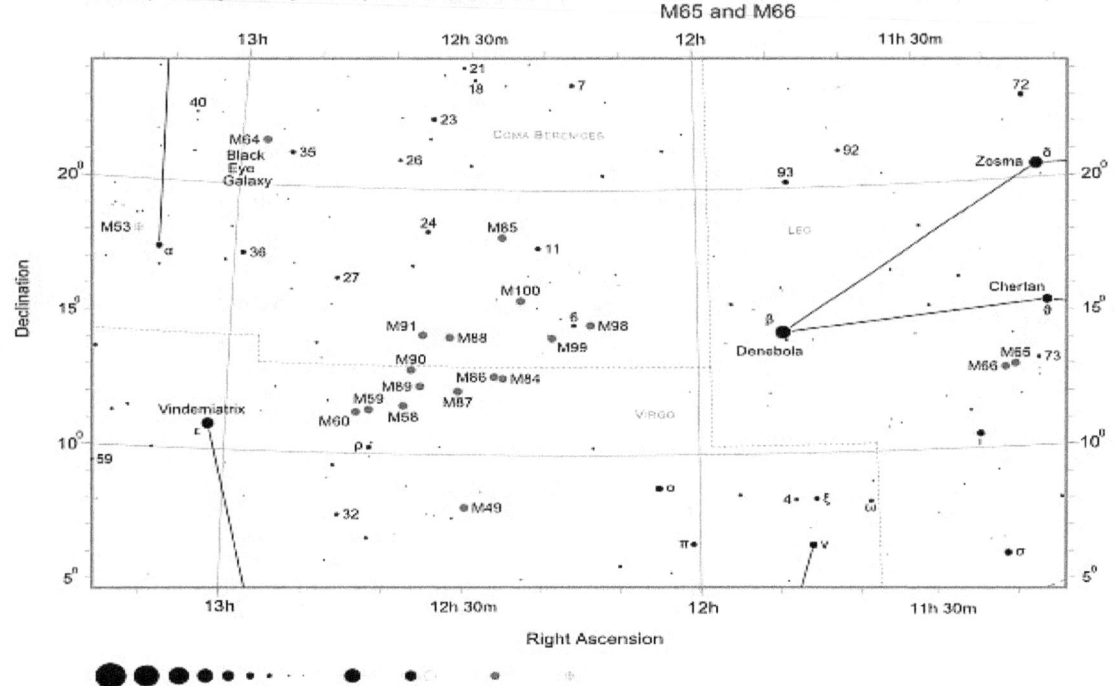

Messier Finder Chart for M49, M58, M59, M60, M84, M85, M86, M87, M88, M89, M90, M91, M98, M99 and M100 Also shown M53, M64 Black Eye Galaxy, M65 and M66

Messier 86 - M86 - Lenticular Galaxy (or Elliptical Galaxy)

M86 is a giant lenticular or elliptical galaxy located in Virgo that's one of the brighter galaxies in the Virgo cluster (mag. +9.3). It lies at the heart of the grouping and forms a conspicuous pair with close neighbor and almost twin, M84. Through a medium sized scope, M86 appears as a bright, elongated patch of light. Also visible in the same low/medium power eyepiece field of view is M84.

It's currently not 100% certain what type of galaxy M86 is, it could be either a type S0 lenticular galaxy or an elliptical galaxy of type E3. The galaxy is unusual in that it's blue shifted and hence moving towards the Milky Way. Due to the expansion of the Universe most galaxies are receding and show red shifts. However, M86 is falling towards the centre of the Virgo cluster, causing it to move towards us at a speed of 244 km/s. This resulting blue shift is the highest of all Messier objects.M86 was discovered by Charles Messier on March 18, 1781. This was an extremely productive night for Messier since he also discovered another seven Virgo cluster members and rediscovered globular cluster M92 in Hercules. The galaxy has an apparent diameter of 8.9 x 5.8 arc minutes and at a distance of 52 million light-years this corresponds to a spatial diameter of 135,000 light-years. It's estimated to contain at least 400 billion stars.

M86 is positioned close to the Virgo-Coma Berenices constellation border and can be found by imagining a line connecting Denebola (β Leo - mag. +2.1) with Vindemiatrix (ε Vir - mag. +2.8). Towards the centre of this line is M86 and positioned 17 arc minutes west of M86 is M84.

The Virgo cluster galaxies are best seen during the months of March, April and May.

M86 is a challenging binocular object. It can be spotted with 7x50 or 10x50 models but requires dark skies and good transparency. Together with M84, both galaxies appear as faint smudges of light in the same field of view. A 80mm (3.1-inch) scope reveals them as small oval shapes with brighter centers. M86 is slightly bigger than M84. Due to their low surface brightness, larger telescopes show the surrounding halos better. With 200mm (8-inch) scopes under dark skies it's possible to spot several more faint galaxies in the same field of view. These include NGC 4435, NGC 4388, NGC 4402 and NGC 4438. The giant elliptical galaxy M87 is located about 1.5 degrees southeast of the M84/M86 pair. M86 has an extensive system of globular clusters, totaling around 3,800.

M86 Data Table

Messier	86
NGC	4406
Object Type	Lenticular galaxy or (Elliptical galaxy)
Classification	S0 (or E3)
Constellation	Virgo
Distance (kly)	52000

Apparent Mag.	9.3
RA (J2000)	12h 26m 12s
DEC (J2000)	12d 56m 47s
Apparent Size (arcmins)	8.9 x 5.8
Radius (light years)	67,500
Number of	>400 Billion

Stars	
Notable Feature	Displays the highest blue shift of all Messier objects

Messier # 086

Date:	Time:

Site:

Temp:	Wind:	Hum:

Clouds:	Moon:

Scope:

EP:	Mag:
NELM:	See/Trans:
Type:	# Stars:
Mag:	Age:

Const:

Notes:

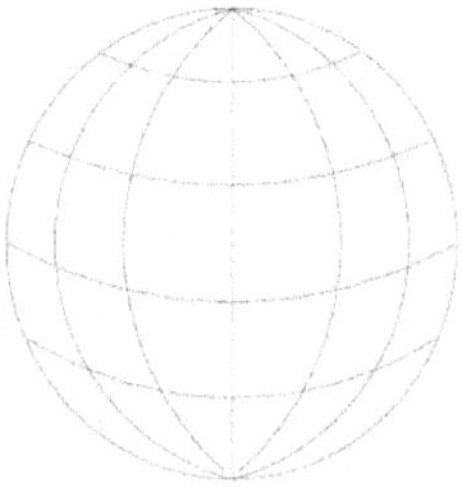

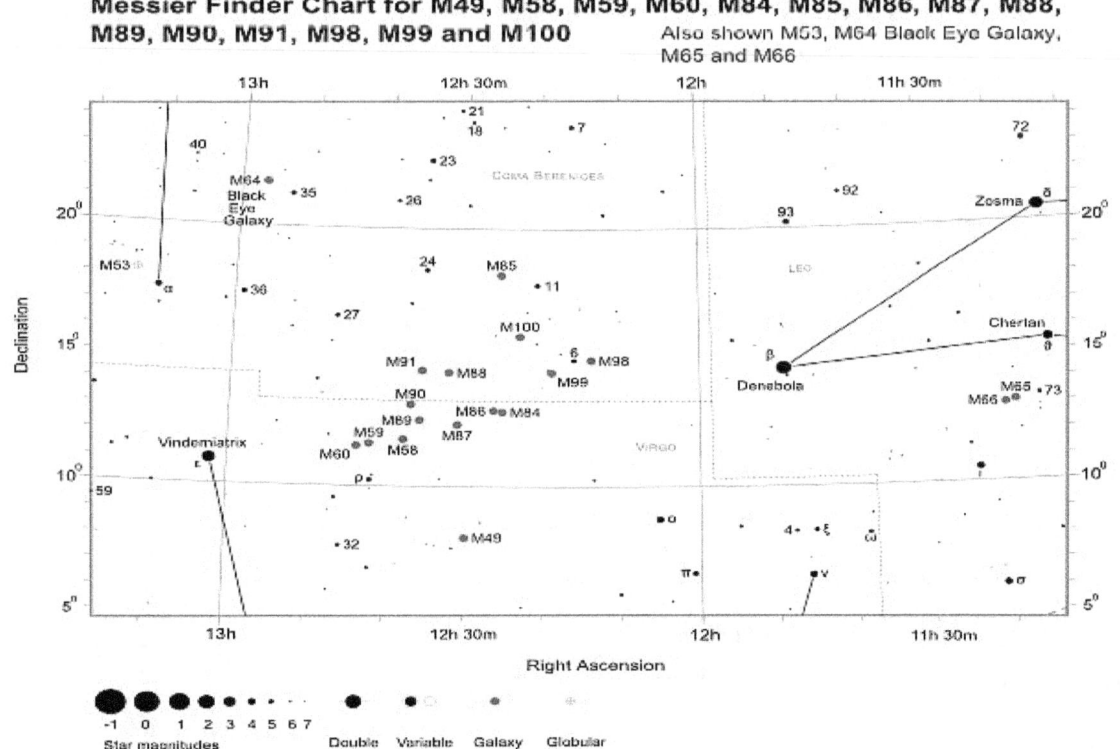

Messier Finder Chart for M49, M58, M59, M60, M84, M85, M86, M87, M88, M89, M90, M91, M98, M99 and M100 Also shown M53, M64 Black Eye Galaxy, M65 and M66

Right Ascension

-1 0 1 2 3 4 5 6 7
Star magnitudes

Double star   Variable stars   Galaxy   Globular cluster

Messier 87 - M87 - Elliptical Galaxy

M87 is a super giant elliptical galaxy that's a prominent member of the Virgo cluster of galaxies. It's one of the largest and most luminous galaxies known and a strong source of radiation, particularly radio and X-ray emissions. At the centre of M87 is a super massive black hole with a jet of extremely energetic plasma extending outwards for at least 5000 light-years. The galaxy is therefore an interesting object for both professional and amateur astronomers alike.

With an apparent magnitude of +8.6, M87 is the second brightest of the Virgo cluster galaxies; only M49at mag. +8.4 is brighter. On dark moonless nights it's visible with 7x50 or 10x50 binoculars, appearing as a faint hazy patch of light. The galaxy was one of eight discovered by Charles Messier on March 18, 1781. On this day he also re-discovered fine globular cluster M92.

M87 lies at the heart of the Virgo cluster. It can be found by imagining a line connecting Denebola (β Leo - mag. +2.1) with Vindemiatrix (ε Vir - mag. +2.8). Just over half way along this line is M87. Faint elliptical galaxy M89 is positioned just over a degree east of M87 with galaxy pair M84/M86 located 1.5 degrees northwest of M87.

The Virgo galaxies are best seen during the months of March, April and May.

Through a 80mm (3.1-inch) scope M87 appears as a fuzzy elliptical ball of light that's brighter towards the centre. Even with larger scopes the galaxy remains essentially featureless although much easier to detect. It has no distinctive dust lanes and diminishes in luminosity with distance from the center. The jet is far too faint to be observed with most backyard scopes, although it has been reportedly observed with extremely large amateur scopes under excellent conditions. It is much easier to image or photograph. Within the same low-power field as M87 are two fainter elliptical galaxies, NGC 4476 and NGC 4378.

M87 spans 8.3 x 6.6 arc minutes of apparent sky. It's located 53.5 million light-years distant, which corresponds to a spatial diameter of 130,000 light years and is estimated to contain a trillion stars. The only supernova recorded for in M87 occurred in February 1919, but was not detected until 1922 when photographic plates were examined. The maximum brightness was estimated at +11.5.

Orbiting M87 are an extremely large number of globular clusters of which at least 12,000 have been identified. For comparison our Milky Way galaxy contains only 200.

The galaxy is also referred to as Virgo A.

M87 Data Table

Messier	87
NGC	4486
Object Type	Elliptical galaxy
Classification	E1
Constellation	Virgo
Distance (kly)	53500

Apparent Mag.	8.7
RA (J2000)	12h 30m 49.3s
DEC (J2000)	12d 23m 26s
Apparent Size (arcmins)	8.3 x 6.6
Radius (light years)	65,000

Number of Stars	1 Trillion
Notable Feature	Contains a spectacular jet of ejected matter

Messier # 087

Date:	Time:

Site:

Temp:	Wind:	Hum:

Clouds:	Moon:

Scope:

EP:	Mag:
NELM:	See/Trans:
Type:	# Stars:
Mag:	Age:
Const:	

Notes:

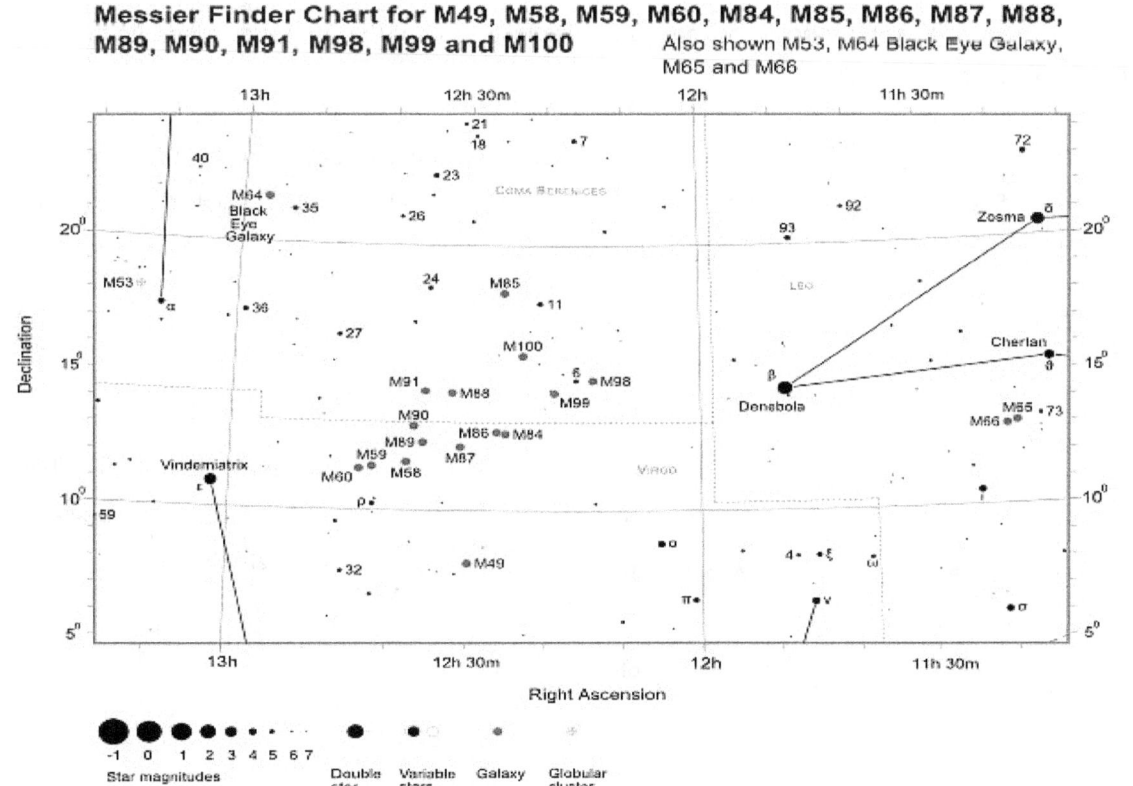

Messier Finder Chart for M49, M58, M59, M60, M84, M85, M86, M87, M88, M89, M90, M91, M98, M99 and M100

Also shown M53, M64 Black Eye Galaxy, M65 and M66

Messier 88 - M88 - Spiral Galaxy

M88, mag. +9.6, is a fine spiral galaxy located in Coma Berenices that's a member of the Virgo cluster of galaxies. It has a reasonably high surface brightness - partly due to its favorable inclination of 30 degrees - and therefore a nice small telescope object. It appears somewhat like a much smaller and fainter version of M31, the spectacular Andromeda Galaxy. M88 is one of the brightest Seyfert galaxies in the sky. These types of galaxies exhibit extremely active quasar like nuclei and are strong emitters of electromagnetic radiation with highly ionized spectral emission lines present. They are named after 20th century American astronomer Carl Seyfert who first identified them. Galaxies M51, M66, M77, M81, M87 and M106 also belong to this class of object.

M88 was one of the eight Virgo cluster galaxies discovered by Messier on his most productive night, March 18, 1781. Messier's description of M88 was of a "nebula without star between two small stars and one star of the sixth magnitude, which appear at the same time as the nebula in the field of the telescope". He also remarked that it was similar in appearance to M58. William Parsons the 3rd Earl of Rosse was the first to recognize the spiral shape and listed M88 as one of 14 "spiral nebulae" discovered to 1850.As with some of the Virgo galaxies, locating M88 can be challenging since there are no bright stars located in the vicinity. The galaxy is positioned about a degree north of the Coma Berenices-Virgo constellation boundary with the general area of sky located midway between stars Denebola (β Leo - mag. +2.1) and Vindemiatrix (ε Vir - mag. +2.8). Tenth magnitude barred spiral galaxy M91 is located just east of M88.

The best time of year to look for the Virgo galaxies is during the months of March, April and May.

M88 is bright enough to be seen with 7x50 or 10x50 binoculars on dark nights and is one of the better Virgo cluster galaxies for small telescopes. A small 80mm (3.1-inch) scope reveals an elongated glow of light with a bright centre surrounded by a large outer envelope of nebulosity. It takes high magnifications well so don't be afraid to push up the power especially when the seeing conditions are good.

In total, the galaxy spans 6.9 x 3.7 arc minutes of apparent sky but of course through the eyepiece it appears smaller. With a medium size 200mm (8-inch) scope, M88 displays subtle changes in brightness especially along the edges. The core is well defined, condensed and bright.

M88 is located 53 million light-years distant, which corresponds to an actual diameter of 105,000 light-years. It's estimated to contain about 400 billion stars. To date, one supernova (SN 1999cl) has been observed in M88, which peaked at magnitude +13.6 and within the range of larger amateur scopes.

M88 Data Table

Messier	88
NGC	4501
Object Type	Spiral galaxy
Classification	SA(rs)b
Constellation	Coma Berenices
Distance (kly)	53000

Apparent Mag.	9.6
RA (J2000)	12h 31m 59s
DEC (J2000)	14d 25m 15s
Apparent Size (arcmins)	6.9 x 3.7
Radius (light	52,500

years)	
Number of Stars	400 Billion
Notable Feature	Member of the Virgo Cluster of galaxies

Messier # 088

Date:	Time:

Site:

Temp:	Wind:	Hum:

Clouds:	Moon:

Scope:

EP:	Mag:
NELM:	See/Trans:
Type:	# Stars:
Mag:	Age:
Const:	

Notes:

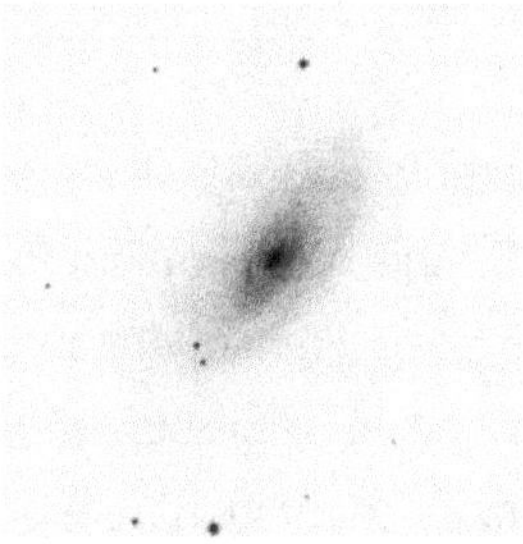

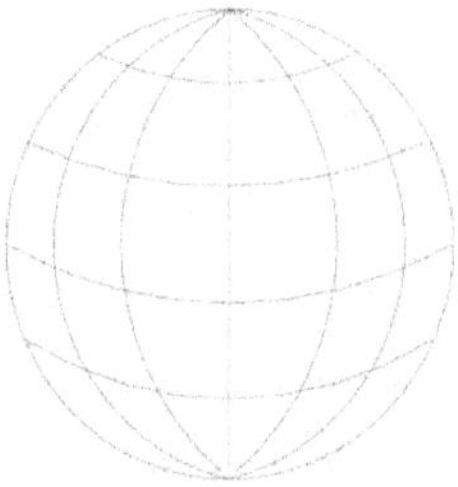

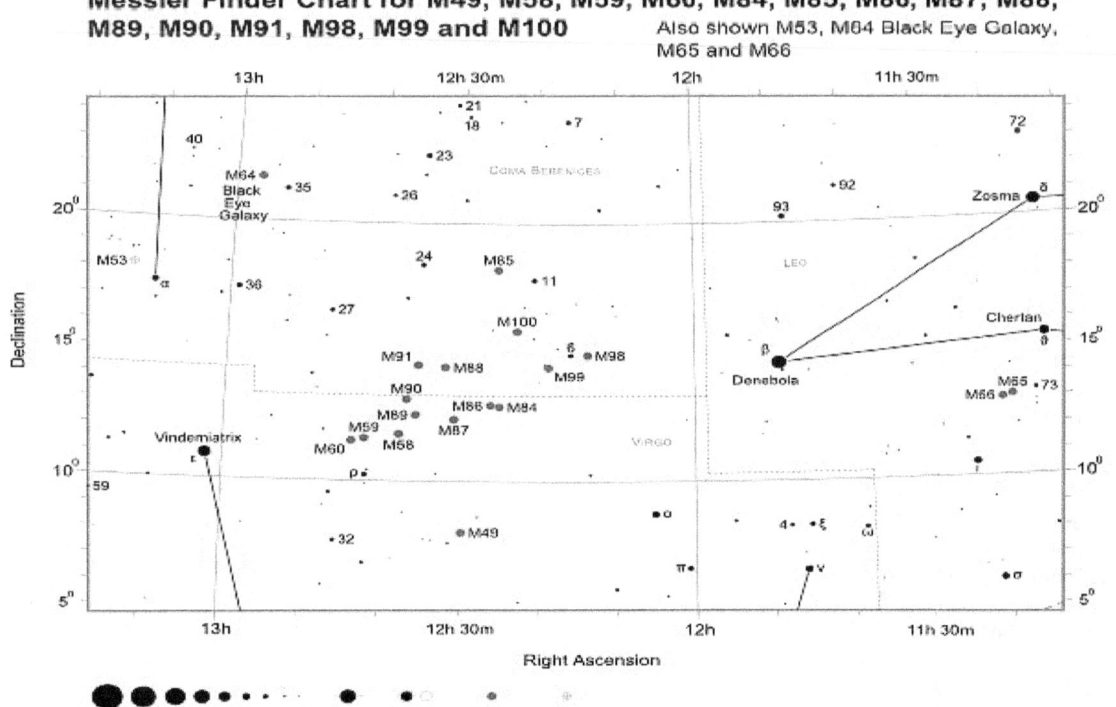

Messier Finder Chart for M49, M58, M59, M60, M84, M85, M86, M87, M88, M89, M90, M91, M98, M99 and M100   Also shown M53, M64 Black Eye Galaxy, M65 and M66

Messier 89 - M89 - Elliptical Galaxy

M89 is another member of the Virgo Cluster of galaxies. It's a small magnitude +10.0 elliptical galaxy (type - E0) discovered by Charles Messier on March 18, 1781. On this bumper night for Messier he also discovered seven other Virgo galaxies and re-discovered globular cluster M92 in Hercules. Recent observations indicate that M89 may be nearly perfectly spherical in shape. This is unusual because all other known ellipticals are elongated. However, it's possible that the spherical nature of M89 is purely a visual affect resulting from its orientation from our perspective.

The galaxy is not as bright as some other group members and therefore a challenging small telescope object. Messier's original discovery observation acknowledges this: "extremely faint and pale and it's not without difficulty that one can distinguish it". The galaxy is best seen with large telescopes but generally featureless and rather unexciting through most amateur instruments.

M89 is located in Virgo just south of the Virgo-Coma Berenices constellation boundary. It's positioned roughly 60% along an imaginary line connecting stars, Denebola (β Leo - mag. +2.1) and Vindemiatrix (ε Vir - mag. +2.8). Slightly brighter spiral galaxy M90 is 0.75 degrees northeast of M89. One degree southeast of M89 is fine barred spiral galaxy M58 with super giant elliptical galaxy M87 located about a degree west of M89.

The Virgo galaxies are best seen during the months of March, April and May.

When viewed through a 200mm (8-inch) telescope M89 appears as a faint, round, small featureless diffuse ball of light. It has a total apparent diameter of about 5 arc-minutes although visually it's about half this size. The view somewhat resembles a distant globular cluster but even larger scopes fail to reveal much more.

M89 is 55 million light-years distant which corresponds to a spatial diameter of 80,000 light years. The galaxy also features a surrounding structure of gas and dust that extends up to 150,000 light-years from the centre. It's estimated to contain 100 billion stars.

M89 Data Table

Messier	89
NGC	4552
Object Type	Elliptical galaxy
Classification	E0
Constellation	Virgo
Distance (kly)	55000
Apparent Mag.	10.0
RA (J2000)	12h 35m 40s
DEC (J2000)	12d 33m 23s
Apparent Size (arcmins)	5.1 x 4.7
Radius (light years)	40,000
Number of Stars	100 Billion

Messier # 089

Date:	Time:

Site:		
Temp:	Wind:	Hum:
Clouds:	Moon:	
Scope:		
EP:	Mag:	
NELM:	See/Trans:	
Type:	# Stars:	
Mag:	Age:	
Const:		

Notes:

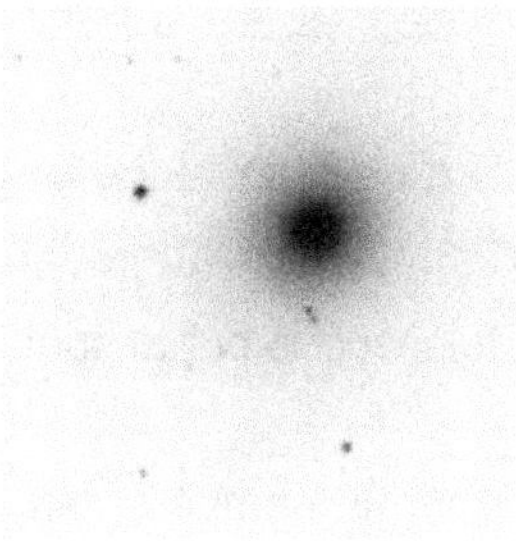

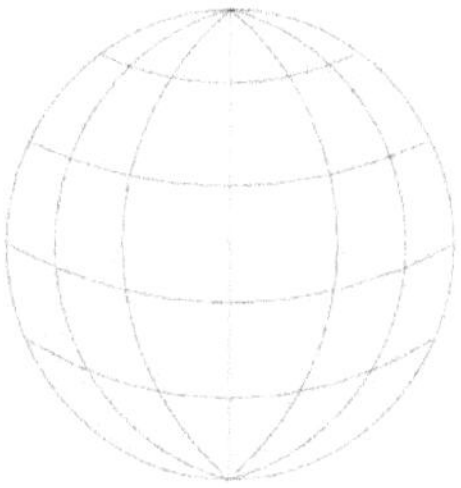

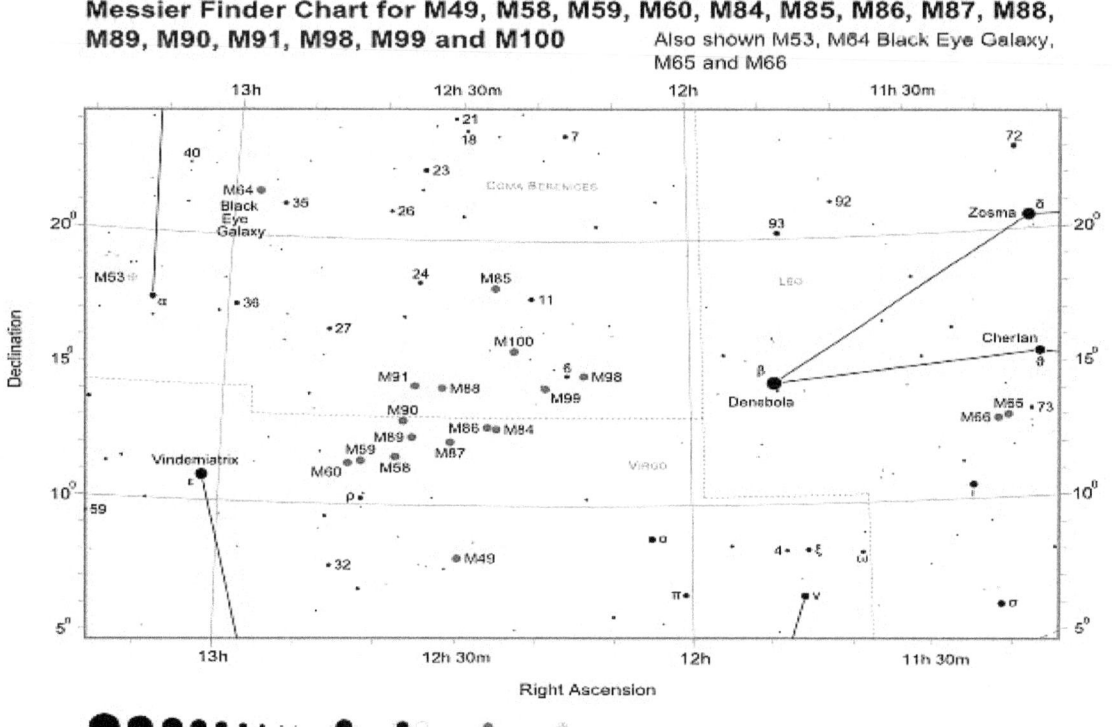

Messier Finder Chart for M49, M58, M59, M60, M84, M85, M86, M87, M88, M89, M90, M91, M98, M99 and M100

Also shown M53, M64 Black Eye Galaxy, M65 and M66

Right Ascension

Declination

Star magnitudes -1 0 1 2 3 4 5 6 7

Double star   Variable stars   Galaxy   Globular cluster

Messier 90 - M90 - Spiral Galaxy

M90 is a spiral galaxy located in Virgo. It's a member of the Virgo Cluster and one of the largest and brightest spirals in the group. With an apparent magnitude of +9.6, it's visible through small scopes as a reasonably bright oval shaped patch of light. M90 appears bright in medium size telescopes but to spot the spiral structure requires a larger amateur scope.

The galaxy was one of eight galaxies, all Virgo members, discovered by Charles Messier on March 18, 1781. It's located about 60 Million light-years distant and is intrinsically large with an actual diameter of 165,000 light-years, more than the Andromeda Galaxy (M31). It's estimated to contain a trillion stars.

M90 is positioned close to the centre of the Virgo cluster and right at the Virgo-Coma Berenices constellation boundary. The centre of the cluster is located roughly halfway along a line connecting stars, Denebola (β Leo - mag. +2.1) and Vindemiatrix (ε Vir - mag. +2.8). In the same area of sky are M84, M86 and M87 with M90 positioned 1.5 degrees northeast of M87. The small elliptical galaxy M89 is 0.75 degrees southwest of M90 with M91 about a degree to the north-northwest of M90. Tenth magnitude spiral galaxy M88 is located 1.5 degrees northwest of M90.

The Virgo galaxies are best seen during the months of March, April and May.

M90 can be spotted with 7x50 or 10x50 binoculars but requires dark skies and excellent seeing conditions. It's easier to see with small scopes; an 80mm (3.1-inch) instrument reveals a reasonably bright oval shaped smudge of light but nothing much else. When viewed through a 200mm (8-inch) scope, M90 has a bright core surrounded by a streak of nebulosity that fades outwards from the centre. On dark nights, larger amateur scopes hint at more detail including the spiral arms. In total, M90 spans 9.5 x 4.4 arc minutes of apparent sky.

The galaxy is one of a few that appear blue shifted. This results from its large velocity within the Virgo Cluster, which means that it's currently moving towards us. Only one Messier galaxy is approaching us faster, M86. This is in contrast to most other galaxies which are red shifted and therefore receding from us.

M90 Data Table

Messier	90	DEC (J2000)	13d 09m 45s
NGC	4569	Apparent Size (arcmins)	9.5 x 4.4
Object Type	Spiral galaxy		
Classification	SAB(rs)ab	Radius (light years)	82,500
Constellation	Virgo		
Distance (kly)	60000	Number of Stars	1 Trillion
Apparent Mag.	9.6	Notable Feature	Very large spiral galaxy belonging to the Virgo Cluster of galaxies
RA (J2000)	12h 36m 50s		

Messier # 090

Date:	Time:	
Site:		
Temp:	Wind:	Hum:
Clouds:	Moon:	
Scope:		
EP:	Mag:	
NELM:	See/Trans:	
Type:	# Stars:	
Mag:	Age:	
Const:		

Notes:

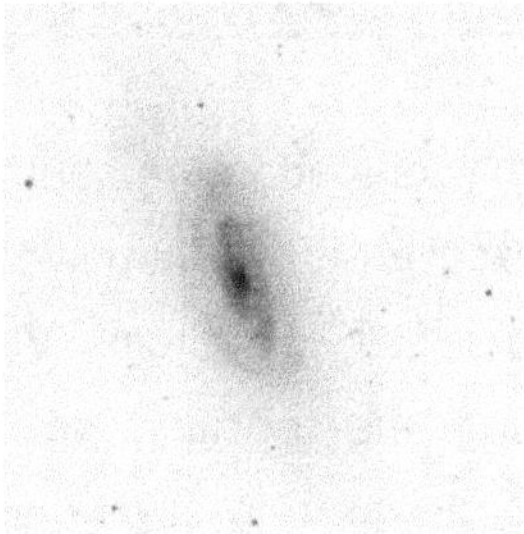

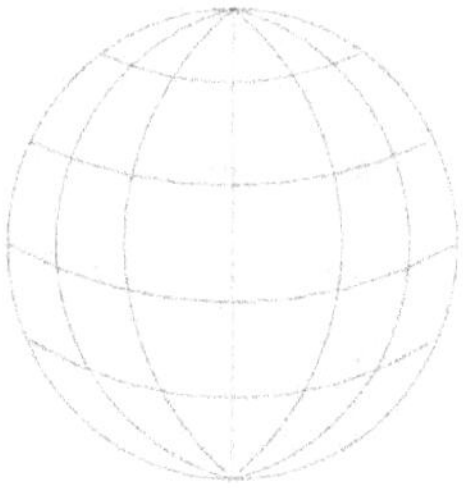

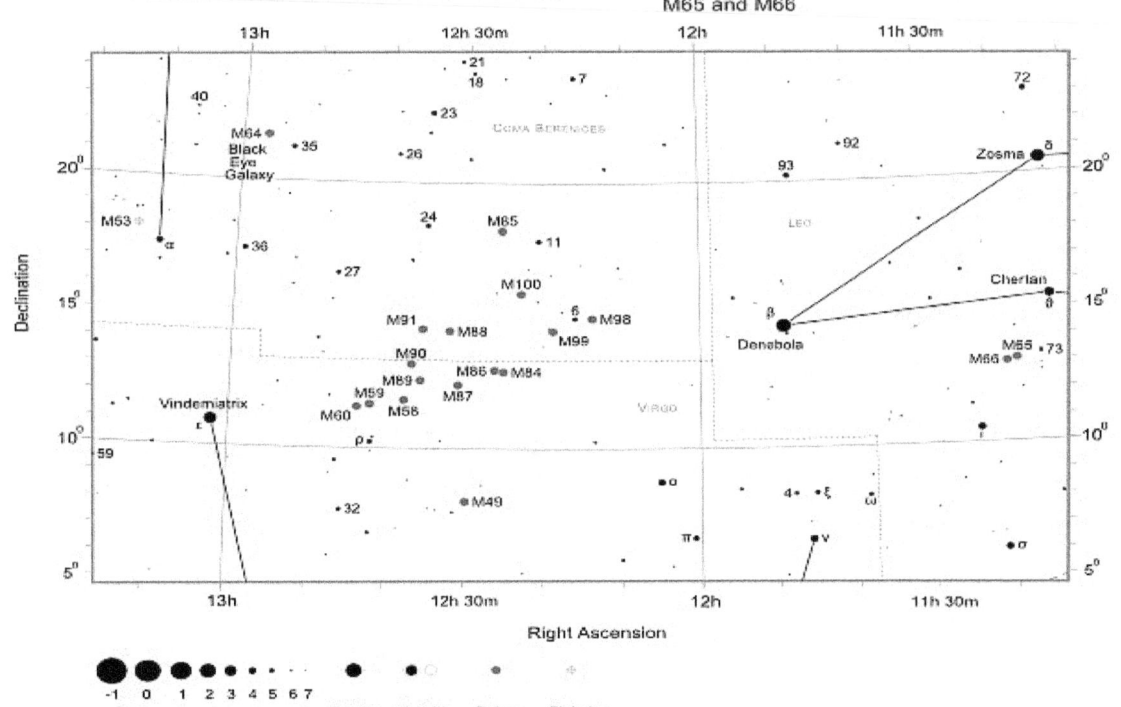

Messier Finder Chart for M49, M58, M59, M60, M84, M85, M86, M87, M88, M89, M90, M91, M98, M99 and M100

Also shown M53, M64 Black Eye Galaxy, M65 and M66

Declination

Right Ascension

Star magnitudes   -1 0 1 2 3 4 5 6 7

Double star   Variable stars   Galaxy   Globular cluster

Messier 91 - M91 - Barred Spiral Galaxy

M91, mag. +10.3, is a barred spiral galaxy located in the southern part of constellation Coma Berenices. It's a member of the Virgo cluster of galaxies and was discovered by Charles Messier on March 18, 1781. This was a productive night for Messier; he discovered eight objects, all of them Virgo cluster galaxies and also rediscovered globular cluster M92 in Hercules. When recording the position of M91, Messier incorrectly referenced its location from galaxy M58 when he meant to use M89. It was only a one degree mistake, however the result meant that M91 was a missing object for almost 200 years!

It was not until 1969 when some astronomy detective work by William Williams solved the mystery of M91. In 1969, he pinpointed the location in the sky after applying Messier's measurements to a starting point of M89 and concluded that the missing object was almost certainly NGC 4548.

Finding M91 can be challenging. The galaxy is located about a degree north of the Coma Berenices-Virgo constellation boundary but there are no bright stars in the vicinity. The general area of sky can be found by imagining a line connecting Denebola (β Leo - mag. +2.1) with Vindemiatrix (ε Vir - mag. +2.8). About 60% of the way along this line is elliptical galaxy M89 (mag. +10.0) with M91 positioned two degrees directly north of it.

M91 and the Virgo cluster galaxies are best seen during the months of March, April and May.

M91 is one of the fainter Messier galaxies and is generally considered as one of the more difficult objects in the list. Through a medium size scope of the order of 150mm (6-inch) or 200mm (8-inch) it appears as a faint ball of light that's brighter towards the middle. In total the galaxy spans 5.4 x 4.3 arc minutes of apparent sky. On dark nights with good transparency the centre bar shape is evident with hints of the spiral structure visible. The spiral arms are better seen with larger amateur scopes.

M91 is about 63 million light years distant and has a spatial diameter of 100,000 light-years. It's estimated to contain 400 billion stars.

M91 Data Table

Messier	91	**DEC (J2000)**	14d 29m 48s
NGC	4548	**Apparent Size (arcmins)**	5.4 x 4.3
Object Type	Barred Spiral galaxy		
Classification	SBb(rs)	**Radius (light years)**	50,000
Constellation	Coma Berenices	**Number of Stars**	400 Billion
Distance (kly)	63000	**Notable Feature**	Member of the Virgo Cluster of galaxies
Apparent Mag.	10.3		
RA (J2000)	12h 35m 27s		

Messier # 091

Date:	Time:
Site:	
Temp:	Wind:	Hum:
Clouds:	Moon:
Scope:	
EP:	Mag:
NELM:	See/Trans:
Type:	# Stars:
Mag:	Age:
Const:	
Notes:	

Messier Finder Chart for M49, M58, M59, M60, M84, M85, M86, M87, M88, M89, M90, M91, M98, M99 and M100

Also shown M53, M64 Black Eye Galaxy, M65 and M66

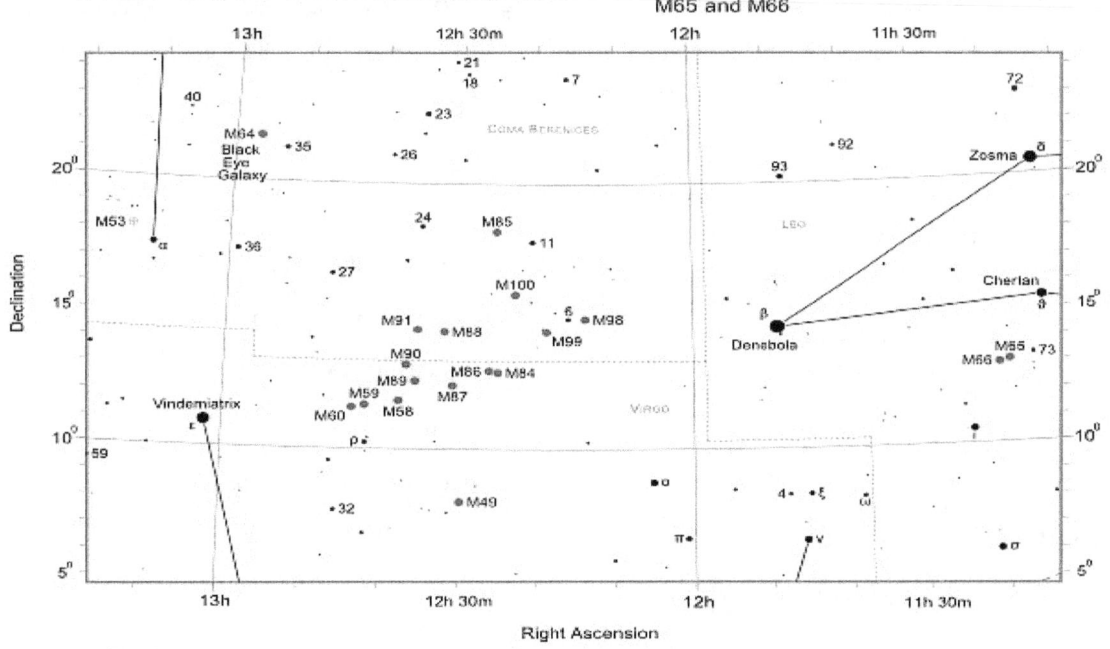

Messier 92 - M92 - Globular Cluster

M92 is a bright globular cluster located in the northern part of the constellation of Hercules. Despite shining at magnitude +6.4 and at the very far limit of naked eye visibility, it's often overlooked by amateur astronomers because of its close proximity to the more spectacular M13.

M92 is located in a relatively blank area of Hercules north of the well-known "Keystone" asterism. It can be found by drawing an imaginary line from the NW corner "Keystone" star Eta Herculis (η Her - mag. +3.5) to star Iota Herculis (ι Her - mag. +3.8). About 60% of the way along this line is M92. When viewed through 10x50 binoculars, the cluster appears distinctly non-stellar, looking like an out of focus star or a hazy patch of light. It has a brighter core that can be seen with direct vision but easier if using averted vision. A small to medium size telescope will start to resolve some of the outer stars. With a 150mm (6-inch) or 200mm (8-inch) telescope, M92 appears slightly oval in shape with a bright core and a smattering of stars in the surrounding halo. Compared with M13, the core of M92 is more dense and compact and hence much more difficult to resolve. This is particularly noticeable in large telescopes of the order of 250mm (12-inch) or greater.

When viewed through a telescope of this size, M92 appears as a large, bright, ball of stars with dozens of bright stars resolved in the halo, across the surface of the cluster and in the dense central core. In total, it has an apparent diameter of 14 arc minutes and is best seen from the Northern Hemisphere during the months of June, July and August.

M92 is one of the original discoveries of Johann Elert Bode, who found it on December 27, 1777 and described it as "A nebula that is more or less round with a pale glow". Charles Messier independently rediscovered it and cataloged it on March 18, 1781. Incidentally, this proved to be a very productive day for Messier as he also cataloged another 8 objects, all of them Virgo Cluster galaxies (M84-M91) at the same time. As with many globulars, it was William Herschel in 1783 who first resolved M92 into stars. Of all the stars in M92, only about 16 variables have been discovered of which 14 of are of RR Lyrae type and one of them is a rare globular eclipsing binary of the W Ursae Majoris type. The cluster is located 26,700 light-years from Earth and has a spatial diameter of 108 light-years.

Due to the effect of precession, the Earth's North Celestial Pole (NCP) occasionally passes within one degree of M92. The last time this occurred was about 12,000 years ago (10,000 BC) and the next time will be in about 14,000 years. So around the year 16,000 AD, M92 will become "Polarissima Borealis", or the "North Cluster" object.

M92 Data Table

Messier	92		**DEC (J2000)**	43d 08m 10s
NGC	6341		**Apparent Size (arcmins)**	14 x 14
Object Type	Globular Cluster		**Radius (light years)**	54
Constellation	Hercules		**Age (years)**	14,200M
Distance (kly)	26.7		**Number of Stars**	250,000
Apparent Mag.	6.4			
RA (J2000)	17h 17m 07s			

Messier # 092

Date:		Time:	
Site:			
Temp:	Wind:		Hum:
Clouds:		Moon:	
Scope:			
EP:		Mag:	
NELM:		See/Trans:	
Type:		# Stars:	
Mag:		Age:	
Const:			
Notes:			

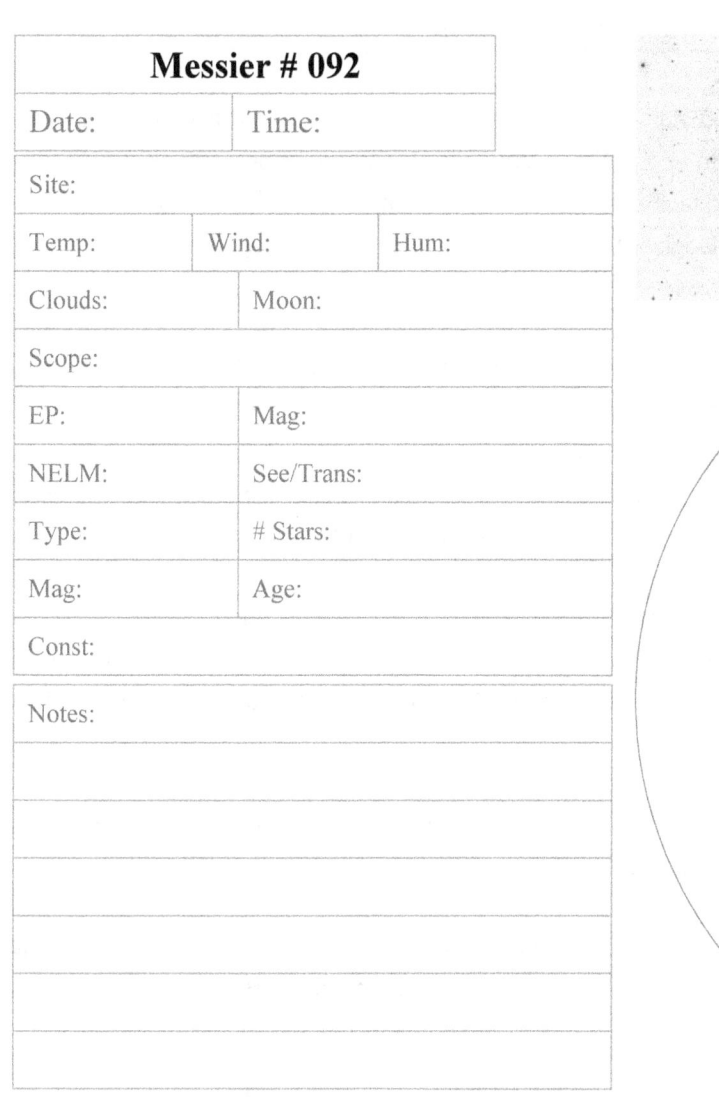

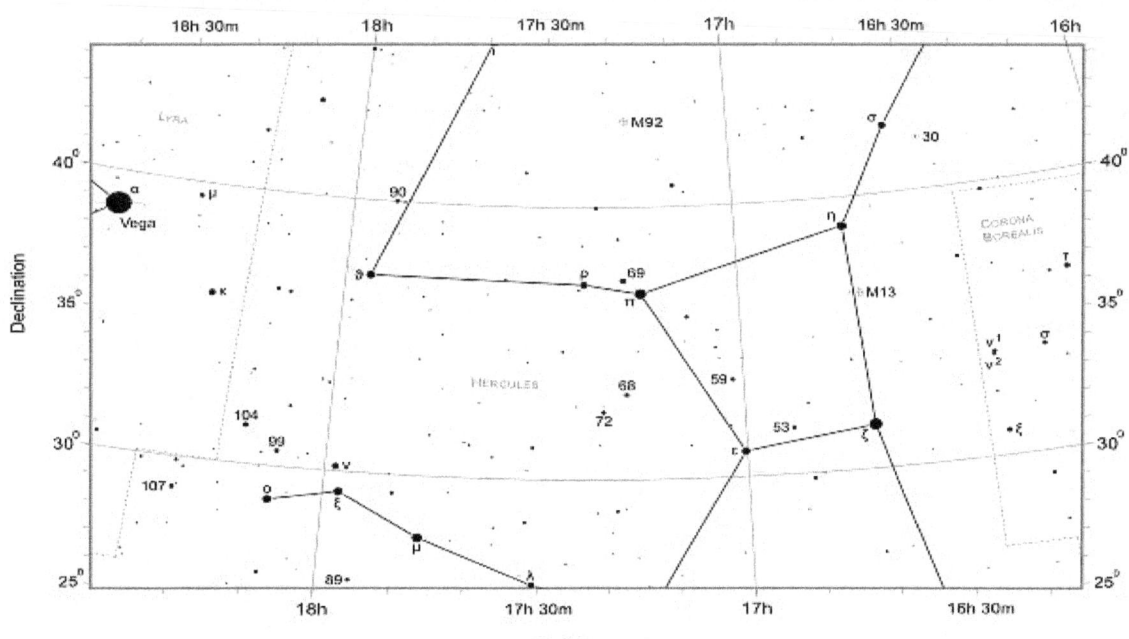

Messier Finder Chart for M13 Great Hercules Globular Cluster and M92

Messier 93 - M93 - Open Cluster

M93, mag. +6.0, is a bright large open cluster of about 80 stars that's located in the southern constellation of Puppis. It has an apparent diameter of 22 arc minutes, corresponding to about 2/3rds that of the full Moon. Under dark skies, M93 is just about visible to the naked eye as a misty patch of light. It's an easy binocular object and a wonderful sight in small telescopes where the brightest stars are shaped like a triangle. The cluster was one of the last deep sky objects discovered by Charles Messier, which he catalogued on March 20, 1781.

Finding M93 is not difficult; it's positioned in western Puppis only a few degrees from the Canis Major border and not far from Sirius (α CMa) the brightest star in the night sky. Sirius shines at mag. -1.46 and can be easily found by connecting the three bright stars of Orion's belt and following the line southwards. Located 8 degrees southeast of Sirius are stars Omicron1 CMa (o^1 CMa - mag. +3.9) and Omicron2 CMa (o^2 CMa - mag. +3.0). Next imagine a line connecting these two stars and extend it eastwards, curving slightly southwards for about 10 degrees to arrive at M93.

A pair of 7x50 or 10x50 binoculars easily shows M93 and the triangular shape formed by the brightest stars. Through a small 80mm (3.1 inch) telescope at medium magnifications, M93 appears large and somewhat dense cluster. Towards the centre of the cluster is an arrowhead or wedge shaped grouping of bright stars. With averted vision, the nebulous background resolves into many fainter stars.

Larger telescopes of aperture 200mm (8-inch) or greater reveal dozens of members of mostly blue giants but some red giant stars that add to the charm of this already dazzling and beautiful cluster.

M93 has a spatial diameter of 20 light years and is located 3600 light years away. It's estimated to be about 100 million years old and is best seen from southern latitudes during the months of December, January and February.

M93 Data Table

Messier	93
NGC	2447
Object Type	Open Cluster
Constellation	Puppis
Distance (kly)	3.6
Apparent Mag.	6.0
RA (J2000)	07h 44m 29s

DEC (J2000)	-23d 51m 11s
Apparent Size (arcmins)	22 x 22
Radius (light years)	10
Age (years)	100M
Number of Stars	80
Other Name	Collinder 160

Messier # 093

Date:	Time:
Site:	
Temp:	Wind:	Hum:
Clouds:	Moon:
Scope:	
EP:	Mag:
NELM:	See/Trans:
Type:	# Stars:
Mag:	Age:
Const:	
Notes:	

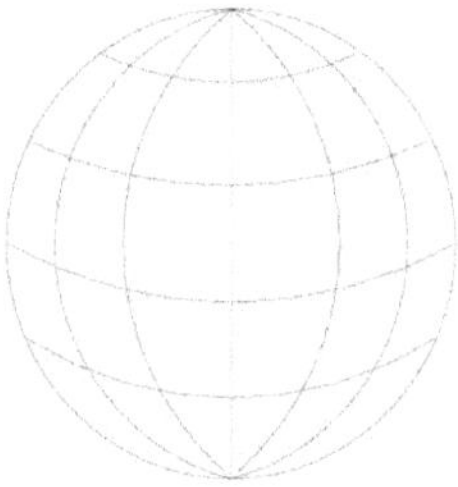

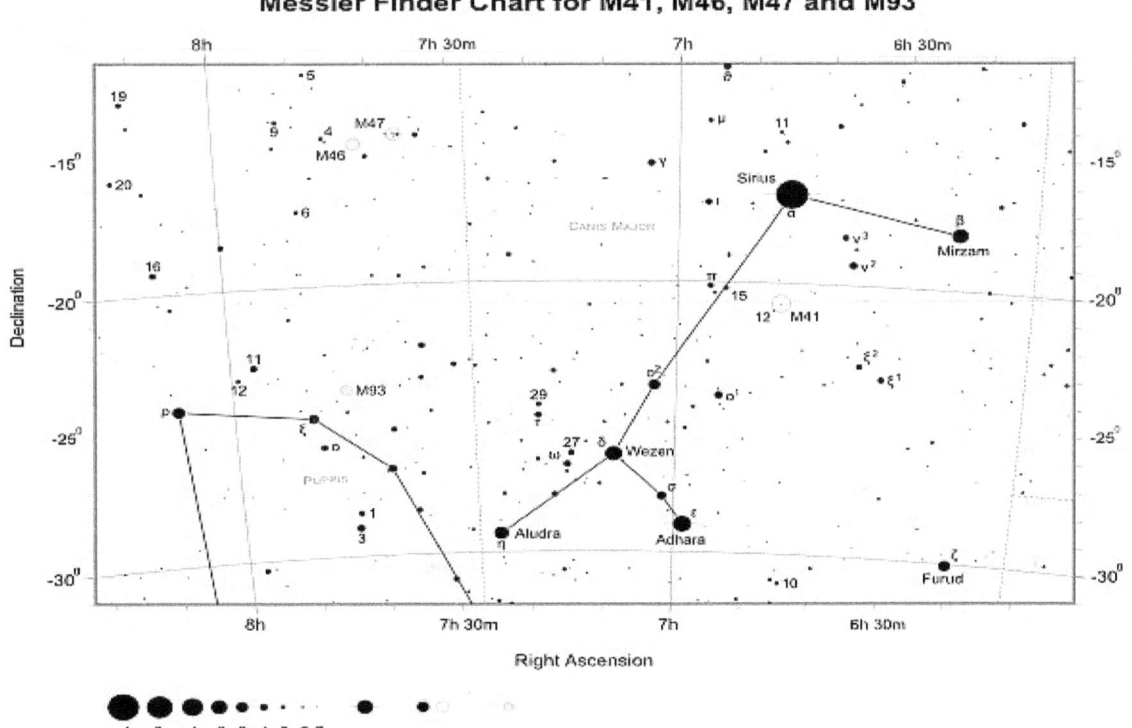

Messier Finder Chart for M41, M46, M47 and M93

Messier 94 - M94 - Cats Eye Galaxy (Spiral Galaxy)

M94 is a nice spiral galaxy located in the constellation of Canes Venatici. It's also known as the Cats Eye Galaxy or the Crocs Eye Galaxy due to its stunning eye-like resemblance. With an apparent magnitude of +8.5, it's a difficult binocular object requiring dark skies and good transparency; at best appearing as only a small faint hazy patch of light.

M94 is one of the nearest galaxies beyond our Local Group of Galaxies. It's located about 16 million light years distance and belongs to the M94 Group, a collection of between 16 and 24 galaxies lying within the Virgo Super cluster. The galaxy was discovered by Pierre Méchain on March 22, 1781 and subsequently confirmed and cataloged two days later by Charles Messier.

The constellation of Canes Venatici is faint but the brightest star Cor Caroli (α CVn - mag. +2.9) can be quite easily found since it's due south of the famous Plough or Big Dipper asterism of Ursa Major. The second brightest star in the constellation is Chara (β CVn - mag. +4.2), located just over 5 degrees northwest of Cor Caroli. M94 is 3 degrees east and a fraction south of Chara. Keep continuing eastwards and you will reach the fine Sunflower Galaxy (M63).

Through any telescope M94 appears unmistakable as a galaxy. A small 80mm (3.1-inch) instrument shows a fuzzy patch with a distinct bright central core. When viewed through a medium size scope of the order of 150mm (6-inch) or 200mm (8-inch) aperture it offers more detail. The galaxy nucleus appears as a brilliant condense point of light that shines through the mist of the surrounding compact nebulosity; somewhat like an eye staring back at you. The halo appears smooth but hints at the spiral nature of the object. Even larger amateur instruments show mottling, a brighter ring around the core and exquisite details.

M94 is a rare galaxy in that it contains two star forming or starburst rings of interstellar material. The inner ring has a diameter of 70 arc seconds, the outer ring 600 arc seconds and both regions are of strong star forming activity.

In total the galaxy spans 11.2 x 9.2 arc minutes of apparent sky, which corresponds to a spatial diameter of 50,000 light-years. It's estimated to contain about 40 billion stars. The galaxy is best seen from the Northern Hemisphere during the months of March, April or May.

M94 Data Table

Messier	94
NGC	4736
Name	Cats Eye Galaxy
Object Type	Spiral galaxy
Classification	(R)SA(r)ab
Constellation	Canes Venatici
Distance (kly)	16,000
Apparent Mag.	8.5

RA (J2000)	12h 50m 53s
DEC (J2000)	41d 07m 12s
Apparent Size (arcmins)	11.2 x 9.2
Radius (light years)	25,000
Number of Stars	40 Billion
Notable Feature	One of the brightest galaxies in the M94 Group

Messier # 094

Date:	Time:
Site:	
Temp:	Wind:	Hum:
Clouds:	Moon:
Scope:	
EP:	Mag:
NELM:	See/Trans:
Type:	# Stars:
Mag:	Age:
Const:	
Notes:	

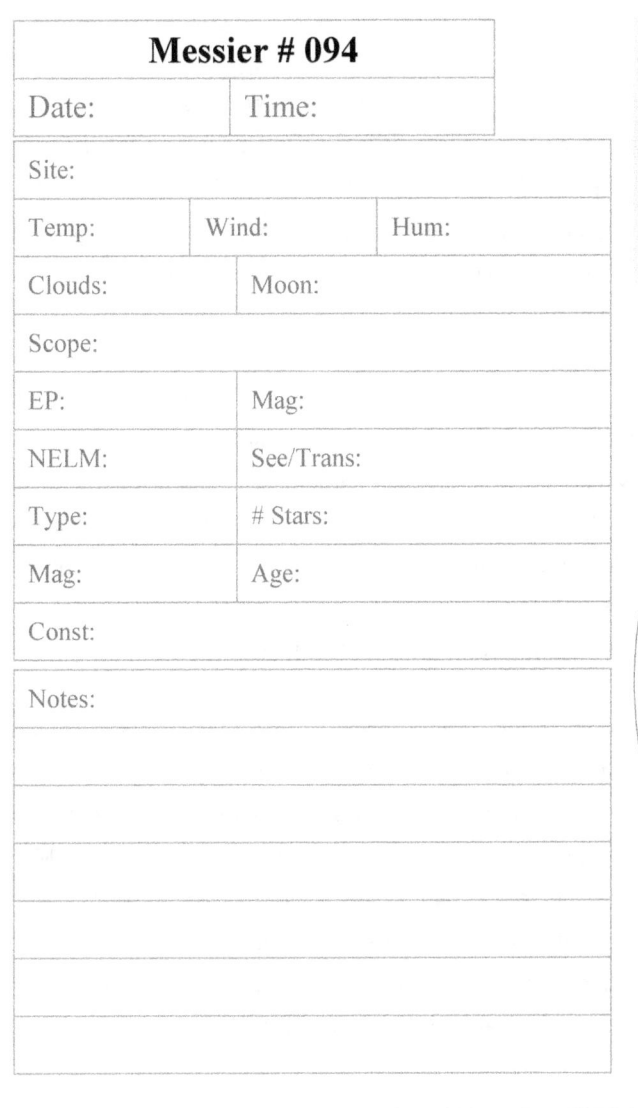

Messier Finder Chart for M51 Whirlpool Galaxy, M63 Sunflower Galaxy, M94, M101 and M106

Also shown M109

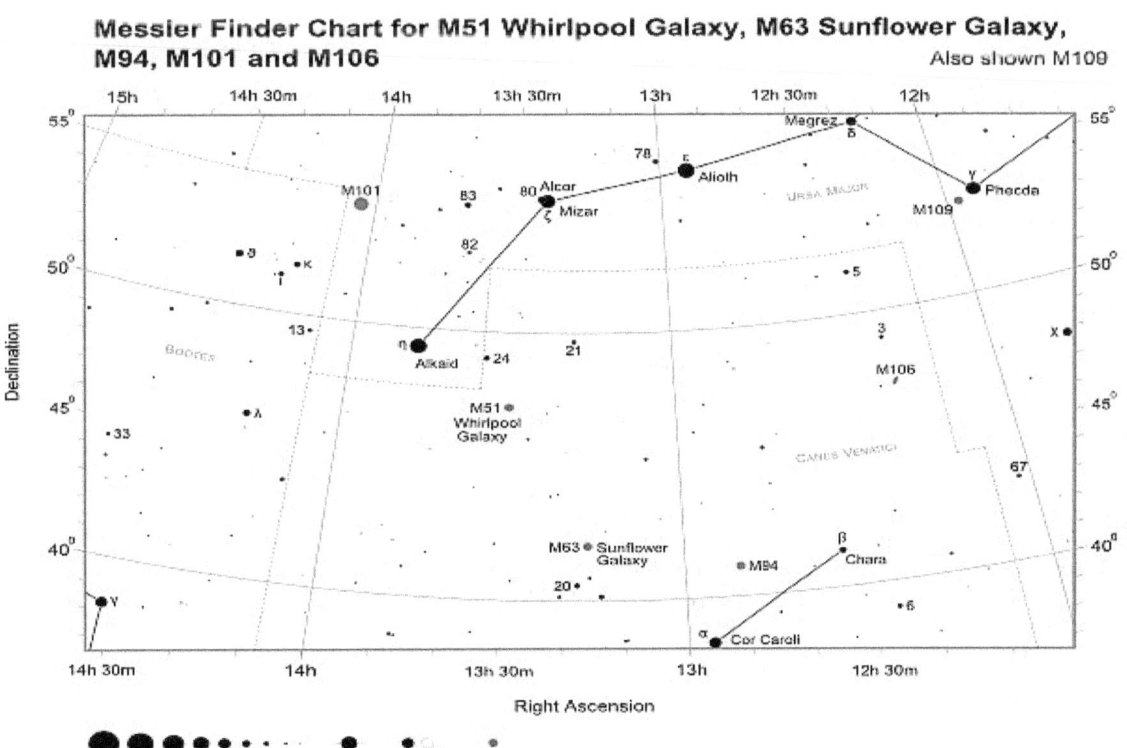

Messier 95 - M95 - Barred Spiral Galaxy

M95 is a barred spiral galaxy about 36 million light-years away in the constellation of Leo. It was discovered by Pierre Méchain on March 20, 1781 - the same night he discovered M96 - and catalogued by Charles Messier four days later. With an apparent magnitude of +10.3 it's visible in small telescopes. Together with M96 and M105, M95 forms a trio of faint gravitationally bound galaxies grouped close together. Of these, M96 is the brightest and the largest. The group is known as the Leo I or M96 group of galaxies, which also contains at least 21 other fainter galaxies and is one of many groups that lie within the Virgo Super cluster.

Charles Messier included M95 and M96 in his catalogue on the March 24, 1781. M105 was not included in the original Messier catalogue but added much later by Helen B. Sawyer Hogg in 1947. The galaxies are best seen during the months of March, April and May. The constellation of Leo the Lion is relatively bright and large. It contains one first magnitude star, Regulus (α Leo - mag. +1.4), which happens to be the brightest star in the surrounding region of sky. About 24 degrees east and two degrees north of Regulus is the third brightest star in Leo, Denebola (β Leo - mag. +2.1). Imagine a line connecting Regulus with Denebola. Just less than half way along this line are M95, M96 and M105. The northernmost member of the trio is M105 with M96 located 50 arc minutes south of M105 and M95 positioned 40 arc minutes west of M96.

These three galaxies are amongst the fainter objects in Messiers catalogue. With large 15x70 or 20x80 binoculars from a dark location they are visible as faint smudges. Each galaxy is of a different type; M95 is a barred spiral galaxy, M96 an intermediate spiral galaxy and M105 an elliptical galaxy. The brightest of the three and easiest to spot is M96. It shines at magnitude +9.6 with an apparent size of 8 x 5 arc minutes. At magnitude +10.3 and spanning 4 x 3 arc minutes, M95 is fainter and smaller than M96. The third member M105 shines at magnitude +9.8 and spans about 5 arc minutes in diameter. In medium sized telescopes of the order of 150mm (6-inch) or 200mm (8-inch), M95 appears as an oval patch of diffuse light with a bright core. On nights of good seeing its possible to notice the central bar structure and surrounding nebulosity. In the same low power telescope field of view is M96 with its bright oval shaped core visible. M105 is the least impressive of the three appearing as a small faint ball of fuzz.

M95 has a diameter of 46,000 light-years. For comparison the diameters of M96 and M105 are 80,000 and 55,000 light-years respectively. On March 16, 2012 a Type II supernova (SN 2012aw) was discovered in M95. It peaked at magnitude +12.7.

M95 Data Table

Messier	95	Apparent Mag.	10.3	Number of Stars	40 Billion
NGC	3351	RA (J2000)	10h 43m 58s		
Object Type	Barred Spiral galaxy	DEC (J2000)	11d 42m 13s	Notable Feature	M95 is a member of the Leo I or M96 group of galaxies
		Apparent Size (arcmins)	4.4 x 3.3		
Classification	SB(r)b				
Constellation	Leo	Radius (light years)	23,000		
Distance (kly)	36,000				

Messier # 095

Date:	Time:
Site:	
Temp:	Wind:	Hum:
Clouds:	Moon:
Scope:	
EP:	Mag:
NELM:	See/Trans:
Type:	# Stars:
Mag:	Age:
Const:	
Notes:	

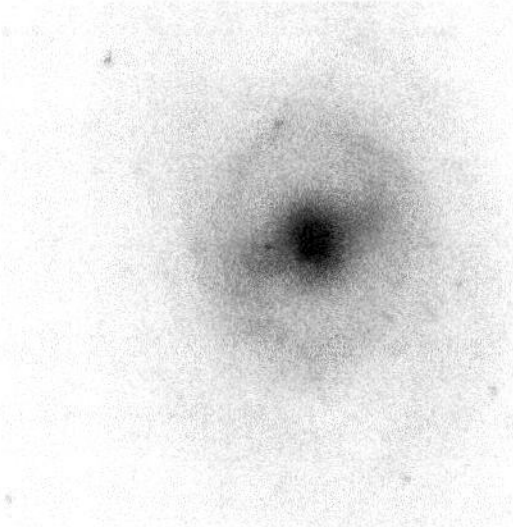

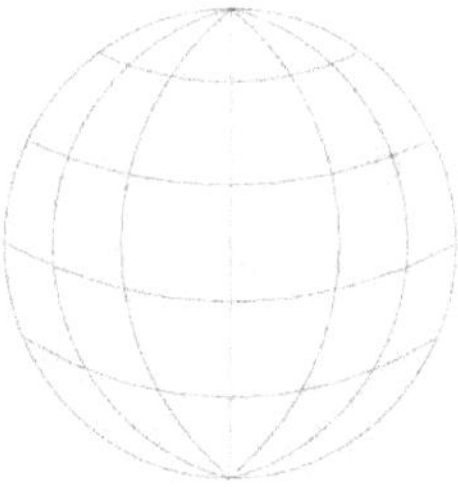

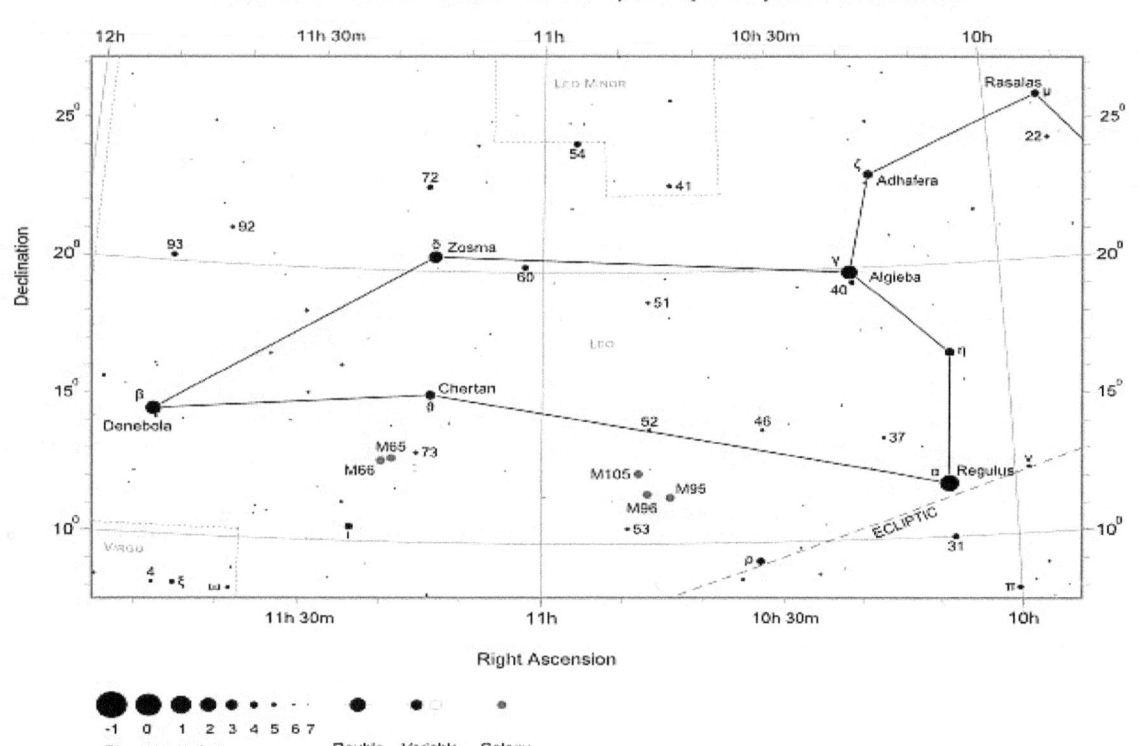

Messier Finder Chart for M65, M66, M95, M96 and M105

Declination / Right Ascension

Star magnitudes   -1 0 1 2 3 4 5 6 7
Double star   Variable stars   Galaxy

Messier 96 - M96 - Intermediate Spiral Galaxy

M96 is an intermediate spiral galaxy 35 million light-years distant in the constellation of Leo. At magnitude +9.6, it's the brightest member of the Leo I or M96 group of galaxies that also contains M95,M105 and at least another 21 fainter galaxies. The grouping is one of many that lie within the Virgo Super cluster. The three galaxies are amongst the fainter objects in Messiers catalogue but all are visible with large 15x70 or 20x80 binoculars from a dark site, appearing as faint smudges of light.M96 is an unusual galaxy in the sense that it has asymmetric arms and a displaced core that were probably caused by gravitational pulling from other nearby galaxies. It was discovered, along with M95, by Pierre Méchain on March 20, 1781. Charles Messier including both items in his catalogue four days later. M105 was not included in the original Messier catalogue but added much later by Helen Sawyer Hogg in 1947.

The galaxies are located in the southern middle section of the relatively large and bright constellation of Leo the Lion, which lies east of Cancer and to the west of Virgo. Leo contains one first magnitude star, Regulus (α Leo - mag. +1.4), which happens to be the brightest star in the surrounding region of sky. About 24 degrees east and two degrees north of Regulus is the third brightest star in Leo, Denebola (β Leo - mag. +2.1). Imagine a line connecting Regulus with Denebola with M95, M96 and M105 located just less than half way along this line. The northernmost member of the trio is M105 with M96 located 50 arc minutes south of M105 and M95 positioned 40 arc minutes west of M96.

They are best seen during the months of March, April and May.

It's difficult to notice much detail in M96 through a small 80mm (3.1-inch) telescope but it's possible, on nights of good seeing, to make out the oval shaped core. Larger apertures of 200mm (8-inch) or greater bring out more, including the bright core surrounded by wispy nebulosity that hints at the true spiral nature of the galaxy. In total, M96 spans 7.8 x 5.2 arcminutes of apparent sky.

In the same field of view as M96 is M95 which displays an oval patch of diffuse light with a bright core. The third galaxy M105 is the least impressive of the three, appearing only as a small faint ball of fuzz. Since each galaxy is of a different type - barred spiral galaxy M95, intermediate spiral galaxy M96 and elliptical galaxy M105 - it's interesting to compare how they appear through the eyepiece.

M96 is 80,000 light-years in diameter while M95 and M105 measure 46,000 and 55,000 light-years in diameter respectively. On May 9, 1998 a type Ia supernova (SN 1998bu) was discovered in M96, peaking at magnitude +11.8.

M96 Data Table

Messier	96
NGC	3368
Object Type	Intermediate Spiral galaxy
Classification	SAB(rs)ab
Constellation	Leo
Distance (kly)	35,000

Apparent Mag.	9.6
RA (J2000)	10h 46m 46s
DEC (J2000)	11d 49m 25s
Apparent Size (arcmins)	7.8 x 5.2
Radius (light years)	40,000

Number of Stars	100 Billion
Notable Feature	M96 is a member of the Leo I or M96 group of galaxies

Messier # 096

Date:	Time:
Site:	
Temp:	Wind:	Hum:
Clouds:	Moon:
Scope:	
EP:	Mag:
NELM:	See/Trans:
Type:	# Stars:
Mag:	Age:
Const:	
Notes:	

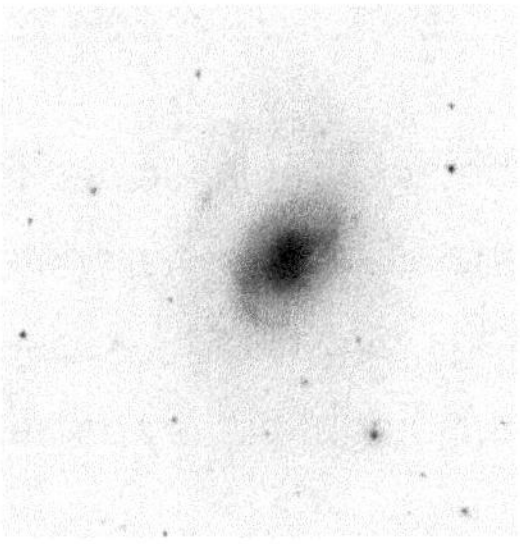

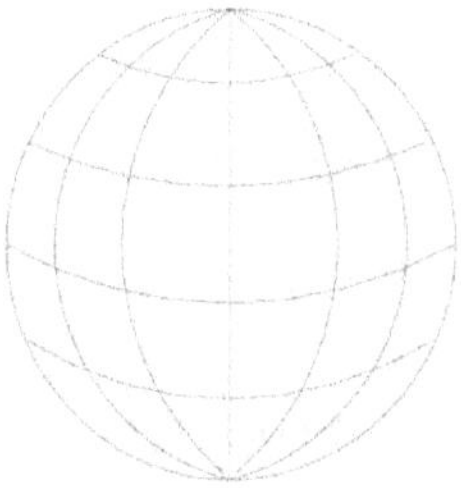

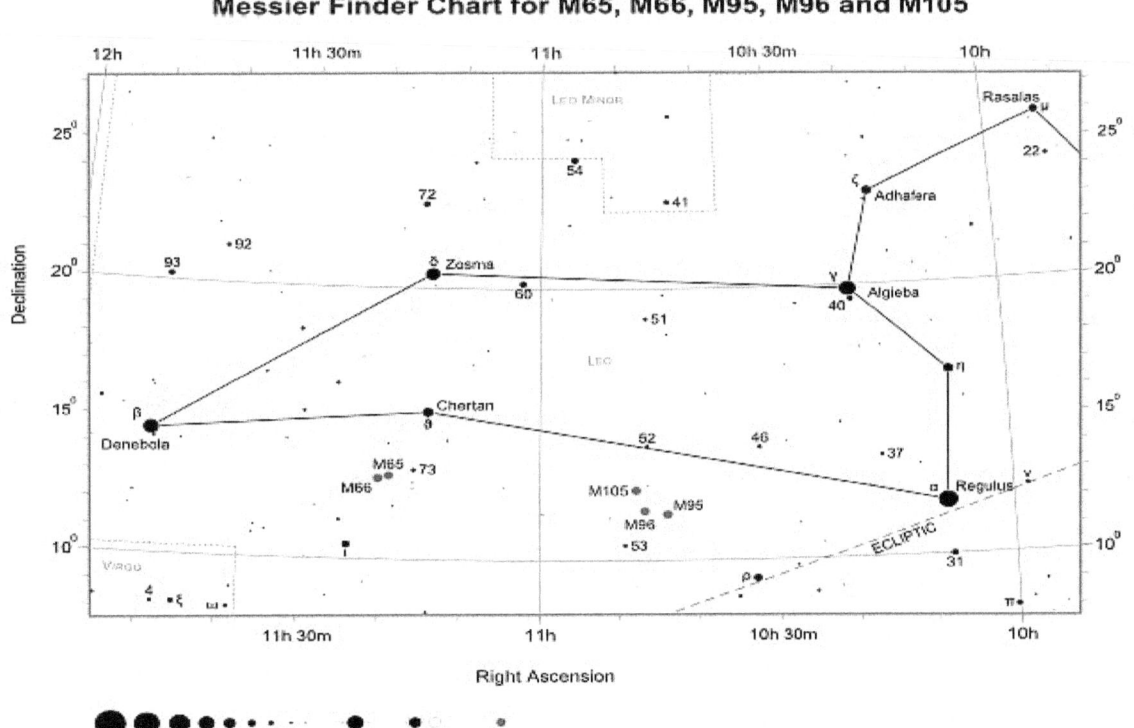

Messier Finder Chart for M65, M66, M95, M96 and M105

Star magnitudes -1 0 1 2 3 4 5 6 7

Double star   Variable stars   Galaxy

Right Ascension

Declination

Messier 97 - M97 - The Owl Nebula (Planetary Nebula)

M97 or "the Owl Nebula" is a famous planetary nebula located in the constellation of Ursa Major. It was discovered by Pierre Mechain on February 16, 1781 and is one of only four planetary nebulae listed in the Messier catalogue. The name Owl Nebula was first coined by William Parsons, 3rd Earl of Rosse who noticed owl-like "eyes" when observing the nebula in 1848. At magnitude +9.9, it's not particularly bright but is spectacular and regarded as one of the more complex examples of its type.

Locating M97 is easy; it's positioned only 2.5 degrees southeast of bright star Merak (β UMa - mag. +2.3), which forms the southwest corner of the bowl of the famous Plough or Big Dipper asterism of Ursa Major. In the same wide field telescope field of view, 50 arc minutes northwest of M97 is the barred spiral galaxy M108 (mag. +10.2).

The Owl Nebula is best seen from Northern Hemisphere latitudes during the months of March, April and May. From latitudes north of 35N it's circumpolar and therefore never sets.

Due to its low surface brightness the Owl Nebula is a challenging object for large binoculars and small telescope observers. It's visible in 20x80 binoculars and 100mm (4-inch) scopes but usually requires very dark skies and excellent observing conditions to be seen. It appears as nothing more than a dim circular disk or fuzzy ball without detail.

The famous eyes consist of two dark patches superimposed on the face of the nebula. Under good conditions a 200mm (8-inch) scope at high power can show the eyes, but normally a 250mm (10-inch) scope is required to see them. An ultra high contrast deep sky or light pollution filter may also help. M97's central star is of only 14th magnitude making it an elusive target in anything less than a 350mm (14-inch) telescope.

M97 has an apparent size of 3.4 x 3.3 arc minutes. At a distance of 2,600 light-years form Earth this equates to an actual diameter of 3 light-years. It's a fantastic deep sky object that's estimated to be 8,000 years old. Although not that bright, it's a worthy object on any observing list.

M97 Data Table

Messier	97
NGC	3587
Name	Owl Nebula
Object Type	Planetary nebula
Constellation	Ursa Major
Distance (kly)	2.6
Apparent Mag.	9.9
RA (J2000)	11h 14m 48s

DEC (J2000)	55d 01m 07s
Apparent Size (arcmins)	3.4 x 3.3
Radius (light years)	1.5
Notable Feature	Owl like "eyes" visible through larger amateur telescopes.

Messier # 097

Date:	Time:
Site:	
Temp:	Wind:	Hum:
Clouds:	Moon:
Scope:	
EP:	Mag:
NELM:	See/Trans:
Type:	# Stars:
Mag:	Age:
Const:	
Notes:	

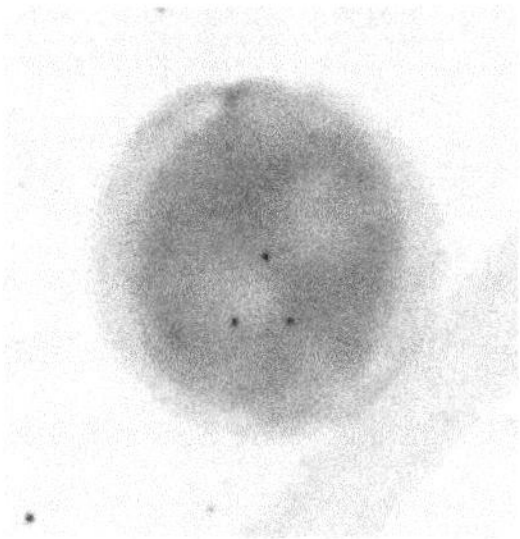

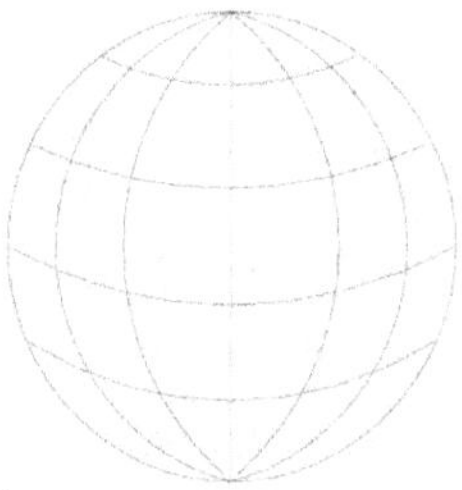

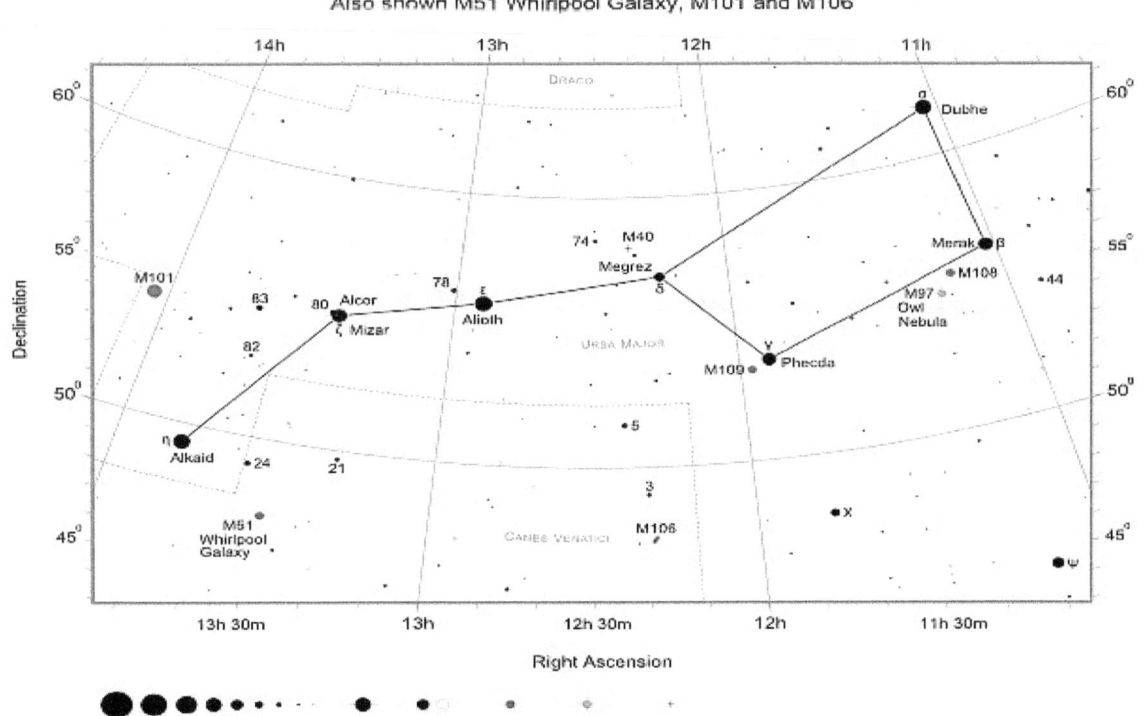

Messier Finder Chart for M40, M97 Owl Nebula, M108 and M109
Also shown M51 Whirlpool Galaxy, M101 and M106

Declination

Right Ascension

Star magnitudes   -1 0 1 2 3 4 5 6 7

Double star   Variable stars   Galaxy   Planetary nebula   Double star

Messier 98 - M98 - Spiral Galaxy

M98, mag. +10.4, is a large beautiful edge-on spiral galaxy that's located in the southern section of the constellation of Coma Berenices. It's one of the fainter Messier objects and to spot it a medium sized amateur telescope or greater is recommended. The galaxy is a member of the Virgo cluster and was discovered by Pierre Mechain on March 15, 1781. On the same night he also discovered M99 andM100. Messier catalogued them shortly afterwards and remarked that M98 was the faintest of the three.

M98 is one of a small group of galaxies that are blue shifted. The vast majority of galaxies are receding from us and display red shifts but due to the movement of M98 within the Virgo cluster, it's currently falling towards us, hence the blue shift. The galaxy is located about 57 Million light-years distant and has an apparent size of 9.8 x 2.8 arc minutes. This corresponds to an actual diameter of 160,000 light-years. It's estimated to contain 1 trillion stars.

Pinpointing the area of sky where M98 is located is easy. It's located 6 degrees east of Denebola (β Leo - mag. +2.1) the third brightest star in Leo. The star 6 Com (mag. +5.1) lies 0.5 degree east of M98 and acts as a perfect marker.

The Virgo cluster galaxies are best seen during the months of March, April and May.

M98 is one of the more difficult Virgo cluster Messier galaxies. Through a 200m (8-inch) scope it appears as a faint, thin, diffuse streak of light with a brighter core. The surface brightness of the galaxy is low, hence it's a tricky object to spot, especially when there is a small amount of light pollution present. For the same reason, it does not take high magnification well. Larger backyard telescopes reveal a slightly curved structure that's brighter on the southern side with a greenish tinge and a distinct nucleus.

About 750 million years ago, M98 may have interacted with the large spiral galaxy NGC 4254. Today, they are separated by a distance of about 1.3 million light-years.

M98 Data Table

Messier	98		DEC (J2000)	14d 54m 00s
NGC	4192		Apparent Size (arcmins)	9.8 x 2.8
Object Type	Spiral galaxy			
Classification	SAB(s)ab		Radius (light years)	80,000
Constellation	Coma Berenices		Number of Stars	1 Trillion
Distance (kly)	57000		Notable Feature	Member of the Virgo Cluster of galaxies
Apparent Mag.	10.4			
RA (J2000)	12h 13m 48s			

Messier # 098

Date:	Time:
Site:	
Temp:	Wind:	Hum:
Clouds:	Moon:
Scope:	
EP:	Mag:
NELM:	See/Trans:
Type:	# Stars:
Mag:	Age:
Const:	
Notes:	

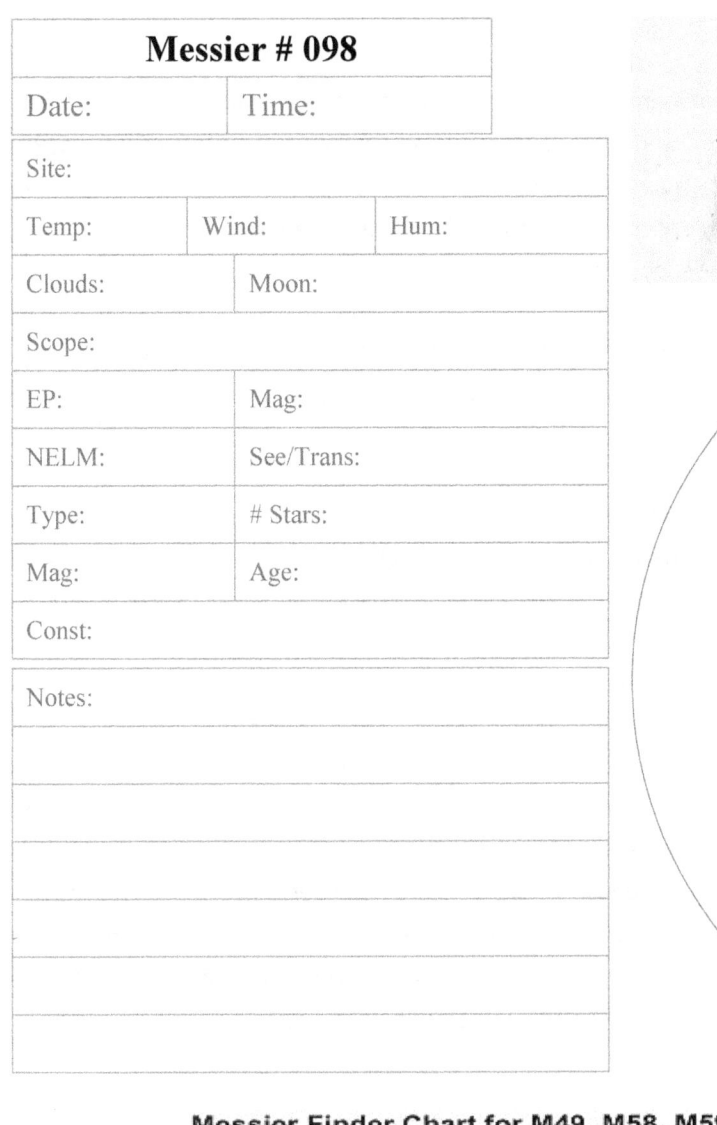

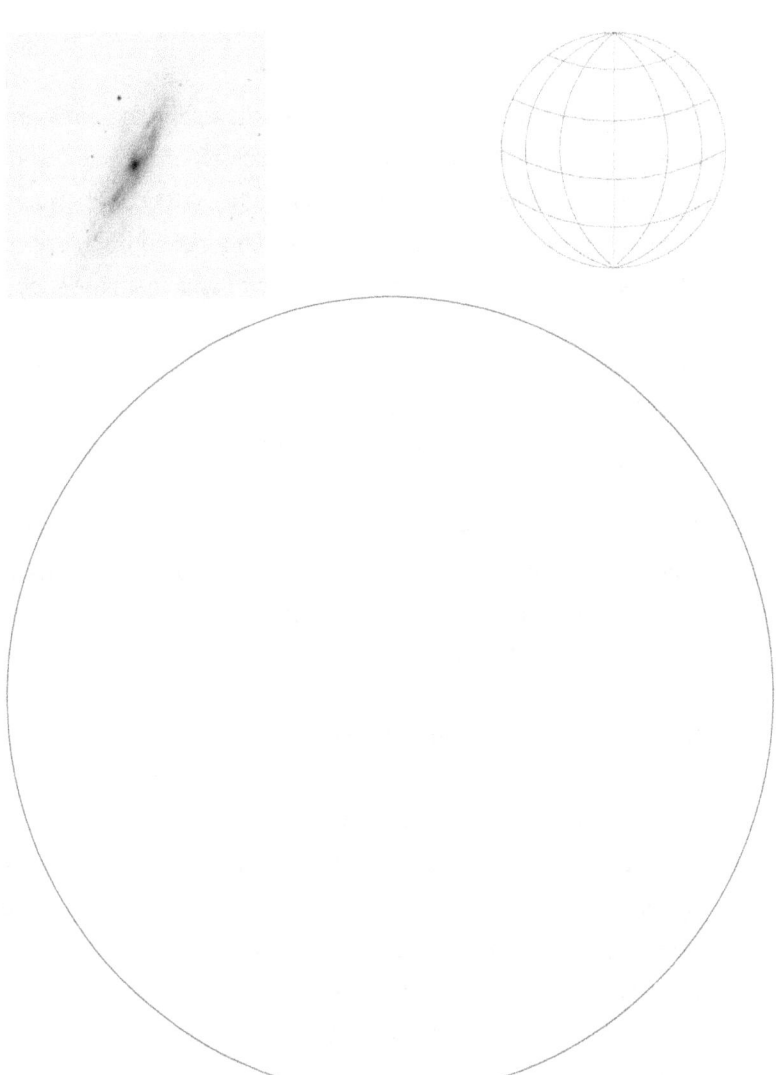

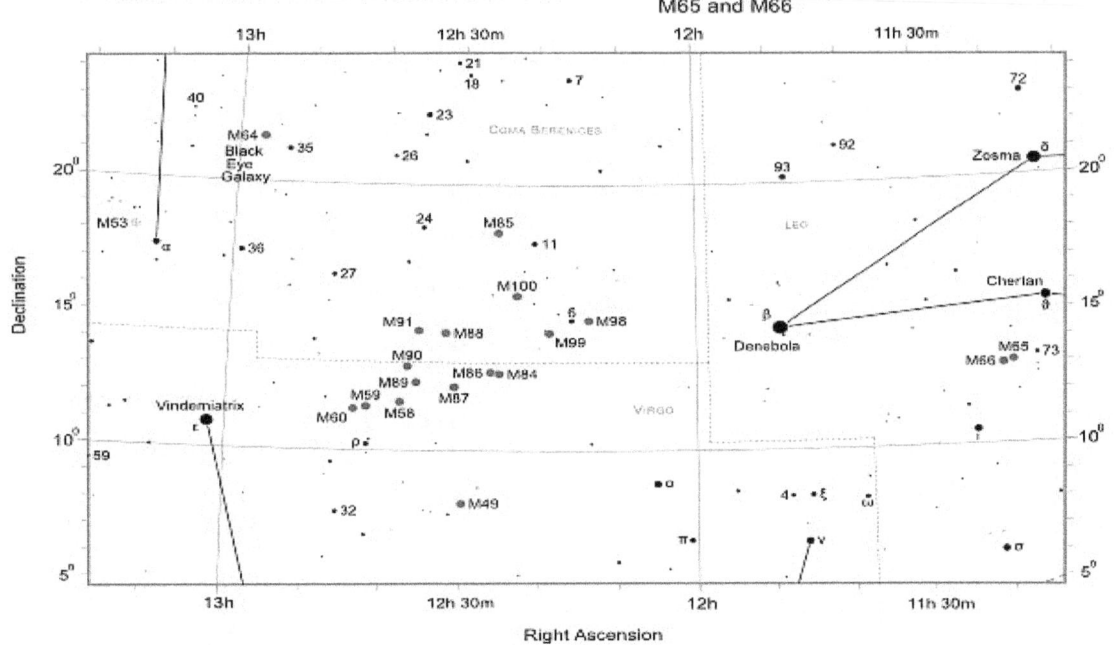

Messier Finder Chart for M49, M58, M59, M60, M84, M85, M86, M87, M88, M89, M90, M91, M98, M99 and M100   Also shown M53, M64 Black Eye Galaxy, M65 and M66

Messier 99 - M99 - Spiral Galaxy

M99 is a magnitude +10.2 spiral galaxy situated in the southern part of the constellation of Coma Berenices. It's a beautiful object that's a member of the Virgo cluster of galaxies and appears almost face-on from our perspective. M99 was discovered by Pierre Mechain on March 15, 1781, on the same night he also discovered M98 and M100. The discoveries were then reported to Charles Messier, who measured the positions before adding them to his catalogue on April 13, 1781. This was just prior to the release of the third and final published edition.

The 3rd Earl of Rosse, William Parsons, first identified the spiral structure of M99 in 1846 using his 72-inch (1.83 m) reflecting telescope at Birr Castle in Ireland. The galaxy was one of the first to have its structure identified. At the time, Rosse was using the World's largest optical telescope.

M99 is located 55 Million light-years from Earth. It covers 5.3 x 4.6 arc minutes of apparent sky, which corresponds to a spatial diameter of 85,000 light-years. The galaxy is positioned 7 degrees east of bright star Denebola (β Leo - mag. +2.1) and just less than 1 degree southeast of star 6 Com (mag. +5.1). Tenth magnitude edge-on spiral galaxy M98 lies 0.5 degrees west of 6 Com.

M99 is visible through small telescopes as a faint roundish glow with a brighter core. Of course aperture helps and it's much easier to detect with a medium sized telescope or larger. Through a 200mm (8-inch) scope the galaxy appears as a fuzzy patch of light with a noticeably brighter centre. On dark nights, the outer haze hints at the spiral structure, which is easily seen at high powers through 250mm (10-inch) scopes. Even larger instruments bring out finer details such as dust bands and subtle details on the spiral surface. It's a testament to the quality of today amateur scopes that the spiral structure can be easily seen, compared to the enormous sized instrument required by Lord Rosse for the same task. While not classified as a starburst galaxy, there is a high activity of star formation occurring in M99. To date, four supernovae have been recorded in the galaxy. They are SN 1967H (mag. +14), SN 1972Q (mag +15.6), SN 1986I (mag +14) and SN 2014L (mag +15.4).

M99 and the other Virgo cluster galaxies are best seen during the months of March, April and May.

M99 Data Table

Messier	99
NGC	4254
Object Type	Spiral galaxy
Classification	SA(s)c
Constellation	Coma Berenices
Distance (kly)	55000
Apparent Mag.	10.2
RA (J2000)	12h 18m 50s

DEC (J2000)	14d 25m 01s
Apparent Size (arcmins)	5.3 x 4.6
Radius (light years)	42,500
Number of Stars	150 Billion
Notable Feature	Three Supernovae have been observed in this galaxy.

Messier # 099

Date:	Time:

Site:	
Temp:	Wind:	Hum:
Clouds:	Moon:
Scope:	
EP:	Mag:
NELM:	See/Trans:
Type:	# Stars:
Mag:	Age:
Const:	

Notes:

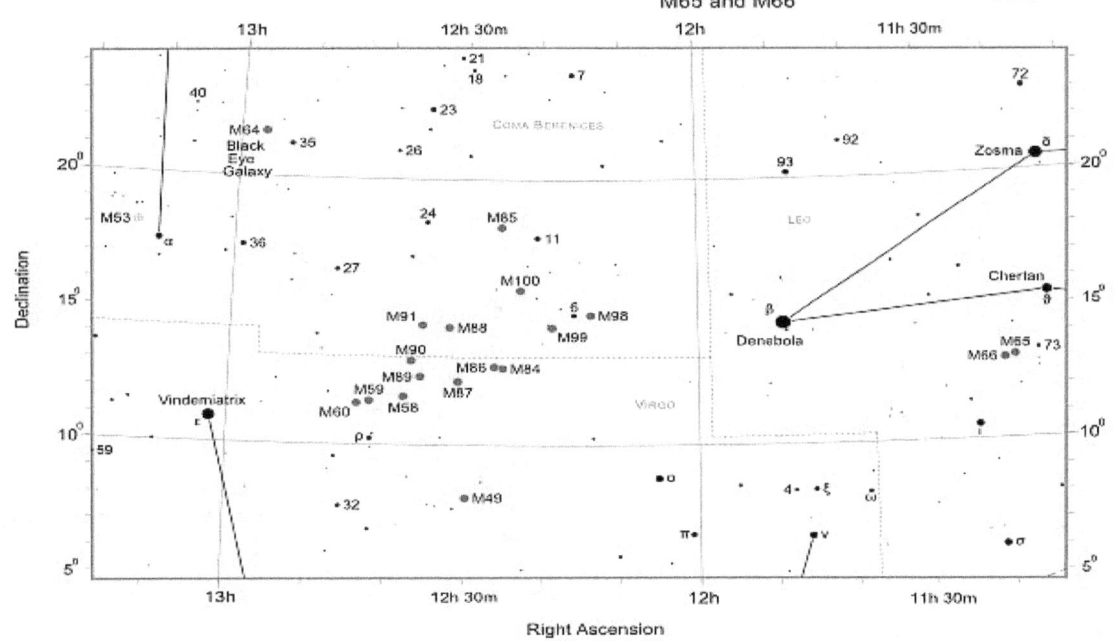

Messier Finder Chart for M49, M58, M59, M60, M84, M85, M86, M87, M88, M89, M90, M91, M98, M99 and M100

Also shown M53, M64 Black Eye Galaxy, M65 and M66

Declination

Right Ascension

Star magnitudes
-1 0 1 2 3 4 5 6 7

Double star   Variable stars   Galaxy   Globular cluster

Messier 100 - M100 - Spiral Galaxy

M100, mag. +9.5, is a spiral galaxy located in the southern part of constellation of Coma Berenices. It's one of the brightest members of the Virgo cluster of galaxies appearing almost face-on from our perspective. M100 exhibits prominent well-defined spiral arms and is therefore regarded as an example of a grand design spiral galaxy; other notable galaxies that fall into this category are M51, M74, M81and M101.

M100 was discovered - along with M98 and M99 - by Pierre Méchain on March 15, 1781. Charles Messier subsequently observed all three objects and added them to his catalogue on April 13, 1781. He described the galaxy as faint without stars. It was not until 1850 that the spiral nature of M100 was first detected. Ango-Irish astronomer William Parsons the 3rd Earl of Rosse was the person to achieve this. He included M100 in a list of 14 spiral nebulae he had observed.

Finding the area of sky where M100 is positioned is not so difficult once one is familiar with the location of Virgo cluster. The centre of the cluster is located close to super giant elliptical galaxy M87 about halfway along a line connecting Denebola (β Leo - mag. +2.1) with Vindemiatrix (ε Vir - mag. +2.8). M100 is positioned towards the northern section of the group, 2 degrees southeast of star 11 Com (mag. +4.7).

The Virgo cluster galaxies are best seen during the months of March, April and May.

At mag. +9.5, M100 is within the range of 7x50 or 10x50 binoculars. However, since the galaxy appears face it suffers from low surface brightness and therefore a difficult object. It's easier to spot through larger 11x80/20x80 binoculars or small telescopes, where it appears as a faint hazy patch of light with an uneven texture. A medium size 200mm (8-inch) scope reveals a bright core surrounding by an envelope of shady nebulosity. With large amateur instruments some dust structure is visible but only on dark nights of excellent seeing. In total, M100 covers 7.5 x 6.1 arc minutes of apparent sky although it appears visually smaller.

The galaxy is located 57.5 Million light-years distant. It has an actual spatial diameter of 125,000 light-years and is estimated to contain 400 Billion stars. Two satellite galaxies - NGC 4323 and NGC 4328 - are present within M100.

To date, five supernovae have been observed in M100. They are SN 1901B (mag. +15.6 - March 1901), SN 1914A (mag. +15.7 - Feb 1914), SN 1959E (mag. +17.5 - Aug 1959), SN 1979C (mag. +11.6 - April, 1979) and SN 2006X (mag. +15.3).

M100 Data Table

Messier	100	DEC (J2000)	15d 49m 21s
NGC	4321	Apparent Size (arcmins)	7.5 x 6.1
Object Type	Spiral galaxy	Radius (light years)	62,500
Classification	SAB (s)6c	Number of Stars	400 Billion
Constellation	Coma Berenices	Notable Feature	One of the first spiral galaxies to be discovered.
Distance (kly)	57500		
Apparent Mag.	9.5		
RA (J2000)	12h 22m 55s		

Messier # 100

Date:	Time:

Site:

Temp:	Wind:	Hum:

Clouds:	Moon:

Scope:

EP:	Mag:
NELM:	See/Trans:
Type:	# Stars:
Mag:	Age:
Const:	

Notes:

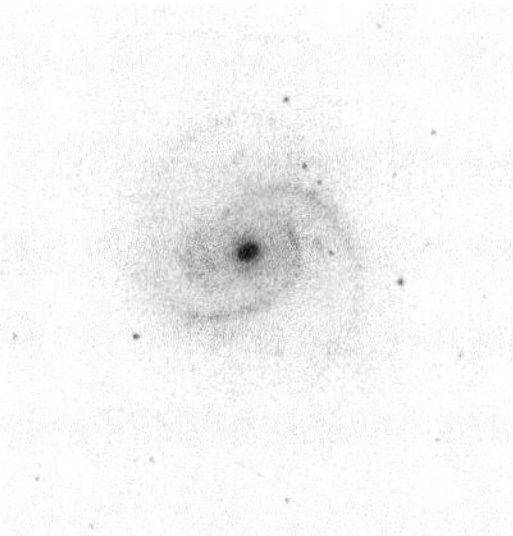

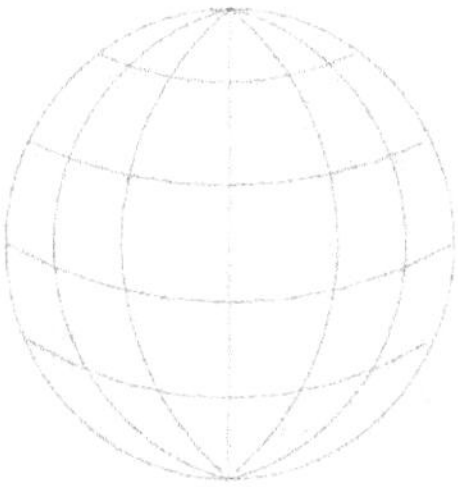

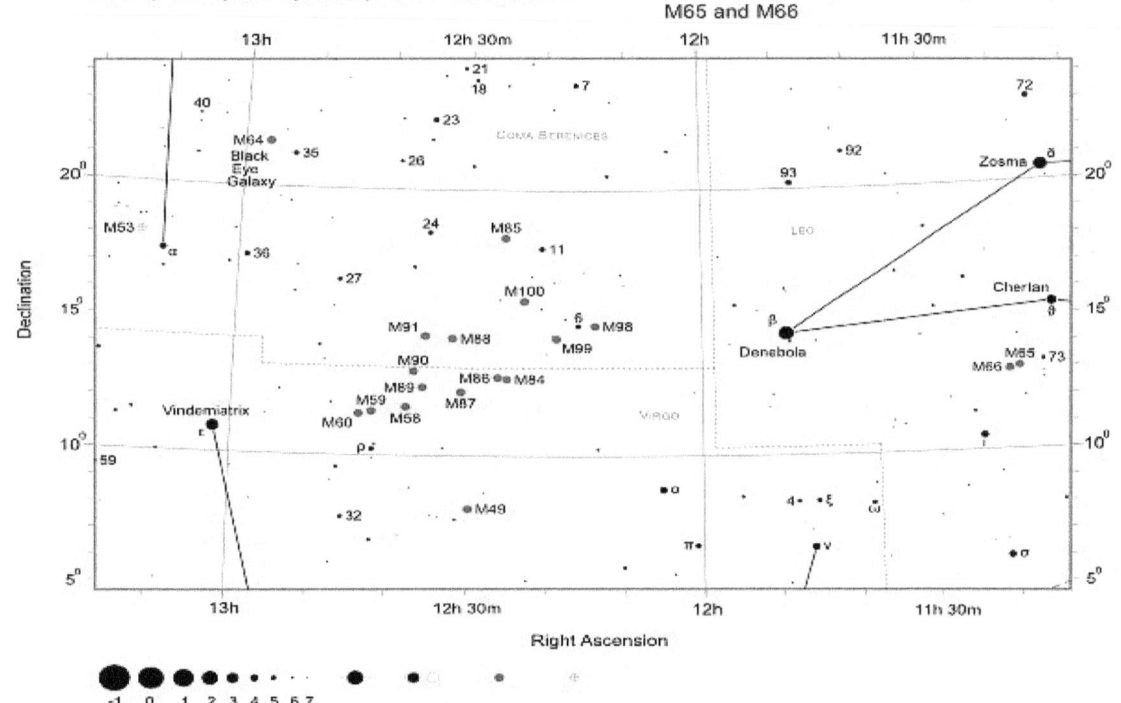

Messier Finder Chart for M49, M58, M59, M60, M84, M85, M86, M87, M88, M89, M90, M91, M98, M99 and M100

Also shown M53, M64 Black Eye Galaxy, M65 and M66

Right Ascension

Declination

Star magnitudes

Double star Variable stars Galaxy Globular cluster

Messier 101 - M101 - The Pinwheel Galaxy (Spiral Galaxy)

M101 is a large face-on spiral galaxy located 22 million light-years away in the constellation of Ursa Major. At mag. +7.9, it can be glimpsed in binoculars or small telescopes from dark sites but suffers from low surface brightness and in bad seeing conditions or light polluted areas, the galaxy can be difficult to spot even with a 200mm (8-inch) scope. It's best seen from the Northern Hemisphere during the months of March, April and May.

M101 is also known as "The Pinwheel Galaxy" and was discovered by Pierre Méchain on March 27, 1781. He described it as "nebula without star, very obscure and pretty large, 6' to 7' in diameter, between the left hand of Bootes and the tail of the great Bear." He communicated this to Charles Messier who verified its position and then included it in his catalogue as one of the final entries.

Locating the part of sky where M101 is positioned is easy since it's close to the handle of the bowl that forms the "Plough" or "Big Dipper" asterism of Ursa Major. The Pinwheel galaxy is located at one corner of an equatorial triangle formed with second magnitude stars Mizar (ζ UMa - mag. +2.2) and Alkaid (η UMa - mag. +1.8). M101 is 5.5 degrees east of Mizar (the celebrated naked eye double star) and 5.5 degrees northeast of Alkaid.

The low surface brightness of M101 is a combination of its large apparent size - 29 x 27 arc minutes - and face-on appearance when viewed from our line of sight. In this case, and also for other similar large face-on galaxies, dark moonless skies are a must! In such seeing conditions M101 can even be spotted with 7x50 or 10x50 binoculars as a large featureless dim patch of light. A small 80mm (3.1-inch) telescope displays a large nebulous haze with a brighter centre. On good nights using a 200mm (8-inch) scope it's possible to observe the bright-condensed core surrounded by a halo of nebulosity, which includes several knotty patches and fills an entire low power field of view. It may be possible to make out a weak spiral shape.

M101 is an extremely large galaxy with a diameter of 180,000 light-years; almost double that of our Milky Way galaxy. It contains about 1 trillion stars and is the brightest member of a group of at least 9 galaxies, called the M101 Group. There have been 4 recorded supernovae in M101; SN 1909A, SN 1951H, SN 1970G and SN 2011fe with the last one discovered on August 24, 2011. This type of supernova was the brightest of the four reaching magnitude +9.9, making it was visible with small telescopes.

M101 Data Table

Messier	101		**Apparent Mag.**	7.9
NGC	5457		**RA (J2000)**	14h 03m 12s
Name	Pinwheel Galaxy		**DEC (J2000)**	54d 20m 55s
Object Type	Spiral galaxy		**Apparent Size (arcmins)**	28.8 x 26.9
Classification	SAB (rs) cd		**Radius (light years)**	90,000
Constellation	Ursa Major		**Number of Stars**	1 Trillion
Distance (kly)	22,000			

Messier # 101

Date:	Time:	
Site:		
Temp:	Wind:	Hum:
Clouds:	Moon:	
Scope:		
EP:	Mag:	
NELM:	See/Trans:	
Type:	# Stars:	
Mag:	Age:	
Const:		

Notes:

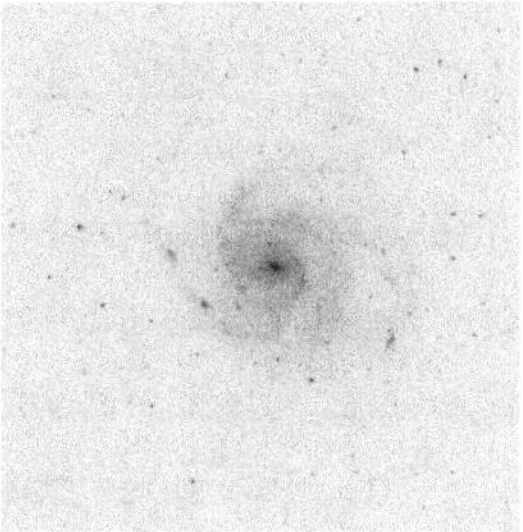

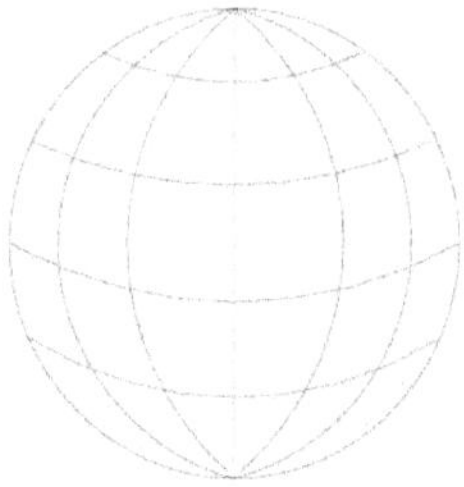

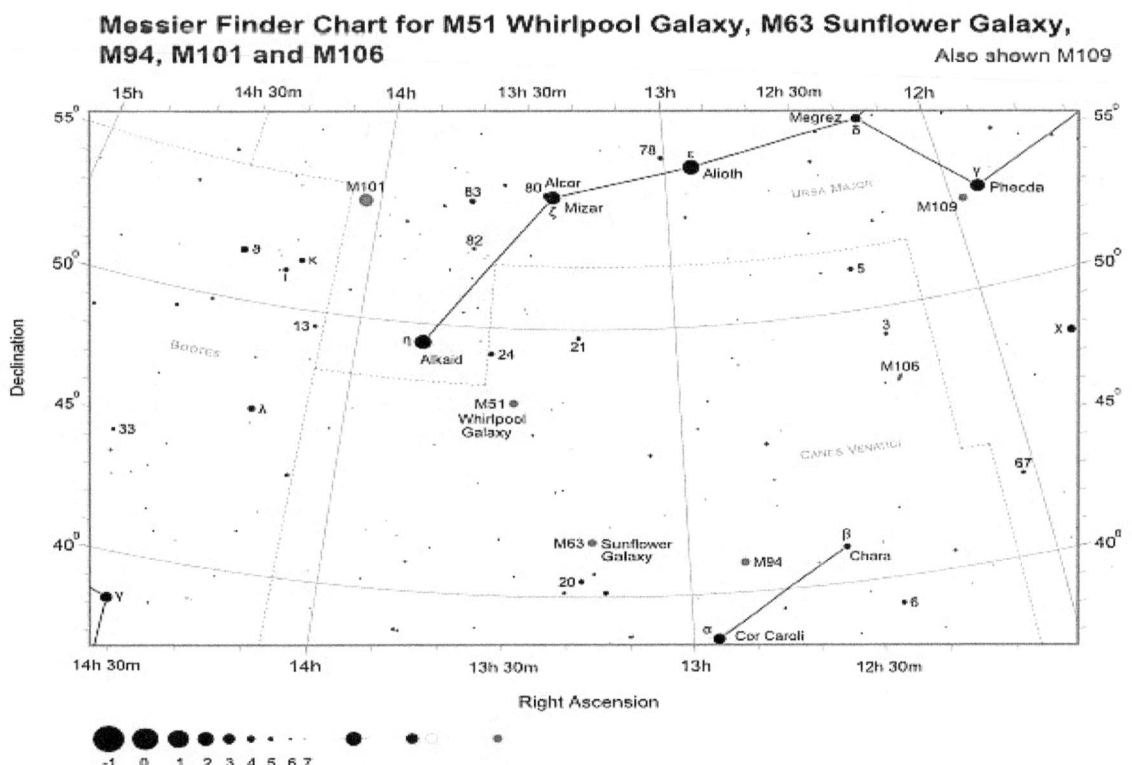

Messier Finder Chart for M51 Whirlpool Galaxy, M63 Sunflower Galaxy, M94, M101 and M106

Also shown M109

Star magnitudes: -1 0 1 2 3 4 5 6 7

Double star   Variable stars   Galaxy

Messier 102 - M102 - NGC 5866 - Spindle Galaxy (Lenticular Galaxy)

M102 is a galaxy catalogued by Charles Messier that hasn't been explicitly identified. It was Pierre Mechain who made the original observation in March 1781 before passing the information onto Messier, who catalogued it without verification. However, Mechain himself believed M102 was an error and wrote a letter on the May 6, 1783 expressing his view that the object was in fact a duplicate entry of M101. The story does not end there, historical evidence based on Messier description of the galaxy combined with its co-ordinates suggest that M102 could well be lenticular galaxy NGC 5866, also known as the Spindle galaxy. A number of other possible candidates have been suggested but it seems both Messier and Mechain have observed NGC 5866 in the past and therefore we list it as the missing item.

The Spindle galaxy (mag. +9.9) is located at the southern edge of the far northern constellation of Draco. It's positioned four degrees southwest of star Iota Draconis (ι Dra - mag. +3.3). Directly west of NGC 5866 are the seven stars that form the famous "Plough" or "Big Dipper" asterism of Ursa Major.

NGC 5866 is a challenging binocular object but easier to spot with small scopes. It's best seen from the Northern Hemisphere during the months of April, May and June. From latitudes of 35N or greater, the galaxy is circumpolar and therefore never sets.

A small 80mm (3.1-inch) telescope under dark skies shows NGC 5866 as a thin saucer shape smudge of nebulosity that spans 4.7 x 1.9 arc minutes. Through a 150mm (6-inch) or 200mm (8-inch) scope the galaxy appears as a halo of greenish tinged light that hints at the dark dust lane. It has a well-defined bright central core with larger scopes revealing more intriguing details. Photographically NGC 5866 is a beautiful sight with the almost edge on dusk disk spectacular.

The Spindle galaxy is located 50 million light-years distant and has a radius of 70,000 light-years. It's estimated to contain about 100 billion stars.

M102 Data Table

Messier	102		**RA (J2000)**	15h 06m 29s
NGC	5866		**DEC (J2000)**	55d 45m 47s
Object Type	Lenticular Galaxy		**Apparent Size (arcmins)**	4.7 x 1.9
Classification	S0		**Radius (light years)**	35,000
Constellation	Draco		**Number of Stars**	100 Billion
Distance (kly)	50000		**Notable Feature**	Disputed Messier object
Apparent Mag.	9.9			

Messier # 102

Date:	Time:

Site:

Temp:	Wind:	Hum:

Clouds:	Moon:

Scope:

EP:	Mag:
NELM:	See/Trans:
Type:	# Stars:
Mag:	Age:
Const:	

Notes:

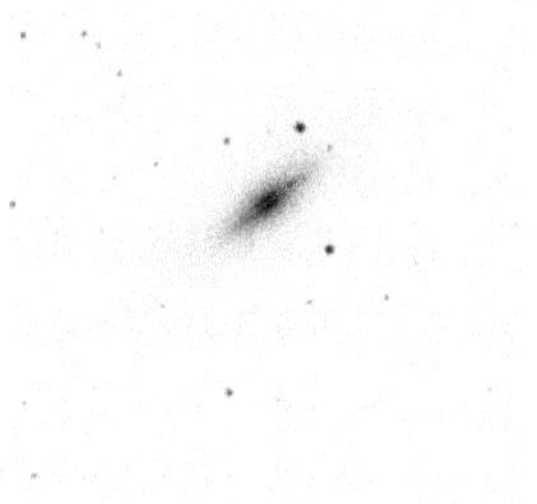

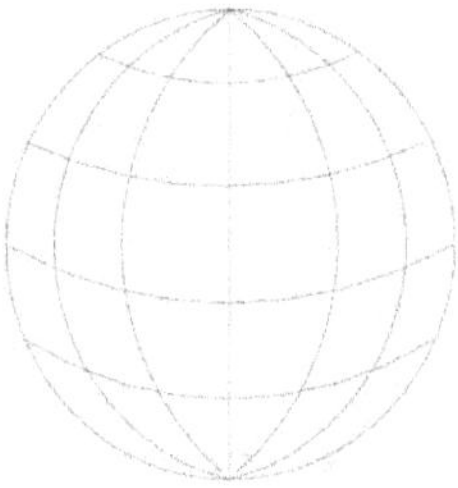

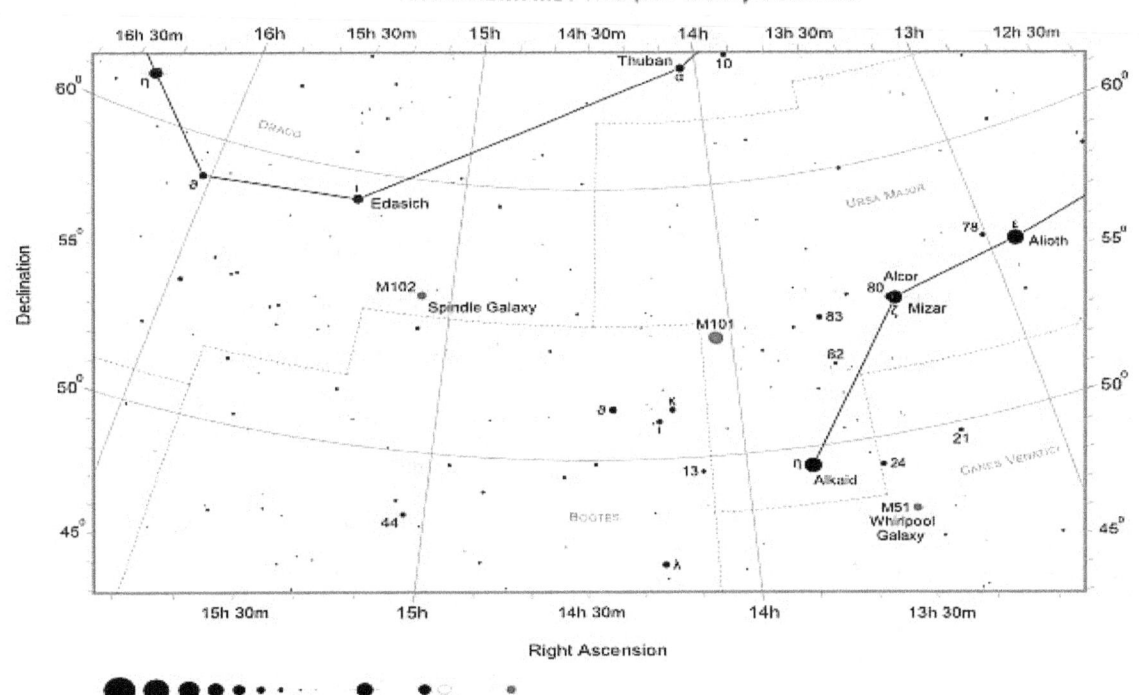

Messier Finder Chart for M102 Spindle Galaxy
Also shown M51 Whirlpool Galaxy and M101

Messier 103 - M103 - Open Cluster

M103 is a small loose but sparkling open cluster of at least 40 stars located amongst the Milky Way star fields of Cassiopeia. At magnitude +7.4, it's beyond naked eye visibility but an easy binocular target, appearing somewhat like a fan shaped hazy patch of light spanning about 6 arc minutes across.

M103 was discovered by Pierre Méchain in either March or April of 1781 and subsequently reported to Charles Messier. Normally, Messier would take the opportunity to observe the newly discovered object himself but on this occasion he didn't, probably due to lack of time or unfavorable observing conditions. This cluster along with M101 and M102 were the last objects added by Messier to his catalogue with objects M104 to M110 added much later during the 20th century.

M103 is easy to find as it's located on the eastern side of the well-known Cassiopeia "W" shape. Its positioned 1 degree east of Ruchbah (δ Cas - mag. 2.7) and almost along the line connecting Ruchbah with epsilon Cas (ε Cas - mag. 3.4). Situated nearby are a number of other open clusters, including NGC 654, NGC 659 and NGC 663. The latter is occasionally confused with M103.Located some 10,000 light years away and spanning 17.5 light years, this wide spaced cluster is easy to find and identify with binoculars where it appears as a fan or wedge-shaped diffuse patch of light. Through a 100mm (4-inch) telescope the brightest four stars of the cluster are resolvable looking somewhat like the Greek letter lambda (λ), with averted vision revealing a nebulous triangle shaped patch of light that extends beyond the brightest stars.

One star in particular, Struve 131 dominates the scene. It's a 7th magnitude multiple star that's easily split in telescopes of 100mm (4-inch) aperture. However, it's not actually a true member of the cluster; it's much closer to us and just happens to be just in the line of sight. The brightest members of M103 are of magnitude +10.5 and at the centre of the cluster lies a prominent red giant star. With larger scopes, the cluster is not so easy to identify due to its looseness and is easy to confuse with other star groups or clusters in the vicinity. Of course, larger telescopes will show many fainter member stars. A 200mm (8-inch) scope at about 100x magnification reveals a dozen or so stars with the brighter stars anchoring the points of the wedge. Still larger sized amateur scopes reveal at least 20 stars.

Its estimated that M103 is about 25 million years old. It's one of the more distant open clusters in the Messier catalogue and is best seen during the Northern Hemisphere winter months when it appears high in the sky. From latitudes greater than 30N it's circumpolar and therefore never sets.

M103 Data Table

Messier	103
NGC	581
Object Type	Open Cluster
Constellation	Cassiopeia
Distance (kly)	10.0
Apparent Mag.	7.4
RA (J2000)	01h 33m 22s

DEC (J2000)	60d 39m 29s
Apparent Size (arcmins)	6.0 x 6.0
Radius (light years)	8.75
Age (years)	25M
Number of Stars	>40
Other Name	Collinder 14

Messier # 103

Date:	Time:
Site:	
Temp:	Wind:	Hum:
Clouds:	Moon:
Scope:	
EP:	Mag:
NELM:	See/Trans:
Type:	# Stars:
Mag:	Age:
Const:	

Notes:

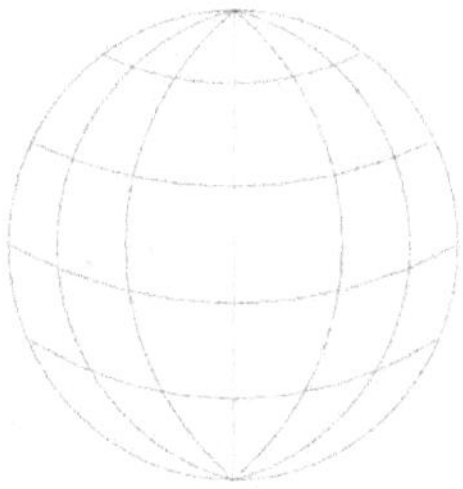

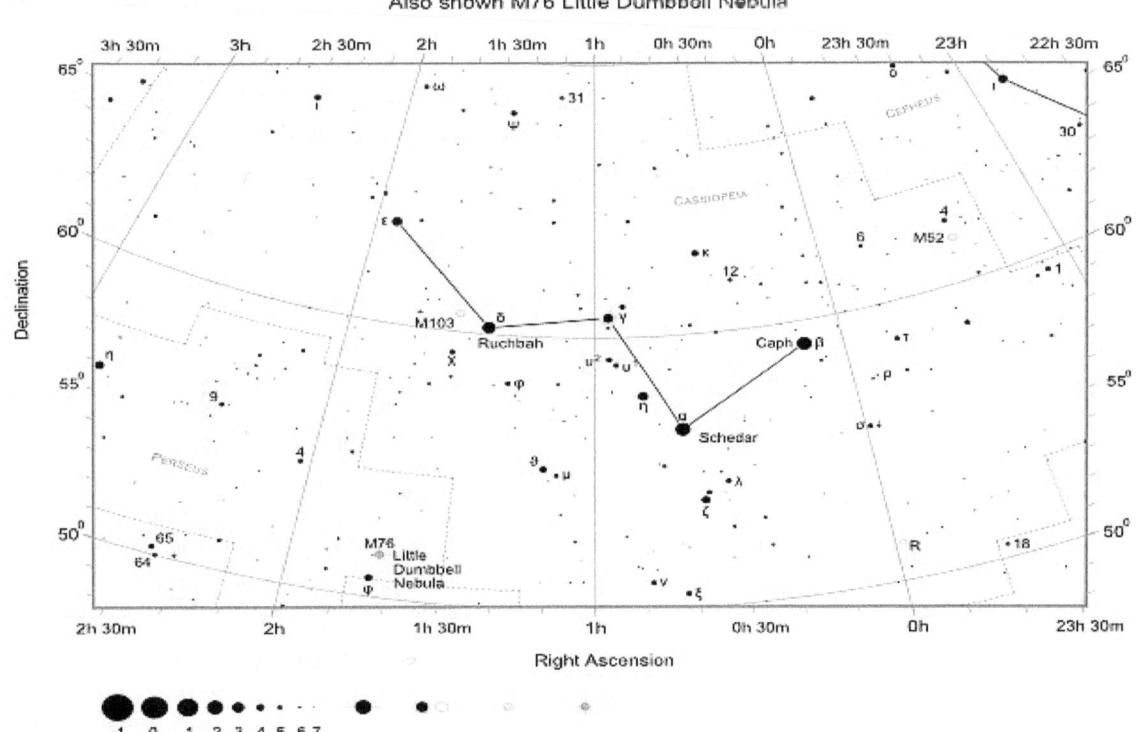

Messier Finder Chart for M52 and M103
Also shown M76 Little Dumbbell Nebula

Messier 104 - M104 - Sombrero Galaxy (Spiral Galaxy)

M104 more commonly known as the Sombrero Galaxy is a spectacular, almost edge-on, spiral galaxy located in Virgo. At magnitude +8.4, the Sombrero appears in binoculars as a small patch of nebulosity. Its most striking feature - visible in medium/large sized amateur scopes - is a ring of thick dust that encapsulates the bulge of the galaxy, giving the appearance of a Sombrero hat. Many astronomers regard M104 as the finest of all galaxies in Virgo. Pierre Méchain discovered M104 on May 11, 1781. A couple of years later he described the galaxy in a letter to Johann Bernoulli and subsequently it was published in the Berliner Astronomisches Jahrbuch (Berlin Astronomy Year Book). Charles Messier made hand-written notes about this and five other objects but none were included in the final published catalogue version. The Sombrero Galaxy was finally added to the "official" catalogue in 1921 with the other five "missing" items (M105 to M109) added a few years later.

M104 was one of the first galaxies to have its spectra and velocity measured by Vesto Slipher in 1912. He noted that the object was red shifted and therefore receding from us, the current accepted rate being 900 km/s. Slipher's red shift calculation of M104 along with similar observations from other galaxies pointed towards an expanding Universe, hence providing a key piece of evidence for the Big Bang Theory. The Sombrero Galaxy is located very close to the constellation boundary between Virgo and Corvus. It's positioned 11.5 degrees directly west of Spica (α Vir - mag. +1.0) and 6 degrees northeast of delta Crv (δ Crv - mag. +2.9). The stars chi Vir (χ Vir - mag. +4.7), psi Vir (ψ Vir - mag. +4.8) and 21 Vir (mag. +5.5) form a faint naked-eye triangle just north of M104.

It's best seen during the months of March, April and May.

Through 7x50 or 10x50 binoculars, M104 is visible under dark skies as a small, round patch of nebulosity. The galaxy is easy to observe with almost any telescope. A 100mm (4-inch) scope under good seeing conditions will hint at a dark band within the structure. When seen through larger 200mm (8-inch) scopes, M104 appears as a flattened saucer with a large bulge bisected by a dark lane. Even larger amateur instruments of the order of 250mm (12-inch) aperture reveal a bright nucleus, extensive halo and the dark dusk lane extending outwards on both sides. On good nights, the Sombrero Galaxy is a spectacular sight.

M104 covers 8.6 x 4.2 arc minutes of apparent sky. It's located 30 million light-years distant, which corresponds to an actual diameter of 75,000 light-years. The galaxy is estimated to contain about 100 billion stars. It contains a reasonable large number of globular clusters, as many as 2,000. M104 is also the dominant member of a small group of galaxies known as the M104 group or NGC 4594 group.

M104 Data Table

Messier	104		**Apparent Mag.**	8.4
NGC	4594		**RA (J2000)**	12h 39m 59s
Name	Sombrero Galaxy		**DEC (J2000)**	-11d 37m 23s
Object Type	Spiral galaxy		**Apparent Size (arcmins)**	8.6 x 4.2
Classification	Sa (s) a		**Radius (light years)**	37,500
Constellation	Virgo		**Number of Stars**	100 Billion
Distance (kly)	30000			

Messier # 104

Date:	Time:

Site:

Temp:	Wind:	Hum:

Clouds:	Moon:

Scope:

EP:	Mag:
NELM:	See/Trans:
Type:	# Stars:
Mag:	Age:
Const:	

Notes:

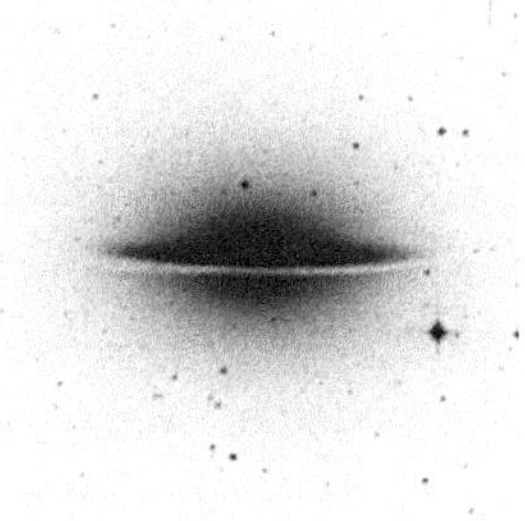

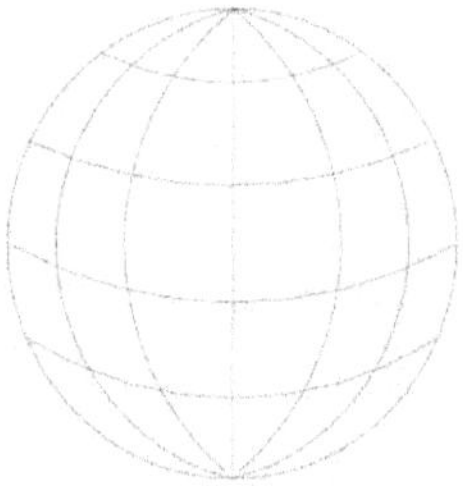

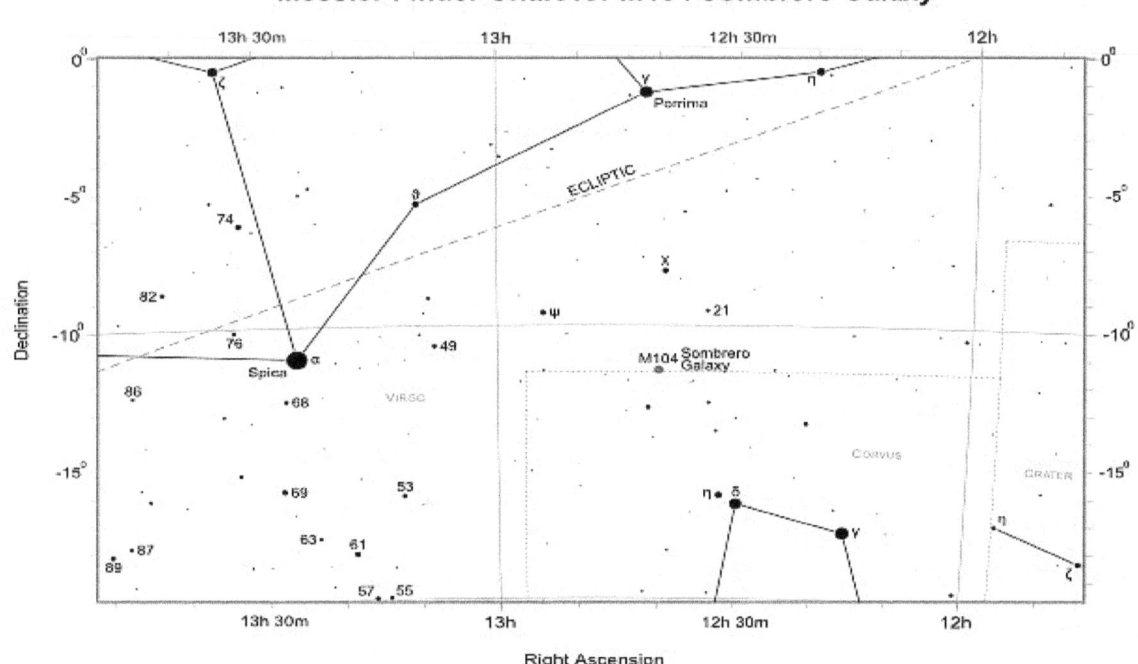

Messier Finder Chart for M104 Sombrero Galaxy

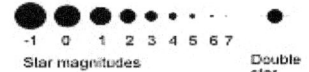

Star magnitudes

-1 0 1 2 3 4 5 6 7

Double star   Variable stars   Galaxy

Messier 105 - M105 - Elliptical Galaxy

M105 (mag. +9.8) is an elliptical galaxy located in the constellation Leo that's visible with small telescopes. It was discovered by Pierre Méchain on March 24, 1781, which was three days before he discovered M101. However, due to unknown reasons the galaxy is one of several not included in Charles Messier's final published list. It was eventually added to the list in 1947 by Helen Sawyer Hogg, together with M106 and M107. William Herschel independently rediscovered M105 on March 11, 1784.

M105 is the brightest elliptical member of the Leo I or M96 group of galaxies. This grouping is one of many that lie within the Virgo Super cluster and includes M95, M96 and at least another 21 fainter members. M105 is located 35 Million light-years distant and is known to contain a super massive black hole at its centre.

Galaxies M95, M96 and M105 are located in the southern section of the middle part of Leo. Imagine a line connecting Regulus (α Leo - mag. +1.4) the brightest star in Leo with Denebola (β Leo - mag. +2.1) the constellations third brightest star. Denebola is positioned about 24 degrees east and a couple of degrees north of Regulus. Located just short of half way along this line are M95, M96 and M105. The northernmost member of the trio is M105 with M96 located 50 arc minutes south of M105 and M95 positioned 40 arc minutes west of M96.

The galaxies are best seen during the months of March, April and May.

Since it's only 10th magnitude in brightness, M105 is a very challenging binocular object. It's easier to spot with large binoculars and when seen through a small 80mm (3.1-inch) scope it appears as a faint diffuse patch of light that's small and round. A larger 200mm (8-inch) scope fairs better with M105 appearing larger and brighter without detail. It has an apparent diameter of about 5 arc minutes but is visually unimpressive when compared with the spirals in Leo. Being an elliptical galaxy, M105 has a high surface brightness, but doesn't show any detail regardless of the size of telescope used.

In the same field of view as M105 - for owners of large scopes - are faint galaxies NGC 3384 and NGC 3389. The three galaxies form a small triangle with NGC 3384 a mag. +10.9 lenticular galaxy and NGC 3389 a mag. +12.4 spiral galaxy, although spotting NGC 3389 requires at least a 250mm (10-inch) scope.

M105 has an actual diameter of 54,000 light-years and is estimated to contain 40 billion stars.

M105 Data Table

Messier	105
NGC	3379
Object Type	Elliptical galaxy
Classification	E1
Constellation	Leo
Distance (kly)	35,000
Apparent Mag.	9.8
RA (J2000)	10h 47m 50s

DEC (J2000)	12d 34m 53s
Apparent Size (arcmins)	5.3 x 4.8
Radius (light years)	27,000
Number of Stars	40 Billion
Notable Feature	Member of the Leo I or M96 group of galaxies

Messier # 105

Date:	Time:

Site:

Temp:	Wind:	Hum:

Clouds:	Moon:

Scope:

EP:	Mag:
NELM:	See/Trans:
Type:	# Stars:
Mag:	Age:
Const:	

Notes:

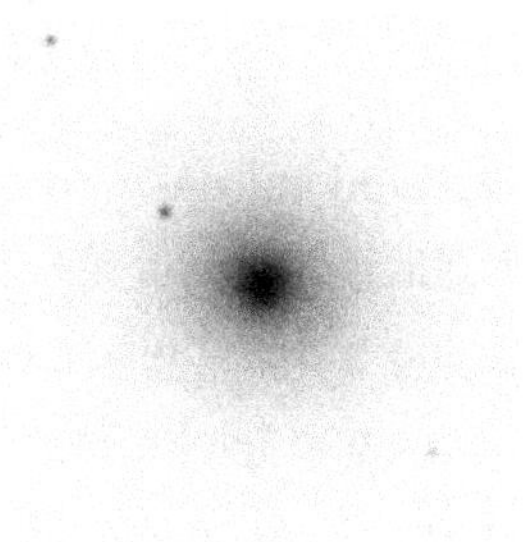

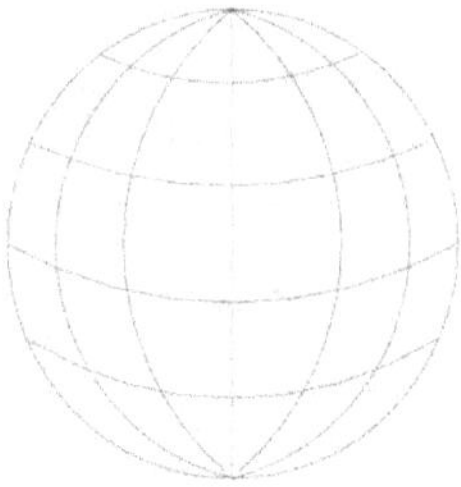

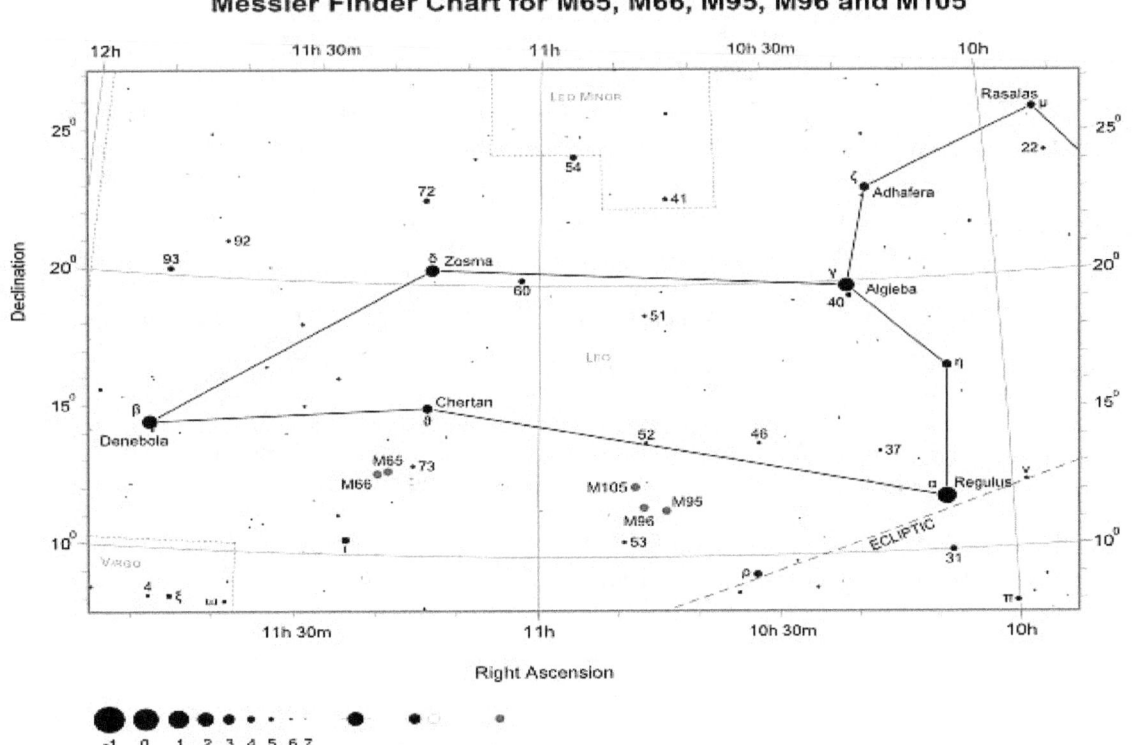

Messier Finder Chart for M65, M66, M95, M96 and M105

Right Ascension

Messier 106 - M106 - Spiral Galaxy

M106, mag. +8.5, is a large spiral galaxy located in Canes Venatici that was discovered by Pierre Méchain in July 1781. He described the galaxy in little detail; referring to it only as a nebula close to star 3 CVn. William Herschel then rediscovered it on March 9, 1788. Since Herschel was using a better telescope than Méchain he was able to see much more detail and noted it as "very brilliant with a bright nucleus and faint milky branches north preceding and south following." Although not one of Messiers original catalogue entries, M106 was included, along with M105 and M107 in 1947 by Helen Sawyer Hogg. It seemed reasonable to assume that Méchain had already intended to add these objects to a future edition.

M106 is one of the brightest examples of a Seyfert type II galaxy and is therefore strong in X-rays and unusual emission lines, which are believed to result from sections of the galaxy falling into the super massive black hole located at the centre. American astronomer Carl Seyfert first identified this class of object in 1943.

The galaxy is located towards the northwestern corner of Canes Venatici; a faint constellation with only one star Cor Caroli (α CVn - mag. +2.9) that's brighter than magnitude +4.0. However, locating M106 is not difficult as the Plough or Big Dipper asterism of Ursa Major is positioned just to the north and can be used as a starting point. Once found, focus on Megrez (δ UMa - mag. +3.2) the faintest star of the Plough. Positioned 5.5 degrees south and slightly east of Megrez is 5 CVn (mag. +4.8). M106 is located just over 4 degrees south of 5 CVn with star 3 CVn (mag. +5.3) positioned along the line connecting the two.

M106 is best seen from the Northern Hemisphere during the months of March, April and May. From southern temperate latitudes it's a difficult object as it never rises very high above the northern horizon.

M106 covers 19 x 7 arc minutes of apparent sky. Despite being spread over such a large area, it has a remarkably high surface brightness and therefore a relatively easy binocular target, appearing as a faint smudge. A 80mm (3.1-inch) telescope reveals it as a diffuse streak of light with a slightly brighter core that looks like a galaxy. A larger 200mm (8-inch) scope enhances the view with more subtle details visible including dusty markings and a faint outer halo of nebulosity. The largest of amateur scopes reveal the galaxy's spiral shape.

Located at a distance of 25 million light-years from Earth, M106 is intrinsically large with a diameter of 135,000 light years. It contains at least 400 billion stars. In August 1981 a 16th magnitude supernova (1981K) was observed.

M106 Data Table

Messier	106
NGC	4258
Object Type	Spiral galaxy
Classification	SAB (s)bc
Constellation	Canes Venatici

Distance (kly)	25,000
Apparent Mag.	8.5
RA (J2000)	12h 18m 57s
DEC (J2000)	47d 18m 15s
Apparent Size (arcmins)	18.6 x 7.2

Radius (light years)	67,500
Number of Stars	>400 Billion
Notable Feature	Seyfert II galaxy

Messier # 106

Date:	Time:

Site:		
Temp:	Wind:	Hum:
Clouds:	Moon:	
Scope:		
EP:	Mag:	
NELM:	See/Trans:	
Type:	# Stars:	
Mag:	Age:	
Const:		

Notes:

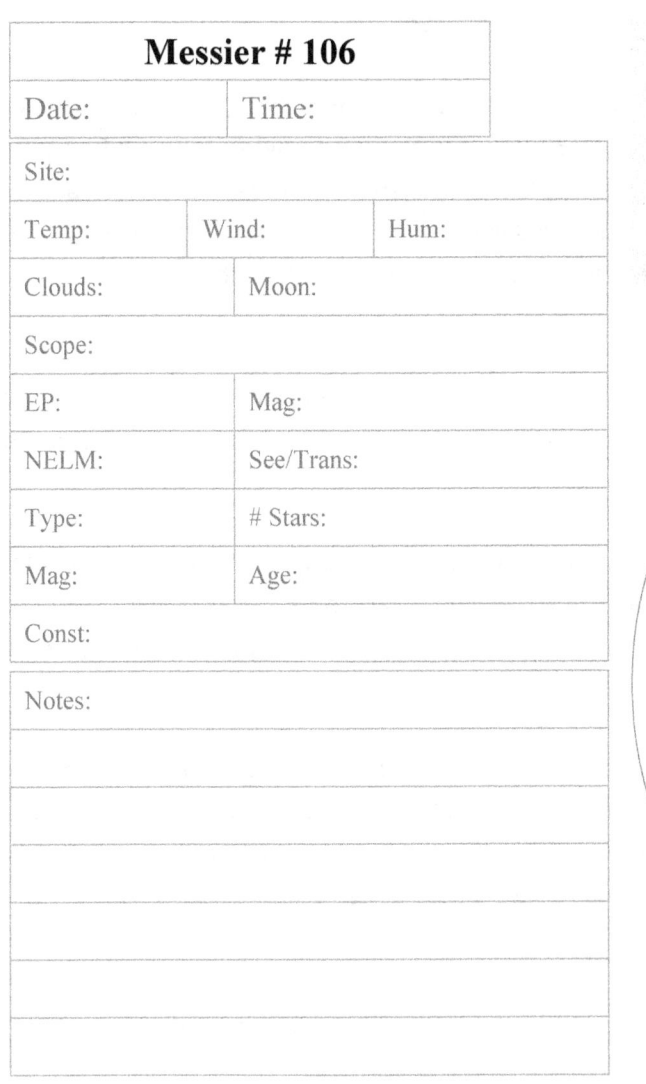

Messier Finder Chart for M51 Whirlpool Galaxy, M63 Sunflower Galaxy, M94, M101 and M106

Also shown M109

Messier 107 - M107 - Globular Cluster

M107 is a loose eight-magnitude globular cluster located in Ophiuchus that's a difficult binocular object but much easier in small telescopes. It was discovered by Pierre Méchain in April, 1782 and then independently re-discovered by William Herschel on May 12, 1793. Herschel was also the first person to resolve M107 into stars. This cluster is one of the additional catalogue items that weren't included in Messier's final version but added much later by Helen Sawyer Hogg in 1947. She also added M105 and M106 since it seems probable that Méchain had also intended to include these items to a future edition of the catalogue.

The globular is located 20,900 light-years from Earth and spans 13 arc minutes of apparent sky, which corresponds to a spatial diameter of 80 light-years. It contains 100,000 stars and has an estimated age of 13.95 billion years, making it one of the oldest known globulars.

Locating M107 is relatively easy as its positioned 2.75 degrees southwest of zeta Oph (ζ Oph - mag. +2.5). This star is located 16 degrees north of Antares (α Sco - mag. +1.0) and can be also found by imagining a line connecting Yed Prior (δ Oph - mag. +2.7) with Yed Posterior (ε Oph - mag. +3.2) and extending it in a south-easterly direction for 9 degrees. The cluster is best observed during the months of May, June and July.

At magnitude +8.0, M107 appears as a faint small diffuse object through a small 80mm (3.1-inch) telescope. There is a subtle increase in brightness from the outer to the inner region but no particularly bright centre. A 150mm (6-inch) scope will start to show the brighter stars around the outer edges but the view is much better with larger telescopes. Instruments of 300mm (12-inch) aperture or greater reveal an incredible site, many stars visible across the entire face of the globular. What's also noticeable is the looseness of this cluster.

At least 25 known variable stars have been identified in M107. It seems to contain some regions obscured by darkness, which is unusual for globular clusters.

M107 Data Table

Messier	107
NGC	6171
Object Type	Globular cluster
Constellation	Ophiuchus
Distance (kly)	20.9
Apparent Mag.	8.0
RA (J2000)	16h 32m 32s
DEC (J2000)	-13d 03m 10s
Apparent Size (arcmins)	13 x 13
Radius (light years)	40
Age (years)	13,950M
Number of Stars	100,000

Messier # 107

Date:	Time:	
Site:		
Temp:	Wind:	Hum:
Clouds:	Moon:	
Scope:		
EP:	Mag:	
NELM:	See/Trans:	
Type:	# Stars:	
Mag:	Age:	
Const:		

Notes:

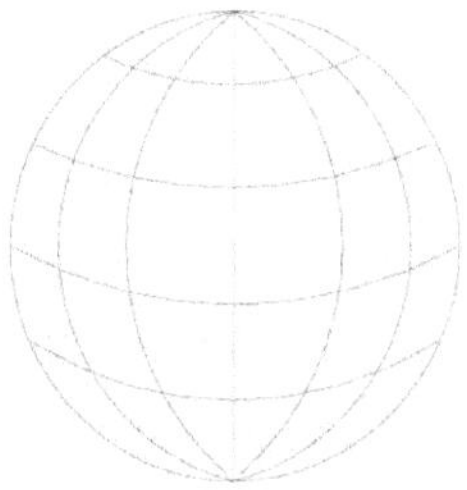

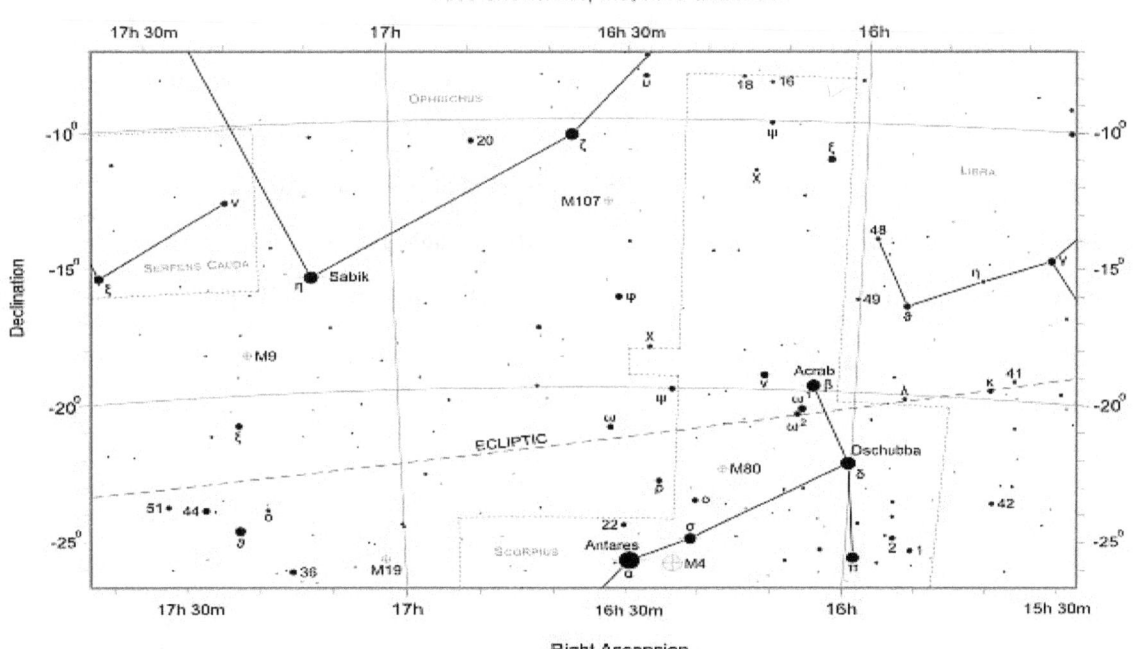

Messier Finder Chart for M107
Also shown M4, M9, M19 and M80

17h 30m
17h
16h 30m
16h
15h 30m
-10°
-15°
-20°
-25°
Declination
Right Ascension
Ophiuchus
Libra
Serpens Cauda
Scorpius
Sabik
M107
M9
M80
M4
M19
Acrab
Dschubba
Antares
ECLIPTIC
Star magnitudes
-1 0 1 2 3 4 5 6 7
Double star
Variable stars
Globular cluster

Messier 108 - M108 - Barred Spiral Galaxy

M108 is a nice edge on barred spiral galaxy located in Ursa Major that was discovered by Pierre Méchain on February 19, 1781. It's not one of the objects included by Messier in his final published catalogue version but was added much later by Owen Gingerich in 1953. This was based on analysis of notes written by Messier and Méchain that referenced M108 suggesting that the object was intended for inclusion in a later version. William Herschel independently rediscovered M108 on April 17, 1789.

Locating M108 is easy since it's positioned only 1.5 degrees southeast of bright Merak (β UMa - mag. +2.3) the southwest corner star of the bowl of the famous Plough or Big Dipper asterism of Ursa Major. Located 50 arc minutes southeast of M108 is the planetary nebula "Owl Nebula" (M97) and both items fit easily in the same wide field telescope field of view.

M108 is best seen from Northern Hemisphere latitudes during the months of March, April and May. For observes located at latitudes greater than 35N, the galaxy is circumpolar and therefore never sets.

M108 shines at apparent magnitude +10.2 and since it's aligned almost face-on the galaxy exhibits a high surface brightness. Therefore, it can be spotted with a small 80mm (3.1-inch) telescope; appearing as a faint strongly elongated streak of light with a slightly brighter central region. A larger 200mm (8-inch) scope reveals a well-defined thin needle structure that displays a mottled dusty complexion with subtle variations in brightness. M108 is a galaxy that can withstand high magnifications and is somewhat similar in appearance to the brighter galaxy M82.

In total, M108 spans 8.6 x 2.4 arc minutes of apparent sky and is located 45 million light-years from Earth. This corresponds to an actual diameter of 110,000 light-years. It's estimated to contain 400 billion stars and the galaxy is believed to be an isolated member of the Ursa Major Cluster of galaxies.

A type II supernova (1969B) was observed in M108 on January 23, 1969, peaking at magnitude +13.9.

M108 Data Table

Messier	108		RA (J2000)	11h 11m 31s
NGC	3556		DEC (J2000)	55d 40m 24s
Object Type	Barred Spiral galaxy		Apparent Size (arcmins)	8.6 x 2.4
Classification	SB(s)cd		Radius (light years)	55,000
Constellation	Ursa Major		Number of Stars	400 Billion
Distance (kly)	45000			
Apparent Mag.	10.2			

Messier # 108

Date:	Time:
Site:	
Temp:	Wind:	Hum:
Clouds:	Moon:
Scope:	
EP:	Mag:
NELM:	See/Trans:
Type:	# Stars:
Mag:	Age:
Const:	
Notes:	

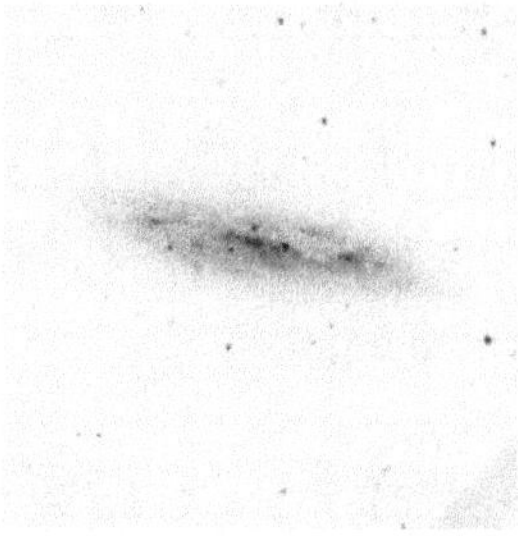

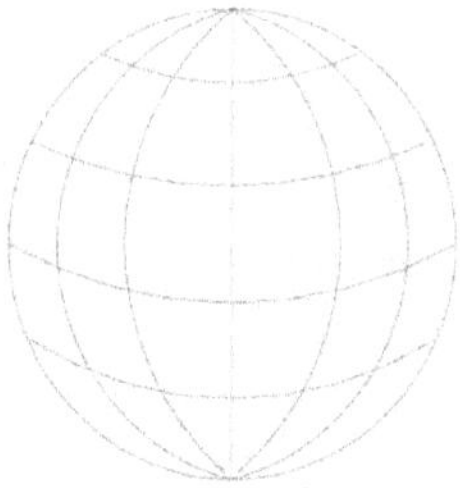

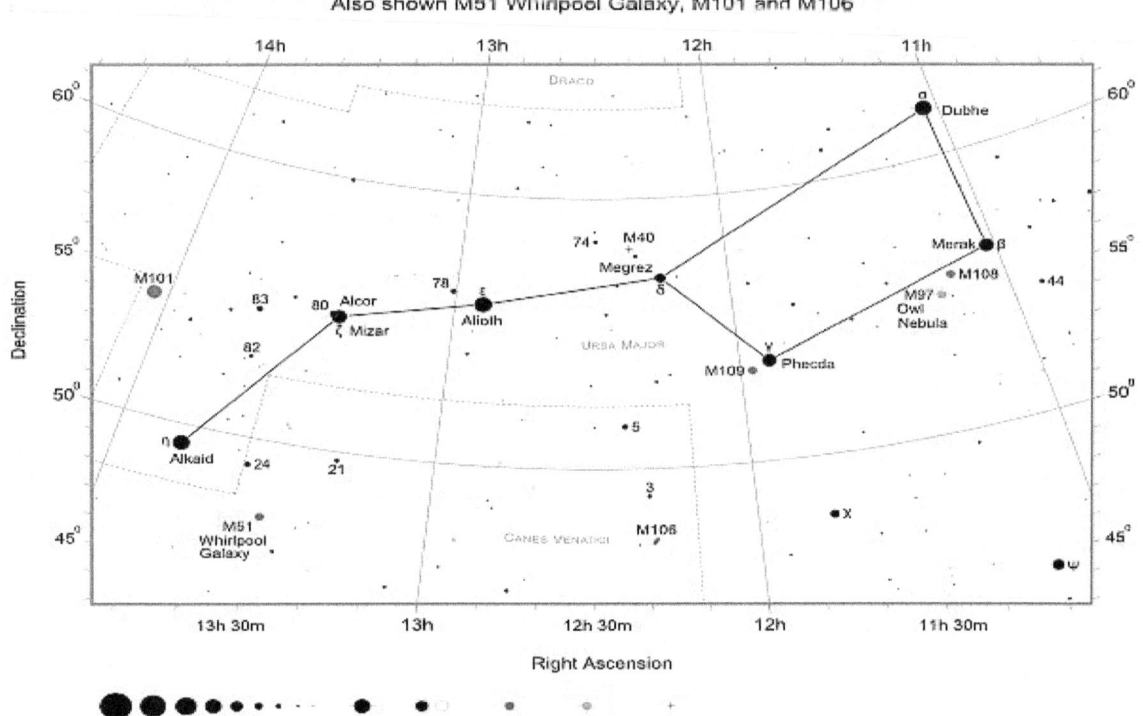

Messier Finder Chart for M40, M97 Owl Nebula, M108 and M109
Also shown M51 Whirlpool Galaxy, M101 and M106

Declination

Right Ascension

Star magnitudes -1 0 1 2 3 4 5 6 7

Double star   Variable stars   Galaxy   Planetary nebula   Double star

Messier 109 - M109 - Barred Spiral Galaxy

M109 (NGC 3992) is a barred spiral galaxy located in the constellation of Ursa Major. It's estimated to be located 83.5 Million light-years from Earth, making it the furthest object in Messier's catalogue. Despite its large distance it's relatively bright; with an apparent magnitude of +10.3 the galaxy is within the range of small to medium sized amateur telescopes.

M109 has a complicated history. In March 1781, Pierre Méchain passed three nebulae he recently found to Charles Messier for confirmation. The first one was to become M97 while the others were recorded by Messier as objects 98 and 99 in a rough draft. However, Messier never assigned positions for these items in the main catalogue and hence they were never included in the final version. Many years later in 1953, American astronomer and historian Owen Gingerich added draft objects 98 and 99 to the "official" Messier catalogue and they became items M108 and M109. The story is further complicated by recent analysis that suggest Méchain may have not originally observed NGC 3992 but instead nearby galaxy NGC 3953. If so, this implies that Messier in fact discovered NGC 3992 and not Méchain. Despite this, it's generally accepted that M109 is identified as NGC 3992.

Finding M109 is easy, it's located only 0.75 degrees to the southeast of Phecda (γ UMa - mag. +2.4) one of the stars of the Plough asterism of Ursa Major. The galaxy is best seen from northern temperate latitudes during the months of March, April and May. From the Southern Hemisphere it never rises very high above the northern horizon.

M109 is visible in large 20x80 binoculars but requires good seeing conditions and is somewhat washed out from the resulting glare due to its close proximity to second magnitude Phecda. Through a 100mm (4-inch) scope the galaxy appears as a faint hazy elongated streak of nebulosity, which is best observed by switching to higher magnifications and moving Phecda outside the field of view. A 150mm (6-inch) telescope reveals a small sharp nucleus surrounded by a mottled nebulosity. An even larger amateur scope shows hints of structure including the bar shaped nucleus. Of course, it's much easier to photograph or image the bar shape than to actually observe it.

In total, M109 measures about 7.6 by 4.7 arc minutes in apparent size. It's an extremely large galaxy with a physical diameter of 180,000 light-years and contains about a trillion stars. On March 17, 1956 a magnitude +12.8 type I supernova (1956A) was observed in M109.

M109 Data Table

Messier	109	**DEC (J2000)**	53d 22m 28s
NGC	3992	**Apparent Size (arcmins)**	7.6 x 4.7
Object Type	Barred Spiral galaxy	**Radius (light years)**	90,000
Classification	SB(rs)bc	**Number of Stars**	1 Trillion
Constellation	Ursa Major	**Notable Feature**	The most distant object in the Messier Catalogue
Distance (kly)	83,500		
Apparent Mag.	10.3		
RA (J2000)	11h 57m 36s		

Messier # 109

Date:	Time:

Site:		
Temp:	Wind:	Hum:
Clouds:	Moon:	
Scope:		
EP:	Mag:	
NELM:	See/Trans:	
Type:	# Stars:	
Mag:	Age:	
Const:		

Notes:

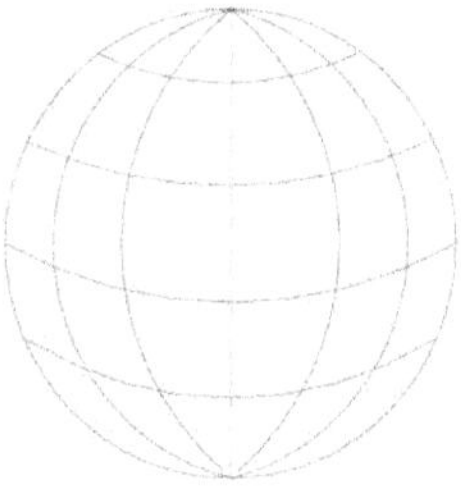

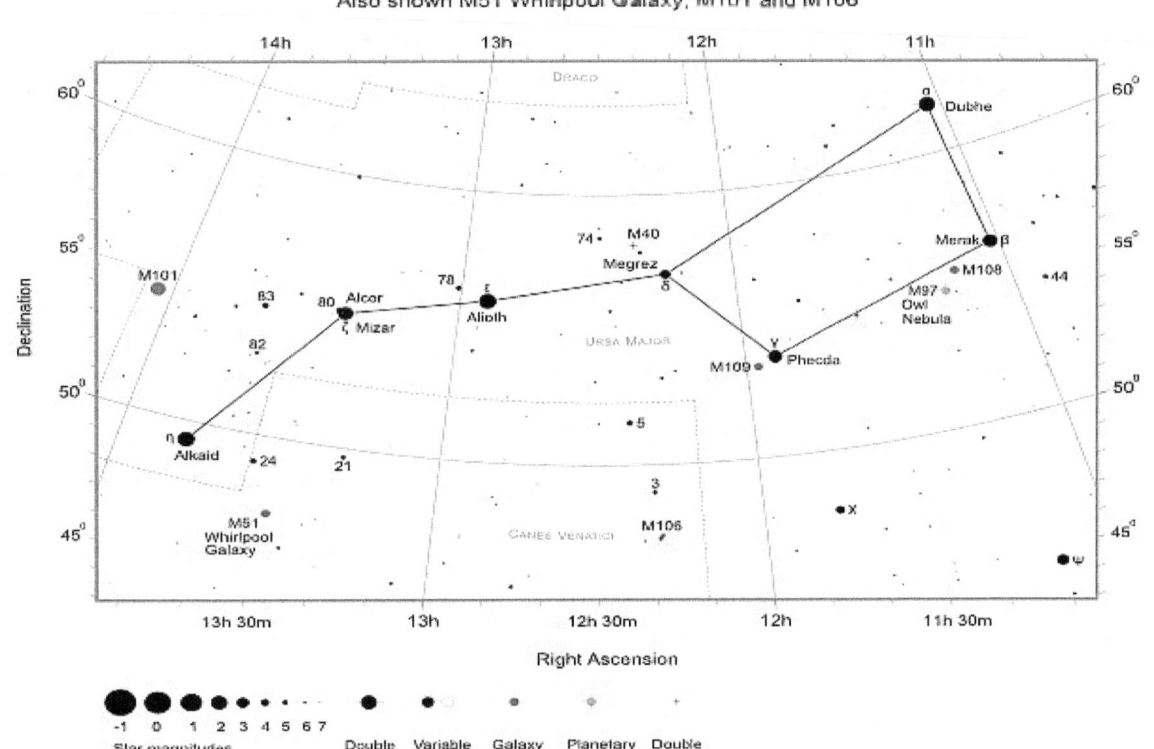

Messier Finder Chart for M40, M97 Owl Nebula, M108 and M109
Also shown M51 Whirlpool Galaxy, M101 and M106

Messier 110 - M110 - Dwarf Elliptical Galaxy

M110 is a dwarf elliptical galaxy located in the constellation of Andromeda. It's one of many satellite galaxies orbiting M31, the famous and spectacular Andromeda galaxy. Of these, at least 14 are dwarf galaxies with M110 being the second brightest of them (after M32). The galaxy is classified as Hubble type E5 and designated as "peculiar" due to unusual dark structures that are probably due to dust clouds.

At magnitude +8.7, M110 is a very challenging binocular object. Although quite large - it covers 22 x 11 arc minutes of apparent sky - it suffers from a low surface brightness and hence even a small amount of light pollution can render it a difficult object to spot with small telescopes. Surprisingly, Charles Messier never included M110 in his famous list. However he depicted it, together with M32 on a drawing of the Andromeda galaxy he made on the August 10, 1773. Caroline Herschel independently discovered the galaxy on August 27, 1783 and much later in 1967, Kenneth Glyn Jones suggested assigning the galaxy a Messier number. Although now commonly known as M110, it's still often referred to in many texts and charts by its New General Catalogue number, NGC 205.

To find M110, first locate the Andromeda Galaxy, which is positioned northeast of the famous "Great Square of Pegasus". Of the four stars of the square, only three of them actually belong to Pegasus. The northeast corner star and brightest of the four at magnitude 2.1, Alpheratz (α And) is part of neighboring Andromeda. Located 7 degrees to the northeast of Alpheratz is δ And (mag. 3.3) and a further 8 degrees to the northeast of δ And is mag. 2.1, Mirach (β And). The Andromeda galaxy is a further 8 degrees to the northwest of Mirach at the end of a line connecting Mirach with μ And and v And. M110 is located 36 arc minutes northwest of the centre of M31.

The galaxies are best seen from the Northern Hemisphere during the months of September, October and November.

In a 80mm (3.1 inch) telescope M110 appears very dim and diffuse. It has a soft, low luminosity without a bright point core (unlike M31 and M32). In a 200mm (8-inch) scope, M110 appears as a large oval nebulosity that's slightly brighter towards the centre. The edges are diffuse.

M110 is located 2.69 million light-years from Earth, which is about 150 million light-years further from us than M31. Its actual diameter is 17,000 light years and the galaxy is estimated to contain 10 billion stars. Surrounding M110 are at least 8 globular clusters, the brightest of them (G73) is of 15th magnitude which is visible in very large amateur telescopes. In 1999, R. Johnson and M. Modjaz of the University of California at Berkeley on behalf of the Lick Observatory Supernova search discovered a nova in M110 at magnitude +18.

M110 Deep Sky Data Table

Messier	110	**Distance (kly)**	2,690	**Radius (light years)**	8,500
NGC	205	**Apparent Mag.**	8.7		
Object Type	Dwarf elliptical galaxy	**RA (J2000)**	00h 40m 22s	**Number of Stars**	10 Billion
		DEC (J2000)	41d 41m 26s	**Notable Feature**	Satellite galaxy of M31
Classification	E5	**Apparent Size (arcmins)**	21.9 x 11.0		
Constellation	Andromeda				

Messier # 110

Date:	Time:
Site:	
Temp:	Wind:	Hum:
Clouds:	Moon:
Scope:	
EP:	Mag:
NELM:	See/Trans:
Type:	# Stars:
Mag:	Age:
Const:	
Notes:	

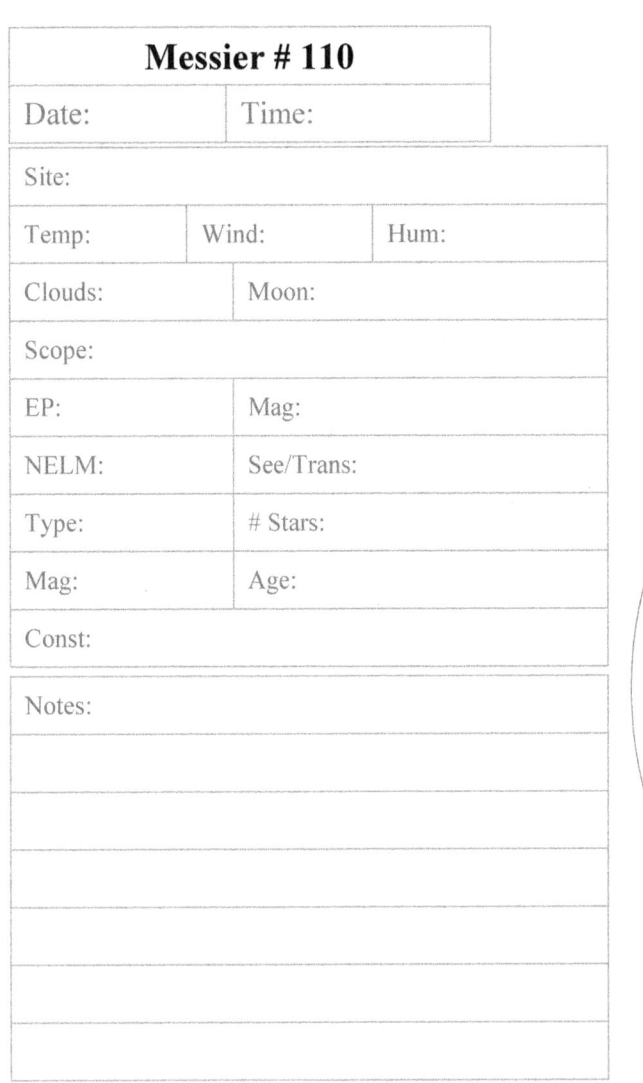

Messier Finder Chart for M31 Andromeda Galaxy, M32, M33 Triangulum Galaxy and M110

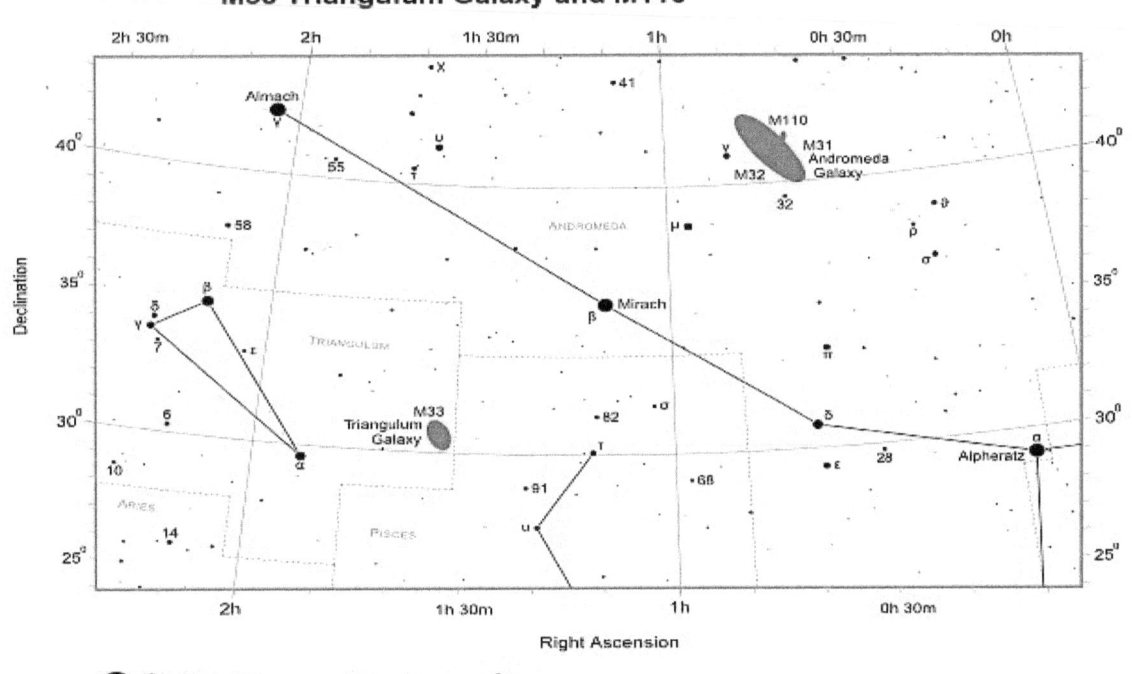

The Complete Messier Object Log

Atlas Info

"Look up and get lost."

- Unknown

The Complete Messier Object Log - Atlas Info

M #	NGC#	Constellation	Type	Name, If Any	Mag .	RA	Dec .	Seen Date
M001	1952	Taurus	Supernova Rem	Crab Nebula	8.4	5h 34.5m	22 01	
M002	7089	Aquarius	Globular Cluster	N/A	6.5	21h 33.5m	-0 49	
M003	5272	Canes Venatici	Globular Cluster	N/A	6.4	13h 42.2m	28 23	
M004	6121	Scorpio	Globular Cluster	N/A	5.9	16h 23.6m	-26 32	
M005	5904	Serpens Caput	Globular Cluster	N/A	5.8	15h 18.6m	2 5	
M006	6405	Scorpius	Open Cluster	Butterfly Cluster	4.2	17h 40.1m	-32 13	
M007	6475	Scorpius	Open Cluster	Ptolemy's Cluster	3.3	17h 53.9m	-34 49	
M008	6523	Sagittarius	Diffuse Nebula	Lagoon Nebula	5.0	18h 3.1m	-24 23	
M009	6333	Ophiuchus	Globular Cluster	N/A	7.9	17h 19.2m	-18 31	
M010	6254	Ophiuchus	Globular Cluster	N/A	6.6	16h 57.1m	-4 6	
M011	6705	Scutum	Open Cluster	Wild Duck Cluster	5.8	18h 51.1m	-6 16	
M012	6218	Ophiuchus	Globular Cluster	N/A	6.6	16h 47.2m	-1 57	
M013	6205	Hercules	Globular Cluster	N/A	5.9	16h 41.7m	36 28	
M014	6402	Ophiuchus	Globular Cluster	N/A	7.6	17h 37.6m	-3 15	
M015	7078	Pegasus	Globular Cluster	N/A	6.4	21h 30m	12 10	
M016	6611	Serpens Claudia	Open Cluster	Eagle Neb Cluster	6.5	18h 18.8m	-13 47	
M017	6618	Sagittarius	Diffuse Nebula	Omega, Swan,	7.0	18h 20.8m	-16 11	
M018	6613	Sagittarius	Open Cluster	N/A	8.0	18h 19.9m	-17 8	
M019	6273	Ophiuchus	Globular Cluster	N/A	8.5	17h 2.6m	-26 16	
M020	6514	Sagittarius	Diffuse Nebula	Trifid Nebula	5.0	18h 2.3m	-23 2	
M021	6531	Sagittarius	Open Cluster	N/A	7.0	18h 4.6m	-22 30	
M022	6656	Sagittarius	Globular Cluster	N/A	6.5	18h 36.4m	-29 54	
M023	6494	Sagittarius	Open Cluster	N/A	6.0	17h 56.8m	-19 1	
M024	6603	Sagittarius	Star Cloud	Milky Way Patch	11.5	18h 18.4m	-18 25	
M025	IC4725	Sagittarius	Open Cluster	N/A	4.9	18h 28.8m	-19 17	
M026	6694	Scutum	Open Cluster	N/A	9.5	18h 45.2m	-9 24	
M027	6853	Vulpecula	Planetary Nebula	Dumbbell Nebula	7.5	19h 59.6m	22 43	
M028	6626	Sagittarius	Globular Cluster	N/A	8.5	18h 24.5m	-24 52	
M029	6913	Cygnus	Open Cluster	N/A	9.0	20h 23.9m	38 32	
M030	7099	Capricornus	Globular Cluster	N/A	8.5	21h 40.4m	-23 11	
M031	0224	Andromeda	Spiral Galaxy	Andromeda Gal.	4.5	0h 42.8m	41 16	

The Complete Messier Object Log - Atlas Info

M #	NGC#	Constellation	Type	Name, If Any	Mag.	RA	Dec.	Seen	Date
M032	0221	Andromeda	Elliptical Galaxy	Satellite of M31	10.0	0h 42.8m	40 52		
M033	0598	Triangulum	Spiral Galaxy	Triangulum, Pinwheel	7.0	1h 33.9m	30 40		
M034	1039	Perseus	Open Cluster	N/A	6.0	2h 42.0m	42 47		
M035	2168	Gemini	Open Cluster	N/A	5.5	6h 8.9m	24 20		
M036	1960	Auriga	Open Cluster	N/A	6.5	5h 36.1	34 08		
M037	2099	Auriga	Open Cluster	N/A	6.0	5h 52.4m	32 33		
M038	1922	Auriga	Open Cluster	N/A	7.0	5h 28.7m	35 50		
M039	7092	Cygnus	Open Cluster	N/A	5.5	21h 32.2m	48 26		
M040	7092	Ursa Major	Double Star	Winecke 4	9.0	12h 20.0m	58 22		
M041	2287	Canis Major	Open Cluster	N/A	5.0	6h 47.0m	-20 44		
M042	1976	Orion	Diffuse Nebula	Great Orion Nebula	5.0	5h 35.3m	-5 23		
M043	1982	Orion	Diffuse Nebula	de Mairan's Nebula	7.0	5h 35.5m	-5 16		
M044	2632	Cancer	Open Cluster	Beehive (Praesepe)	4.0	8h 40.1m	19 59		
M045	1432	Taurus	Open Cluster	Pleiades, 7 Sisters	1.4	3h 47.0m	24 07		
M046	2437	Puppis	Open Cluster	N/A	6.5	7h 41.8m	-14 49		
M047	2422	Puppis	Open Cluster	N/A	4.5	7h 36.6m	-14 30		
M048	2548	Hydra	Open Cluster	N/A	5.5	8h 13.8m	-5 48		
M049	4472	Virgo	Elliptical Galaxy	N/A	10.0	12h 29.8m	8 1		
M050	2323	Monocerus	Open Cluster	N/A	7.0	7h 3.2m	-8 20		
M051	5194	Ursa Major	Spiral Galaxy	Whirlpool Galaxy	8.0	13h 30.0m	47 11		
M052	7654	Cassiopeia	Open Cluster	N/A	8.0	23h 24.2m	61 35		
M053	5024	Coma Berenices	Globular Cluster	N/A	8.5	13h 12.9m	18 10		
M054	6715	Sagittarius	Globular Cluster	N/A	8.5	18h 55.1m	-30 29		
M055	6809	Sagittarius	Globular Cluster	N/A	7.0	19h 40.0m	-30 58		
M056	6779	Lyra	Globular Cluster	N/A	9.5	19h 16.6m	30 11		
M057	6720	Lyra	Planetary Nebula	Ring Nebula	9.5	18h 53.6m	33 2		
M058	4579	Virgo	Spiral Galaxy	N/A	11.0	12h 37.8m	11 50		
M059	4621	Virgo	Elliptical Galaxy	N/A	11.5	12h 42.1m	11 39		
M060	4649	Virgo	Elliptical Galaxy	N/A	10.5	12h 43.7m	11 34		
M061	4303	Virgo	Spiral Galaxy	N/A	10.5	12h 22.0m	4 29		
M062	6266	Ophiuchus	Globular Cluster	N/A	8.0	17h 1.2m	-30 7		

The Complete Messier Object Log - Atlas Info

M #	NGC#	Constellation	Type	Name, If Any	Mag.	RA	Dec.	Seen Date
M063	5055	Canes Venatici	Spiral Galaxy	Sunflower Galaxy	8.5	13h 15.8m	42 2	
M064	4826	Coma Berenices	Spiral Galaxy	Blackeye Galaxy	9.0	12h 56.7m	21 41	
M065	3623	Leo	Spiral Galaxy	N/A	10.5	11h 18.9m	13 6	
M066	3627	Leo	Spiral Galaxy	N/A	10.0	11h 20.2m	13 0	
M067	2628	Cancer	Open Cluster	N/A	7.5	8h 50.4m	11 49	
M068	4590	Hydra	Globular Cluster	N/A	9.0	12h 39.5m	-26 45	
M069	6637	Sagittarius	Globular Cluster	N/A	9.0	18h 34.4m	-32 21	
M070	6681	Sagittarius	Globular Cluster	N/A	9.0	18h 43.2m	-32 18	
M071	6838	Sagittarius	Globular Cluster	N/A	8.5	19h 53.8m	18 47	
M072	6981	Aquarius	Globular Cluster	N/A	10.0	20h 53.5m	-12 32	
M073	6994	Aquarius	Group/Asterism	N/A	9.0	20h 59.0m	-12 3	
M074	0628	Pisces	Spiral Galaxy	N/A	10.5	1h 36.6m	15 48	
M075	6864	Sagittarius	Globular Cluster	N/A	9.5	20h 6.1m	-21 55	
M076	0650	Perseus	Planetary Nebula	Little DBl, Butterfly	12.0	1h 42.4m	51 34	
M077	1068	Cetus	Spiral Galaxy	N/A	10.5	2h 42.7m	-9 2	
M078	2068	Orion	Diffuse Nebula	N/A	8.0	5h 46.8m	0 4	
M079	1904	Lepus	Globular Cluster	N/A	8.5	5h 24.5m	-24 33	
M080	6093	Scorpius	Globular Cluster	N/A	8.5	16h 17.0m	-22 59	
M081	3031	Ursa Major	Spiral Galaxy	Bode's Galaxy	8.5	9h 55.6m	69 4	
M082	3034	Ursa Major	Irregular Galaxy	Cigar Galaxy	9.5	9h 55.9m	69 41	
M083	5236	Hydra	Spiral Galaxy	Small Pinwheel Gal.	8.5	13h 37.1m	-29 52	
M084	4374	Virgo	Lenticular Galaxy	N/A	11.0	12h 25.1m	12 54	
M085	4382	Coma Berenices	Lenticular Galaxy	N/A	10.5	12h 25.5m	18 12	
M086	4406	Virgo	Lenticular Galaxy	N/A	11.0	12h 26.3m	12 57	
M087	4486	Virgo	Elliptical Galaxy	Virgo A	11.0	12h 30.9m	12 24	
M088	4501	Coma Berenices	Spiral Galaxy	N/A	11.0	12h 32.1m	14 26	
M089	4552	Virgo	Elliptical Galaxy	N/A	11.5	12h 35.7m	12 34	
M090	4569	Virgo	Spiral Galaxy	N/A	11.0	12h 36.9m	13 10	
M091	4548	Coma Berenices	Spiral Galaxy	N/A	11.5	12h 35.5m	14 30	
M092	6341	Hercules	Globular Cluster	N/A	7.5	17h 17.1m	43 8	
M093	2447	Puppis	Open Cluster	N/A	6.5	7h 44.6m	-23 52	

The Complete Messier Object Log - Atlas Info

M #	NGC#	Constellation	Type	Name, If Any	Mag.	RA	Dec.	Seen	Date
M094	4736	Canes Venatici	Spiral Galaxy	N/A	9.5	12h 50.9m	41 8		
M095	3351	Leo	Spiral Galaxy	N/A	11.0	10h 43.9m	11 42		
M096	3368	Leo	Spiral Galaxy	N/A	10.5	10h 46.7m	11 49		
M097	3587	Ursa Major	Planetary Neb.	Owl Nebula	12.0	11h 14.8m	55 1		
M098	4192	Coma Berenices	Spiral Galaxy	N/A	11.0	12h 13.9m	14 55		
M099	4254	Coma Berenices	Spiral Galaxy	N/A	10.5	12h 18.9m	14 26		
M100	4321	Coma Berenices	Spiral Galaxy	N/A	10.5	12h 23.0m	15 50		
M101	5457	Ursa Major	Spiral Galaxy	Pinwheel Galaxy	8.5	14h 3.3m	54 22		
M102	5866	Draco	Lenticular Gal.	Spindle Galaxy	10.5	15h 6.5m	55 45		
M103	0581	Cassiopeia	Open Cluster	N/A	7.0	1h 33.2m	60 42		
M104	4594	Virgo	Spiral Galaxy	Sombrero Galaxy	9.5	12h 39.9m	-11 37		
M105	3379	Leo	Elliptical Galaxy	N/A	11.0	10h 47.8m	12 35		
M106	4258	Ursa Major	Spiral Galaxy	N/A	9.5	12h 18.9m	47 19		
M107	6171	Ophiuchus	Globular Cluster	N/A	10.0	16h 32.5m	-13 3		
M108	3556	Ursa Major	Spiral Galaxy	N/A	11.0	11h 11.6m	55 41		
M109	3992	Ursa Major	Spiral Galaxy	N/A	11.0	11h 57.6m	53 23		
M110	0205	Andromeda	Elliptical Galaxy	Satellite of M31	10.0	0h 40.4m	41 41		

The Complete
Messier Object Log

Personal Viewing Notes

"Mortal as I am, I know that I am born for a day. But when I follow, at my pleasure, the serried multitude of the stars in their circular course, my feet no longer touch the earth."

- Ptolemy

The Complete Messier Object Log
Personal Viewing Notes

#	Constellation	Type	Name, if any	Notes
M001	Taurus	Supernova remnant	Crab Nebula	
M002	Aquarius	Globular Cluster	N/A	
M003	Canes Venatici	Globular Cluster	N/A	
M004	Scorpio	Globular Cluster	N/A	
M005	Serpens Caput	Globular Cluster	N/A	
M006	Scorpius	Open Cluster	Butterfly Cluster	
M007	Scorpius	Open Cluster	Ptolemy's Cluster	
M008	Sagittarius	Diffuse Nebula	Lagoon Nebula	
M009	Ophiuchus	Globular Cluster	N/A	
M010	Ophiuchus	Globular Cluster	N/A	
M011	Scutum	Open Cluster	Wild Duck Cluster	
M012	Ophiuchus	Globular Cluster	N/A	
M013	Hercules	Globular Cluster	N/A	
M014	Ophiuchus	Globular Cluster	N/A	
M015	Pegasus	Globular Cluster	N/A	
M016	Serpens Claudia	Open Cluster	Eagle Neb Cluster	
M017	Sagittarius	Diffuse Nebula	Omega, Swan,	
M018	Sagittarius	Open Cluster	N/A	
M019	Ophiuchus	Globular Cluster	N/A	
M020	Sagittarius	Diffuse Nebula	Trifid Nebula	
M021	Sagittarius	Open Cluster	N/A	
M022	Sagittarius	Globular Cluster	N/A	
M023	Sagittarius	Open Cluster	N/A	
M024	Sagittarius	Star Cloud	Milky Way Patch	
M025	Sagittarius	Open Cluster	N/A	
M026	Scutum	Open Cluster	N/A	
M027	Vulpecula	Planetary Nebula	Dumbbell Nebula	
M028	Sagittarius	Globular Cluster	N/A	
M029	Cygnus	Open Cluster	N/A	
M030	Capricornus	Globular Cluster	N/A	
M031	Andromeda	Spiral Galaxy	Andromeda Galaxy	
M032	Andromeda	Elliptical Galaxy	Satellite of M31	
M033	Triangulum	Spiral Galaxy	Triangulum	
M034	Perseus	Open Cluster	N/A	
M035	Gemini	Open Cluster	N/A	
M036	Auriga	Open Cluster	N/A	
M037	Auriga	Open Cluster	N/A	
M038	Auriga	Open Cluster	N/A	
M039	Cygnus	Open Cluster	N/A	
M040	Ursa Major	Double Star	Winecke 4	

The Complete Messier Object Log
Personal Viewing Notes

#	Constellation	Type	Name, if any	Notes
M041	Canis Major	Open Cluster	N/A	
M042	Orion	Diffuse Nebula	Great Orion Nebula	
M043	Orion	Diffuse Nebula	de Mairan's Nebula	
M044	Cancer	Open Cluster	Beehive Cluster	
M045	Taurus	Open Cluster	Pleiades, Subaru,	
M046	Puppis	Open Cluster	N/A	
M047	Puppis	Open Cluster	N/A	
M048	Hydra	Open Cluster	N/A	
M049	Virgo	Elliptical Galaxy	N/A	
M050	Monocerus	Open Cluster	N/A	
M051	Ursa Major	Spiral Galaxy	Whirlpool Galaxy	
M052	Cassiopeia	Open Cluster	N/A	
M053	Coma Berenices	Globular Cluster	N/A	
M054	Sagittarius	Globular Cluster	N/A	
M055	Sagittarius	Globular Cluster	N/A	
M056	Lyra	Globular Cluster	N/A	
M057	Lyra	Planetary Nebula	Ring Nebula	
M058	Virgo	Spiral Galaxy	N/A	
M059	Virgo	Elliptical Galaxy	N/A	
M060	Virgo	Elliptical Galaxy	N/A	
M061	Virgo	Spiral Galaxy	N/A	
M062	Ophiuchus	Globular Cluster	N/A	
M063	Canes Venatici	Spiral Galaxy	Sunflower Galaxy	
M064	Coma Berenices	Spiral Galaxy	Blackeye Galaxy	
M065	Leo	Spiral Galaxy	N/A	
M066	Leo	Spiral Galaxy	N/A	
M067	Cancer	Open Cluster	N/A	
M068	Hydra	Globular Cluster	N/A	
M069	Sagittarius	Globular Cluster	N/A	
M070	Sagittarius	Globular Cluster	N/A	
M071	Sagittarius	Globular Cluster	N/A	
M072	Aquarius	Globular Cluster	N/A	
M073	Aquarius	Group/Asterism	N/A	
M074	Pisces	Spiral Galaxy	N/A	
M075	Sagittarius	Globular Cluster	N/A	
M076	Perseus	Planetary Nebula	Little Dumbbell, Cork,	
M077	Cetus	Spiral Galaxy	N/A	
M078	Orion	Diffuse Nebula	N/A	
M079	Lepus	Globular Cluster	N/A	
M080	Scorpius	Globular Cluster	N/A	

The Complete Messier Object Log
Personal Viewing Notes

#	Constellation	Type	Name, if any	Notes
M081	Ursa Major	Spiral Galaxy	Bode's Galaxy	
M082	Ursa Major	Irregular Galaxy	Cigar Galaxy	
M083	Hydra	Spiral Galaxy	Small Pinwheel	
M084	Virgo	Lenticular (S0) Galaxy	N/A	
M085	Coma Berenices	Lenticular (S0) Galaxy	N/A	
M086	Virgo	Lenticular (S0) Galaxy	N/A	
M087	Virgo	Elliptical Galaxy	Virgo A	
M088	Coma Berenices	Spiral Galaxy	N/A	
M089	Virgo	Elliptical Galaxy	N/A	
M090	Virgo	Spiral Galaxy	N/A	
M091	Coma Berenices	Spiral Galaxy	N/A	
M092	Hercules	Globular Cluster	N/A	
M093	Puppis	Open Cluster	N/A	
M094	Canes Venatici	Spiral Galaxy	N/A	
M095	Leo	Spiral Galaxy	N/A	
M096	Leo	Spiral Galaxy	N/A	
M097	Ursa Major	Planetary Nebula	Owl Nebula	
M098	Coma Berenices	Spiral Galaxy	N/A	
M099	Coma Berenices	Spiral Galaxy	N/A	
M100	Coma Berenices	Spiral Galaxy	N/A	
M101	Ursa Major	Spiral Galaxy	Pinwheel Galaxy	
M102	Draco	Lenticular (S0) Galaxy	Spindle Galaxy	
M103	Cassiopeia	Open Cluster	N/A	
M104	Virgo	Spiral Galaxy	Sombrero Galaxy	
M105	Leo	Elliptical Galaxy	N/A	
M106	Ursa Major	Spiral Galaxy	N/A	
M107	Ophiuchus	Globular Cluster	N/A	
M108	Ursa Major	Spiral Galaxy	N/A	
M109	Ursa Major	Spiral Galaxy	N/A	
M110	Andromeda	Elliptical Galaxy	Satellite of M31	

The Complete Messier Object Log

By Constellation

"No sight is more provocative of awe than is the night sky."

- Llewelyn Powys

The Complete Messier Object Log
By Constellation

Andromeda

Messier #	NGC#	Type	Name, If Any	Mag.	Bino Diff.	Seen	Date
M031	0224	Spiral Galaxy	Andromeda Ga.	4.5	E	☐	
M032	0221	Elliptical Galaxy	Satellite of M31	10.0	E2	☐	
M110	0205	Elliptical Galaxy	Satellite of M31	10.0	C2	☐	

Aquarius

Messier #	NGC#	Type	Name, If Any	Mag.	Bino Diff.	Seen	Date
77M002	7089	Globular Cluster	N/A	6.5	E	☐	
M072	6981	Globular Cluster	N/A	10.0	T2	☐	
M073	6994	Group/Asterism	N/A	9.0	C2	☐	

Auriga

Messier #	NGC#	Type	Name, If Any	Mag.	Bino Diff.	Seen	Date
M036	1960	Open Cluster	N/A	6.5	E	☐	
M037	2099	Open Cluster	N/A	6.0	E	☐	
M038	1922	Open Cluster	N/A	7.0	E	☐	

Cancer

Messier #	NGC#	Type	Name, If Any	Mag.	Bino Diff.	Seen	Date
M044	2632	Open Cluster	Beehive Cluster	4.0	E	☐	
M067	2628	Open Cluster	N/A	7.5	E	☐	

Canes Venatici

Messier #	NGC#	Type	Name, If Any	Mag.	Bino Diff.	Seen	Date
M003	5272	Globular Cluster	N/A	6.4	E	☐	
M063	5055	Spiral Galaxy	Sunflower Gal.	8.5	T	☐	
M094	4736	Spiral Galaxy	N/A	9.5	T	☐	

Canis Major

Messier #	NGC#	Type	Name, If Any	Mag.	Bino Diff.	Seen	Date
M041	2287	Open Cluster	N/A	5.0	E	☐	

difficulty - E=easy, T=tough, C=challenging / 7x50. E2=easy, T2=tough, C2=challenging / 11x80.

The Complete Messier Object Log
By Constellation

Capricornus

Messier #	NGC#	Type	Name, If Any	Mag.	Bino Diff.	Seen	Date
M030	7099	Globular Cluster	N/A	8.5	T	☐	

Cassiopeia

Messier #	NGC#	Type	Name, If Any	Mag.	Bino Diff.	Seen	Date
M052	7654	Globular Cluster	N/A	8.0	E	☐	
M103	0581	Open Cluster	N/A	7.0	E	☐	

Cetus

Messier #	NGC#	Type	Name, If Any	Mag.	Bino Diff.	Seen	Date
M077	1068	Spiral Galaxy	N/A	10.5	C2	☐	

Coma Berenices

Messier #	NGC#	Type	Name, If Any	Mag.	Bino Diff.	Seen	Date
M053	5024	Globular Cluster	N/A	8.5	T	☐	
M064	4826	Spiral Galaxy	Blackeye Gal.	9.0	T	☐	
M085	4382	Lenticular	N/A	10.5	T2	☐	
M088	4501	Spiral Galaxy	N/A	11.0	T2	☐	
M091	4548	Spiral Galaxy	N/A	11.5	C2	☐	
M098	4192	Spiral Galaxy	N/A	11.0	C2	☐	
M099	4254	Spiral Galaxy	N/A	10.5	T2	☐	
M100	4321	Spiral Galaxy	N/A	10.5	C	☐	

Cygnus

Messier #	NGC#	Type	Name, If Any	Mag.	Bino Diff.	Seen	Date
M029	6913	Open Cluster	N/A	9.0	E	☐	
M039	7092	Open Cluster	N/A	5.5	E	☐	

Draco

Messier #	NGC#	Type	Name, If Any	Mag.	Bino Diff.	Seen	Date
M102	5866	Lenticular Galaxy	Spindle Galaxy	10.5	C2	☐	

difficulty - E=easy, T=tough, C=challenging / 7x50. E2=easy, T2=tough, C2=challenging / 11x80.

The Complete Messier Object Log
By Constellation

Gemini

Messier #	NGC#	Type	Name, If Any	Mag.	Bino Diff.	Seen	Date
M035	2168	Open Cluster	N/A	5.5	E	☐	

Hercules

Messier	# NGC#	Type	Name, If Any	Mag.	Bino Diff.	Seen	Date
M013	6205	Globular Cluster	N/A	5.9	E	☐	
M092	6341	Globular Cluster	N/A	7.5	E	☐	

Hydra

Messier	# NGC#	Type	Name, If Any	Mag.	Bino Diff.	Seen	Date
M048	2548	Open Cluster	N/A	5.5	E	☐	
M068	4590	Globular Cluster	N/A	9.0	C	☐	
M083	5236	Spiral Galaxy	Sm Pinwheel Gal.	8.5	T	☐	

Leo

Messier	# NGC#	Type	Name, If Any	Mag.	Bino Diff.	Seen	Date
M065	3623	Spiral Galaxy	N/A	10.5	C	☐	
M066	3627	Spiral Galaxy	N/A	10.0	C	☐	
M095	3351	Spiral Galaxy	N/A	11.0	T2	☐	
M096	3368	Spiral Galaxy	N/A	10.5	T2	☐	
M105	3379	Ellipt. Galaxy	N/A	11.0	T2	☐	

Lepus

Messier	# NGC#	Type	Name, If Any	Mag.	Bino Diff.	Seen	Date
M079	1904	Globular Cluster	N/A	8.5	T2	☐	

Lyra

Messier	# NGC#	Type	Name, If Any	Mag.	Bino Diff.	Seen	Date
M056	6779	Globular Cluster	N/A	9.5	C	☐	
M057	6720	Planetary Neb	Ring Nebula	9.5	?	☐	

difficulty - E=easy, T=tough, C=challenging / 7x50. E2=easy, T2=tough, C2=challenging / 11x80.

The Complete Messier Object Log
By Constellation

Monocerus

Messier #	NGC#	Type	Name, If Any	Mag.	Bino Diff.	Seen	Date
M050	2323	Open Cluster	N/A	7.0	E	☐	

Ophiuchus

Messier #	NGC#	Type	Name, If Any	Mag.	Bino Diff.	Seen	Date
M009	6333	Globular Cluster	N/A	7.9	T2	☐	
M010	6254	Globular Cluster	N/A	6.6	E	☐	
M012	6218	Globular Cluster	N/A	6.6	E	☐	
M014	6402	Globular Cluster	N/A	7.6	T	☐	
M019	6273	Globular Cluster	N/A	8.5	T	☐	
M062	6266	Globular Cluster	N/A	8.0	T	☐	
M107	6171	Globular Cluster	N/A	10.0	T2	☐	

Orion

Messier #	NGC#	Type	Name, If Any	Mag.	Bino Diff.	Seen	Date
M042	1976	Diffuse Nebula	Great Orion Neb	5.0	E	☐	
M043	1982	Diffuse Nebula	de Mairan's Neb	7.0	?	☐	
M078	2068	Diffuse Nebula	N/A	8.0	T	☐	

Pegasus

Messier #	NGC#	Type	Name, If Any	Mag.	Bino Diff.	Seen	Date
M015	7078	Globular Cluster	N/A	6.4	E	☐	

Perseus

Messier #	NGC#	Type	Name, If Any	Mag.	Bino Diff.	Seen	Date
M034	1039	Open Cluster	N/A	6.0	E	☐	
M076	0650	Planetary Nebula	Little Dumbbell, Butterfly	12.0	C2	☐	

Pisces

Messier #	NGC#	Type	Name, If Any	Mag.	Bino Diff.	Seen	Date
M074	0628	Spiral Galaxy	N/A	10.5	C2	☐	

difficulty - E=easy, T=tough, C=challenging / 7x50. E2=easy, T2=tough, C2=challenging / 11x80.

The Complete Messier Object Log
By Constellation

Puppis

Messier	# NGC#	Type	Name, If Any	Mag.	Bino Diff.	Seen	Date
M046	2437	Open Cluster	N/A	6.5	E	☐	
M047	2422	Open Cluster	N/A	4.5	E	☐	
M093	2447	Open Cluster	N/A	6.5	E	☐	

Sagitta

Messier	# NGC#	Type	Name, If Any	Mag.	Bino Diff.	Seen	Date
M071	6838	Globular Cluster	N/A	8.5	C	☐	

Sagittarius

Messier	# NGC#	Type	Name If Any	Mag	Bino Diff.	Seen	Date
M008	6523	Diffuse Nebula	Lagoon Nebula	5.0	E	☐	
M017	6618	Diffuse Nebula	Omega, Swan,	7.0	E	☐	
M018	6613	Open Cluster	N/A	8.0	E	☐	
M020	6514	Diffuse Nebula	Trifid Nebula	5.0	C2	☐	
M021	6531	Open Cluster	N/A	7.0	?	☐	
M022	6656	Globular Cluster	N/A	6.5	E	☐	
M023	6494	Open Cluster	N/A	6.0	E	☐	
M024	6603	Star Cloud	Milky Way Patch	11.5	E	☐	
M025	IC4725	Open Cluster	N/A	4.9	E	☐	
M028	6626	Globular Cluster	N/A	8.5	T	☐	
M054	6715	Globular Cluster	N/A	8.5	C	☐	
M055	6809	Globular Cluster	N/A	7.0	E	☐	
M069	6637	Globular Cluster	N/A	9.0	T2	☐	
M070	6681	Globular Cluster	N/A	9.0	T2	☐	
M075	6864	Globular Cluster	N/A	9.5	C	☐	

Scorpio

Messier	# NGC#	Type	Name If Any	Mag.	Bino Diff.	Seen	Date
M004	6121	Globular Cluster	N/A	5.9	E	☐	

difficulty - E=easy, T=tough, C=challenging / 7x50. E2=easy, T2=tough, C2=challenging / 11x80.

The Complete Messier Object Log
By Constellation

Scorpius

Messier #	NGC#	Type	Name, If Any	Mag	Bino Diff.	Seen	Date
M006	6405	Open Cluster	Butterfly Cluster	4.2	E	☐	
M007	6475	Open Cluster	Ptolemy's Cluster	3.3	E	☐	
M080	6093	Globular Cluster	N/A	8.5	T	☐	

Scutum

Messier #	NGC#	Type	Name, If Any	Mag	Bino Diff.	Seen	Date
M011	6705	Open Cluster	Wild Duck Cluster	5.8	E	☐	
M026	6694	Open Cluster	N/A	9.5	C	☐	

Serpens Caput

Messier #	NGC#	Type	Name, If Any	Mag	Bino Diff.	Seen	Date
M005	5904	Globular Cluster	N/A	5.8	E	☐	

Serpens Claudia

Messier #	NGC#	Type	Name, If Any	Mag	Bino Diff.	Seen	Date
M016	6611	Open Cluster	Eagle Neb Cluster	6.5	E	☐	

Taurus

Messier #	NGC#	Type	Name, If Any	Mag	Bino Diff.	Seen	Date
M001	1952	Supernova remnant	Crab Nebula	8.4	C	☐	
M045	1432	Open Cluster	Pleiades, Seven Sis	1.4	E	☐	

Triangulum

Messier #	NGC#	Type	Name, If Any	Mag	Bino Diff.	Seen	Date
M033	0598	Spiral Galaxy	Triang also Pinwheel	7.0	T2	☐	

Ursa Major

Messier #	NGC#	Type	Name, If Any	Mag	Bino Diff.	Seen	Date
M040	7092	Double Star	Winecke 4	9.0	T	☐	
M051	5194	Spiral Galaxy	Whirlpool Galaxy	8.0	C	☐	

difficulty - E=easy, T=tough, C=challenging / 7x50. E2=easy, T2=tough, C2=challenging / 11x80.

The Complete Messier Object Log
By Constellation

Ursa Major(cont)

Messier	# NGC#	Type	Name, If Any	Mag	Bino Diff.	Seen	Date
M081	3031	Spiral Galaxy	Bode's Galaxy	8.5	T	☐	
M082	3034	Irregular Gal	Cigar Galaxy	9.5	T	☐	
M097	3587	Planetary Neb	Owl Nebula	12.0	C	☐	
M101	5457	Spiral Galaxy	Pinwheel Gal	8.5	C	☐	
M106	4258	Spiral Galaxy	N/A	9.5	C	☐	
M108	3556	Spiral Galaxy	N/A	11.0	T2	☐	
M109	3992	Spiral Galaxy	N/A	11.0	T2	☐	

Virgo

Messier	# NGC#	Type	Name, If Any	Mag	Bino Diff.	Seen	Date
M049	4472	Elliptical Galaxy	N/A	10.0	T	☐	
M058	4579	Spiral Galaxy	N/A	11.0	T2	☐	
M059	4621	Elliptical Galaxy	N/A	11.5	T2	☐	
M060	4649	Elliptical Galaxy	N/A	10.5	E2	☐	
M061	4303	Spiral Galaxy	N/A	10.5	E2	☐	
M084	4374	Lenticular Gal	N/A	11.0	T2	☐	
M086	4406	Lenticular Gal	N/A	11.0	T2	☐	
M087	4486	Elliptical Gal	Virgo A	11.0	C2	☐	
M089	4552	Elliptical Gal	N/A	11.5	T2	☐	
M090	4569	Spiral Galaxy	N/A	11.0	T2	☐	
M104	4594	Spiral Galaxy	Sombrero Gal	9.5	C	☐	

Vulpecula

Messier	# NGC#	Type	Name, If Any	Mag	Bino Diff.	Seen	Date
M027	6853	Planetary Nebula	Dumbbell Neb	7.5	E	☐	

difficulty - E=easy, T=tough, C=challenging / 7x50. E2=easy, T2=tough, C2=challenging / 11x80.

The Complete
Messier Object Log

By Season

"When I, sitting, heard the astronomer, where he lectured with such applause and held unaccountable, I became tired & sick; Till rising and gliding out, I wandered off by myself, in the mystical moist night air, and from time to time, looked up in perfect silence at the stars."

-Walt Whitman

The Complete Messier Object Log - By Season

Early Spring

Messier #	NGC #	Constellation	RA	Dec	Mag.	Size	Seen?	Date
M044	2632	Cancer	8h 40.1m	19 59	4.0	95.0	☐	
M067	2628	Cancer	8h 50.4m	11 49	7.5	30.0	☐	
M003	5272	Canes Venatici	13h 42.2m	28 23	6.4	16.2	☐	
M063	5055	Canes Venatici	13h 15.8m	42 2	8.5	10.0x6.0	☐	
M094	4736	Canes Venatici	12h 50.9m	41 8	9.5	7.0x3.0	☐	
M048	2548	Hydra	8h 13.8m	-5 48	5.5	54.0	☐	
M065	3623	Leo	11h 18.9m	13 6	10.5	8.0x1.5	☐	
M066	3627	Leo	11h 20.2m	13 0	10.0	8.0x2.5	☐	
M095	3351	Leo	10h 43.9m	11 42	11.0	4.4x3.3	☐	
M096	3368	Leo	10h 46.7m	11 49	10.5	6.0x4.0	☐	
M105	3379	Leo	10h 47.8m	12 35	11.0	2.0	☐	
M040	7092	Ursa Major	12h 20.0m	58 22	9.0	0.8	☐	
M051	5194	Ursa Major	13h 30.0m	47 11	8.0	11.0x7.0	☐	
M081	3031	Ursa Major	9h 55.6m	69 4	8.5	21.0x10.0	☐	
M101	5457	Ursa Major	14h 3.3m	54 22	8.5	22.0	☐	
M106	4258	Ursa Major	12h 18.9m	47 19	9.5	19.0x8.0	☐	
M108	3556	Ursa Major	11h 11.6m	55 41	11.0	8.0x1.0	☐	
M109	3992	Ursa Major	11h 57.6m	53 23	11.0	7.0x4.0	☐	

Late Spring

Messier #	NGC #	Constellation	RA	Dec	Mag.	Size	Seen?	Date
M053	5024	Coma Berenices	13h 12.9m	18 10	8.5	12.6	☐	
M064	4826	Coma Berenices	12h 56.7m	21 41	9.0	9.3x5.4	☐	
M085	4382	Coma Berenices	12h 25.5m	18 12	10.5	7.1x5.2	☐	
M088	4501	Coma Berenices	12h 32.1m	14 26	11.0	7.0x4.0	☐	
M091	4548	Coma Berenices	12h 35.5m	14 30	11.5	5.4x4.4	☐	
M098	4192	Coma Berenices	12h 13.9m	14 55	11.0	9.5x3.2	☐	
M100	4321	Coma Berenices	12h 23.0m	15 50	10.5	7.0x6.0	☐	
M068	4590	Hydra	12h 39.5m	-26 45	9.0	12.0	☐	
M082	3034	Ursa Major	9h 55.9m	69 41	9.5	9.x4.0	☐	
M097	3587	Ursa Major	11h 14.8m	55 1	12.0	3.4x3.3	☐	
M049	4472	Virgo	12h 29.8m	8 1	10.0	9.0x7.5	☐	
M058	4579	Virgo	12h 37.8m	11 50	11.0	5.5x4.5	☐	
M059	4621	Virgo	12h 42.1m	11 39	11.5	5.0x3.5	☐	
M060	4649	Virgo	12h 43.7m	11 34	10.5	7.0x6.0	☐	
M061	4303	Virgo	12h 22.0m	4 29	10.5	6.0x5.5	☐	
M084	4374	Virgo	12h 25.1m	12 54	11.0	5.0	☐	
M086	4406	Virgo	12h 26.3m	12 57	11.0	7.5x5.5	☐	
M087	4486	Virgo	12h 30.9m	12 24	11.0	7.0	☐	
M089	4552	Virgo	12h 35.7m	12 34	11.5	4.0	☐	
M090	4569	Virgo	12h 36.9m	13 10	11.0	9.5x4.5	☐	
M104	4594	Virgo	12h 39.9m	-11 37	9.5	9.0x4.0	☐	

The Complete Messier Object Log - By Season

Mid-Summer

Messier #	NGC #	Constellation	RA	Dec	Mag.	Size	Seen?	Date
M102	5866	Draco	15h 6.5m	55 45	10.5	5.2x2.3	☐	
M013	6205	Hercules	16h 41.7m	36 28	5.9	16.6	☐	
M092	6341	Hercules	17h 17.1m	43 8	7.5	11.2	☐	
M083	5236	Hydra	13h 37.1m	-29 52	8.5	11.0x10.0	☐	
M009	6333	Ophiuchus	17h 19.2m	-18 31	7.9	9.3	☐	
M010	6254	Ophiuchus	16h 57.1m	-4 6	6.6	15.1	☐	
M012	6218	Ophiuchus	16h 47.2m	-1 57	6.6	14.5	☐	
M014	6402	Ophiuchus	17h 37.6m	-3 15	7.6	11.7	☐	
M019	6273	Ophiuchus	17h 2.6m	-26 16	8.5	13.5	☐	
M062	6266	Ophiuchus	17h 1.2m	-30 7	8.0	14.1	☐	
M107	6171	Ophiuchus	16h 32.5m	-13 3	10.0	10.0	☐	
M004	6121	Scorpio	16h 23.6m	-26 32	5.9	26.3	☐	
M006	6405	Scorpius	17h 40.1m	-32 13	4.2	15.0	☐	
M007	6475	Scorpius	17h 53.9m	-34 49	3.3	80.0	☐	
M080	6093	Scorpius	16h 17.0m	-22 59	8.5	8.9	☐	
M005	5904	Serpens Caput	15h 18.6m	2 5	5.8	17.4	☐	

Late Summer

Messier #	NGC #	Constellation	RA	Dec	Mag.	Size	Seen?	Date
M008	6523	Sagittarius	18h 3.1m	-24 23	5.0	35.0x50.0	☐	
M017	6618	Sagittarius	18h 20.8m	-16 11	7.0	11.0	☐	
M018	6613	Sagittarius	18h 19.9m	-17 8	8.0	9.0	☐	
M020	6514	Sagittarius	18h 2.3m	-23 2	5.0	28.0	☐	
M021	6531	Sagittarius	18h 4.6m	-22 30	7.0	13.0	☐	
M022	6656	Sagittarius	18h 36.4m	-29 54	6.5	24.0	☐	
M023	6494	Sagittarius	17h 56.8m	-19 1	6.0	27.0	☐	
M024	6603	Sagittarius	18h 18.4m	-18 25	11.5	5.0	☐	
M025	IC4725	Sagittarius	18h 28.8m	-19 17	4.9	40.0	☐	
M028	6626	Sagittarius	18h 24.5m	-24 52	8.5	11.2	☐	
M054	6715	Sagittarius	18h 55.1m	-30 29	8.5	9.1	☐	
M055	6809	Sagittarius	19h 40.0m	-30 58	7.0	19.0	☐	
M069	6637	Sagittarius	18h 34.4m	-32 21	9.0	7.1	☐	
M070	6681	Sagittarius	18h 43.2m	-32 18	9.0	7.8	☐	
M075	6864	Sagittarius	20h 6.1m	-21 55	9.5	6.0	☐	
M011	6705	Scutum	18h 51.1m	-6 16	5.8	14.0	☐	
M026	6694	Scutum	18h 45.2m	-9 24	9.5	15.0	☐	
M016	6611	Serpens Claudia	18h 18.8m	-13 47	6.5	7.0	☐	

Fall/Early Winter

Messier #	NGC #	Constellation	RA	Dec	Mag.	Size	Seen?	Date
M002	7089	Aquarius	21h 33.5m	-0 49	6.5	12.9	☐	
M072	6981	Aquarius	20h 53.5m	-12 32	10.0	5.9	☐	

The Complete Messier Object Log - By Season

Messier	# NGC #	Constellation	RA	Dec	Mag.	Size	Seen?	Date
M073	6994	Aquarius	20h 59.0m	-12 3	9.0	2.8	☐	
M030	7099	Capricornus	21h 40.4m	-23 11	8.5	11.0	☐	
M077	1068	Cetus	2h 42.7m	-9 2	10.5	7.0x6.0	☐	
M099	4254	Coma Berenices	12h 18.9m	14 26	10.5	5.4x4.8	☐	
M029	6913	Cygnus	20h 23.9m	38 32	9.0	7.0	☐	
M039	7092	Cygnus	21h 32.2m	48 26	5.5	32.0	☐	
M056	6779	Lyra	19h 16.6m	30 11	9.5	7.1	☐	
M057	6720	Lyra	18h 53.6m	33 2	9.5	1.4x1.0	☐	
M015	7078	Pegasus	21h 30m	12 10	6.4	12.3	☐	
M074	0628	Pisces	1h 36.6m	15 48	10.5	10.2x9.5	☐	
M071	6838	Sagittarius	19h 53.8m	18 47	8.5	7.2	☐	
M027	6853	Vulpecula	19h 59.6m	22 43	7.5	8.0x5.6	☐	

Winter

Messier	# NGC #	Constellation	RA	Dec	Mag.	Size	Seen?	Date
M031	0224	Andromeda	0h 42.8m	41 16	4.5	178.0	☐	
M032	0221	Andromeda	0h 42.8m	40 52	10.0	8.0x6.0	☐	
M110	0205	Andromeda	0h 40.4m	41 41	10.0	17.0x10.0	☐	
M036	1960	Auriga	5h 36.1	34 08	6.5	12.0	☐	
M037	2099	Auriga	5h 52.4m	32 33	6.0	24.0	☐	
M038	1922	Auriga	5h 28.7m	35 50	7.0	21.0	☐	
M041	2287	Canis Major	6h 47.0m	-20 44	5.0	38.0	☐	
M052	7654	Cassiopeia	23h 24.2m	61 35	8.0	13.0	☐	
M103	0581	Cassiopeia	1h 33.2m	60 42	7.0	6.0	☐	
M035	2168	Gemini	6h 8.9m	24 20	5.5	28.0	☐	
M079	1904	Lepus	5h 24.5m	-24 33	8.5	8.7	☐	
M050	2323	Monocerus	7h 3.2m	-8 20	7.0	16.0	☐	
M042	1976	Orion	5h 35.3m	-5 23	5.0	85.0x60.0	☐	
M043	1982	Orion	5h 35.5m	-5 16	7.0	20.0x15.0	☐	
M078	2068	Orion	5h 46.8m	0 4	8.0	8.0x6.0	☐	
M034	1039	Perseus	2h 42.0m	42 47	6.0	35.0	☐	
M076	0650	Perseus	1h 42.4m	51 34	12.0	2.7x1.8	☐	
M046	2437	Puppis	7h 41.8m	-14 49	6.5	27.0	☐	
M047	2422	Puppis	7h 36.6m	-14 30	4.5	30.0	☐	
M093	2447	Puppis	7h 44.6m	-23 52	6.5	22.0	☐	
M001	1952	Taurus	5h 34.5m	22 01	8.4	6.0x4.0	☐	
M045	1432	Taurus	3h 47.0m	24 07	1.4	110.0	☐	
M033	0598	Triangulum	1h 33.9m	30 40	7.0	73.0x45.0	☐	

The Complete Messier Object Log

Messier Marathon

"I've loved the stars too fondly to be fearful of the night."

- Sarah Williams

The Complete Messier Object Log -
Messier Marathon Info

Search #	M #	NGC #	Constellation	RA	Dec	Mag.	Size	Season	Seen?
1	M077	1068	Cetus	2h 42.7m	-9 2	10.5	7.0x6.0	Fall/Early Winter	☐
2	M074	0628	Pisces	1h 36.6m	15 48	10.5	10.2x9.5	Fall/Early Winter	☐
3	M033	0598	Triangulum	1h 33.9m	30 40	7.0	73.0x45.0	Winter	☐
4	M031	0224	Andromeda	0h 42.8m	41 16	4.5	178.0	Winter	☐
5	M032	0221	Andromeda	0h 42.8m	40 52	10.0	8.0x6.0	Winter	☐
6	M110	0205	Andromeda	0h 40.4m	41 41	10.0	17.0x10.0	Winter	☐
7	M052	7654	Cassiopeia	23h 24.2m	61 35	8.0	13.0	Winter	☐
8	M103	0581	Cassiopeia	1h 33.2m	60 42	7.0	6.0	Winter	☐
9	M076	0650	Perseus	1h 42.4m	51 34	12.0	2.7x1.8	Winter	☐
10	M034	1039	Perseus	2h 42.0m	42 47	6.0	35.0	Winter	☐
11	M045	1432	Taurus	3h 47.0m	24 07	1.4	110.0	Winter	☐
12	M079	1904	Lepus	5h 24.5m	-24 33	8.5	8.7	Winter	☐
13	M042	1976	Orion	5h 35.3m	-5 23	5.0	85.0x60.0	Winter	☐
14	M043	1982	Orion	5h 35.5m	-5 16	7.0	20.0x15.0	Winter	☐
15	M078	2068	Orion	5h 46.8m	0 4	8.0	8.0x6.0	Winter	☐
16	M001	1952	Taurus	5h 34.5m	22 01	8.4	6.0x4.0	Winter	☐
17	M035	2168	Gemini	6h 8.9m	24 20	5.5	28.0	Winter	☐
18	M037	2099	Auriga	5h 52.4m	32 33	6.0	24.0	Winter	☐
19	M036	1960	Auriga	5h 36.1	34 08	6.5	12.0	Winter	☐
20	M038	1922	Auriga	5h 28.7m	35 50	7.0	21.0	Winter	☐
21	M041	2287	Canis Major	6h 47.0m	-20 44	5.0	38.0	Winter	☐
22	M093	2447	Puppis	7h 44.6m	-23 52	6.5	22.0	Winter	☐
23	M047	2422	Puppis	7h 36.6m	-14 30	4.5	30.0	Winter	☐
24	M046	2437	Puppis	7h 41.8m	-14 49	6.5	27.0	Winter	☐
25	M050	2323	Monocerus	7h 3.2m	-8 20	7.0	16.0	Winter	☐
26	M048	2548	Hydra	8h 13.8m	-5 48	5.5	54.0	Early Spring	☐
27	M044	2632	Cancer	8h 40.1m	19 59	4.0	95.0	Early Spring	☐
28	M067	2628	Cancer	8h 50.4m	11 49	7.5	30.0	Early Spring	☐
29	M095	3351	Leo	10h 43.9m	11 42	11.0	4.4x3.3	Early Spring	☐
30	M096	3368	Leo	10h 46.7m	11 49	10.5	6.0x4.0	Early Spring	☐
31	M105	3379	Leo	10h 47.8m	12 35	11.0	2.0	Early Spring	☐
32	M065	3623	Leo	11h 18.9m	13 6	10.5	8.0x1.5	Early Spring	☐
33	M066	3627	Leo	11h 20.2m	13 0	10.0	8.0x2.5	Early Spring	☐
34	M081	3031	Ursa Major	9h 55.6m	69 4	8.5	21.0x10.0	Early Spring	☐
35	M082	3034	Ursa Major	9h 55.9m	69 41	9.5	9.x4.0	Late Spring	☐
36	M097	3587	Ursa Major	11h 14.8m	55 1	12.0	3.4x3.3	Late Spring	☐
37	M108	3556	Ursa Major	11h 11.6m	55 41	11.0	8.0x1.0	Early Spring	☐
38	M109	3992	Ursa Major	11h 57.6m	53 23	11.0	7.0x4.0	Early Spring	☐
39	M040	7092	Ursa Major	12h 20.0m	58 22	9.0	0.8	Early Spring	☐
40	M106	4258	Ursa Major	12h 18.9m	47 19	9.5	19.0x8.0	Early Spring	☐
41	M094	4736	Canes Venatici	12h 50.9m	41 8	9.5	7.0x3.0	Early Spring	☐
42	M063	5055	Canes Venatici	13h 15.8m	42 2	8.5	10.0x6.0	Early Spring	☐
43	M051	5194	Ursa Major	13h 30.0m	47 11	8.0	11.0x7.0	Early Spring	☐

The Complete Messier Object Log - Messier Marathon Info

Search #	M #	NGC #	Constellation	RA	Dec	Mag.	Size	Season	Seen?
44	M101	5457	Ursa Major	14h 3.3m	54 22	8.5	22.0	Early Spring	☐
45	M102	5866	Draco	15h 6.5m	55 45	10.5	5.2x2.3	Mid-Summer	☐
46	M053	5024	Coma Berenices	13h 12.9m	18 10	8.5	12.6	Late Spring	☐
47	M064	4826	Coma Berenices	12h 56.7m	21 41	9.0	9.3x5.4	Late Spring	☐
48	M003	5272	Canes Venatici	13h 42.2m	28 23	6.4	16.2	Early Spring	☐
49	M098	4192	Coma Berenices	12h 13.9m	14 55	11.0	9.5x3.2	Late Spring	☐
50	M099	4254	Coma Berenices	12h 18.9m	14 26	10.5	5.4x4.8	Fall/Early Winter	☐
51	M100	4321	Coma Berenices	12h 23.0m	15 50	10.5	7.0x6.0	Late Spring	☐
52	M085	4382	Coma Berenices	12h 25.5m	18 12	10.5	7.1x5.2	Late Spring	☐
53	M084	4374	Virgo	12h 25.1m	12 54	11.0	5.0	Late Spring	☐
54	M086	4406	Virgo	12h 26.3m	12 57	11.0	7.5x5.5	Late Spring	☐
55	M087	4486	Virgo	12h 30.9m	12 24	11.0	7.0	Late Spring	☐
56	M089	4552	Virgo	12h 35.7m	12 34	11.5	4.0	Late Spring	☐
57	M090	4569	Virgo	12h 36.9m	13 10	11.0	9.5x4.5	Late Spring	☐
58	M088	4501	Coma Berenices	12h 32.1m	14 26	11.0	7.0x4.0	Late Spring	☐
59	M091	4548	Coma Berenices	12h 35.5m	14 30	11.5	5.4x4.4	Late Spring	☐
60	M058	4579	Virgo	12h 37.8m	11 50	11.0	5.5x4.5	Late Spring	☐
61	M059	4621	Virgo	12h 42.1m	11 39	11.5	5.0x3.5	Late Spring	☐
62	M060	4649	Virgo	12h 43.7m	11 34	10.5	7.0x6.0	Late Spring	☐
63	M049	4472	Virgo	12h 29.8m	8 1	10.0	9.0x7.5	Late Spring	☐
64	M061	4303	Virgo	12h 22.0m	4 29	10.5	6.0x5.5	Late Spring	☐
65	M104	4594	Virgo	12h 39.9m	-11 37	9.5	9.0x4.0	Late Spring	☐
66	M068	4590	Hydra	12h 39.5m	-26 45	9.0	12.0	Late Spring	☐
67	M083	5236	Hydra	13h 37.1m	-29 52	8.5	11.0x10.0	Mid-Summer	☐
68	M005	5904	Serpens Caput	15h 18.6m	2 5	5.8	17.4	Mid-Summer	☐
69	M013	6205	Hercules	16h 41.7m	36 28	5.9	16.6	Mid-Summer	☐
70	M092	6341	Hercules	17h 17.1m	43 8	7.5	11.2	Mid-Summer	☐
71	M057	6720	Lyra	18h 53.6m	33 2	9.5	1.4x1.0	Fall/Early Winter	☐
72	M056	6779	Lyra	19h 16.6m	30 11	9.5	7.1	Fall/Early Winter	☐
73	M029	6913	Cygnus	20h 23.9m	38 32	9.0	7.0	Fall/Early Winter	☐
74	M039	7092	Cygnus	21h 32.2m	48 26	5.5	32.0	Fall/Early Winter	☐
75	M027	6853	Vulpecula	19h 59.6m	22 43	7.5	8.0x5.6	Fall/Early Winter	☐
76	M071	6838	Sagittarius	19h 53.8m	18 47	8.5	7.2	Fall/Early Winter	☐
77	M107	6171	Ophiuchus	16h 32.5m	-13 3	10.0	10.0	Mid-Summer	☐
78	M012	6218	Ophiuchus	16h 47.2m	-1 57	6.6	14.5	Mid-Summer	☐
79	M010	6254	Ophiuchus	16h 57.1m	-4 6	6.6	15.1	Mid-Summer	☐
80	M014	6402	Ophiuchus	17h 37.6m	-3 15	7.6	11.7	Mid-Summer	☐
81	M009	6333	Ophiuchus	17h 19.2m	-18 31	7.9	9.3	Mid-Summer	☐
82	M004	6121	Scorpio	16h 23.6m	-26 32	5.9	26.3	Mid-Summer	☐
83	M080	6093	Scorpius	16h 17.0m	-22 59	8.5	8.9	Mid-Summer	☐
84	M019	6273	Ophiuchus	17h 2.6m	-26 16	8.5	13.5	Mid-Summer	☐
85	M062	6266	Ophiuchus	17h 1.2m	-30 7	8.0	14.1	Mid-Summer	☐
86	M006	6405	Scorpius	17h 40.1m	-32 13	4.2	15.0	Mid-Summer	☐

The Complete Messier Object Log - Messier Marathon Info

Search #	M #	NGC #	Constellation	RA	Dec	Mag.	Size	Season	Seen?
87	M007	6475	Scorpius	17h 53.9m	-34 49	3.3	80.0	Mid-Summer	☐
88	M011	6705	Scutum	18h 51.1m	-6 16	5.8	14.0	Late Summer	☐
89	M026	6694	Scutum	18h 45.2m	-9 24	9.5	15.0	Late Summer	☐
90	M016	6611	Serpens Claudia	18h 18.8m	-13 47	6.5	7.0	Late Summer	☐
91	M017	6618	Sagittarius	18h 20.8m	-16 11	7.0	11.0	Late Summer	☐
92	M018	6613	Sagittarius	18h 19.9m	-17 8	8.0	9.0	Late Summer	☐
93	M024	6603	Sagittarius	18h 18.4m	-18 25	11.5	5.0	Late Summer	☐
94	M025	IC4725	Sagittarius	18h 28.8m	-19 17	4.9	40.0	Late Summer	☐
95	M023	6494	Sagittarius	17h 56.8m	-19 1	6.0	27.0	Late Summer	☐
96	M021	6531	Sagittarius	18h 4.6m	-22 30	7.0	13.0	Late Summer	☐
97	M020	6514	Sagittarius	18h 2.3m	-23 2	5.0	28.0	Late Summer	☐
98	M008	6523	Sagittarius	18h 3.1m	-24 23	5.0	35.0x50.0	Late Summer	☐
99	M028	6626	Sagittarius	18h 24.5m	-24 52	8.5	11.2	Late Summer	☐
100	M022	6656	Sagittarius	18h 36.4m	-29 54	6.5	24.0	Late Summer	☐
101	M069	6637	Sagittarius	18h 34.4m	-32 21	9.0	7.1	Late Summer	☐
102	M070	6681	Sagittarius	18h 43.2m	-32 18	9.0	7.8	Late Summer	☐
103	M054	6715	Sagittarius	18h 55.1m	-30 29	8.5	9.1	Late Summer	☐
104	M055	6809	Sagittarius	19h 40.0m	-30 58	7.0	19.0	Late Summer	☐
105	M075	6864	Sagittarius	20h 6.1m	-21 55	9.5	6.0	Late Summer	☐
106	M015	7078	Pegasus	21h 30m	12 10	6.4	12.3	Fall/Early Winter	☐
107	M002	7089	Aquarius	21h 33.5m	-0 49	6.5	12.9	Fall/Early Winter	☐
108	M072	6981	Aquarius	20h 53.5m	-12 32	10.0	5.9	Fall/Early Winter	☐
109	M073	6994	Aquarius	20h 59.0m	-12 3	9.0	2.8	Fall/Early Winter	☐
110	M030	7099	Capricornus	21h 40.4m	-23 11	8.5	11.0	Fall/Early Winter	☐

"Listen To The Stars"

Audio Book

Introduction

The following is the introduction to the audio book of Listen To The Stars.

Hello and welcome to the first in a series of Astronomy Recording Audio Books. These recordings will hopefully enrich your enjoyment of Astronomy as you view distant Stars, Galaxy's & Nebulas.

The first in the series will feature the famous Messier Objects. These objects are some of the most popular and sought after objects to be viewed with telescopes or binoculars. Whether you are a beginner or seasoned astronomer, this information will add a whole new dimension in your viewing session.

Yes, you can use just a good pair of binoculars to view many of the objects depicted in this recording; and an expensive telescope is not necessary to enjoy this experience.

These objects will have their Messier Object # and magnitude in the beginning of each Chapter, so that you can skip over ones you may feel are too difficult to obtain depending on your equipment and location.

Many amateur astronomers enjoy reading the facts and history of their visual objects. It adds an extra perspective of enjoyment when seeing alone is insufficient. To do this, one has to either read about the object on a computer monitor, a smart phone , or usually in a book. And this is usually done at different times away from the telescope.
Some actually will read this information during their viewing session. But this requires the use of a red light so that you can retain your night vision. This, as you know, can become difficult a times. Add to this, the back and forth from EP to book to EP to book... you get the idea.
But what if you could LISTEN to the interesting facts and history of these celestial objects while never taking your eye from the view.

Think about it... You were able to hunt down an elusive object or your revisiting a favorite. You turn on your audio player and now you can listen to all of the interesting history of this famous object you are viewing…while never taking your attention away from the eye piece.

To hear facts, such as, "…You are looking at a stellar nursery where stars are born. Or, "…It contains only a dozen or so variable stars and is estimated to be 13.7 billion years old." Another example, "…there is enough dust/debris, in the nebula, to create 30-40k earths!"

And maybe this one will capture your attention… "It took the light you are looking at over 65 million years to reach your telescope. It left that object while dinosaurs were still roaming the Earth."

There is something very astonishing to read that an object you are looking at is over a BILLION years old. It will give you the sense of viewing "live" history. It will fully immerse you into the view.

So sit back, relax, locate a Messier Object and listen to it's history and beautiful spatial background music… and just "Enjoy The Views"

Note - Most of these recordings fall between 3-6 minutes so there is plenty of time if you would like to change to different eye pieces and/or filters.

As every attempt has been made for excellent audio quality, different headsets could render different audio quality.

"Listen To The Stars" Audio Book

Available on Amazon and iTunes.